Astronomers' Universe

Series editor

Martin Beech, Campion College, The University of Regina, Regina,
Saskatchewan, Canada

More information about this series at http://www.springer.com/series/6960

Steven J. Dick

Classifying the Cosmos

How We Can Make Sense of the Celestial Landscape

 Springer

Steven J. Dick
Ashburn, VA, USA

ISSN 1614-659X   ISSN 2197-6651   (electronic)
Astronomers' Universe
ISBN 978-3-030-10379-8   ISBN 978-3-030-10380-4   (eBook)
https://doi.org/10.1007/978-3-030-10380-4

Library of Congress Control Number: 2019930431

Cover caption: © Mark Garlick / Science Photo Library

This Springer imprint is published by the registered company Springer Nature Switzerland AG
The registered company address is: Gewerbestrasse 11, 6330 Cham, Switzerland

What's so amazing
That keeps us stargazing?
And what do we think we might see?

– Kermit the Frog, "Rainbow Connection"

To the Past
James E. Dick & Elizabeth Grieshaber Dick

The Present
Terry, Gregory & Jenna, Anthony & Elizabeth, Mary Lou,
Karen, Chris & Mark

And the Future
Adeline, Benjamin, John, James, Daniel

The Hubble Ultra Deep Field, released in 2014 with its visible, near-infrared, and ultraviolet components, contains more than 10,000 galaxies stretching back in time almost to the Big Bang. It covers an area of the Southern Hemisphere sky equivalent to looking through a straw. How to understand these objects, and indeed all the objects in the universe? Classification is always the first step. Credit: NASA, ESA, H. Teplitz and M. Rafelski (IPAC/Caltech), A. Koekemoer (STScI), R. Windhorst (Arizona State University), and Z. Levay (STScI)

Acknowledgments

The development of the Three Kingdom system for astronomy has had a long gestation. I began work on the idea shortly after finishing *The Biological Universe* in 1995, the kind of release from a 15-year project that urges the mind in new directions. I had attended numerous meetings of the American Astronomical Society and the International Astronomical Union (IAU), where a zoo of increasingly exotic astronomical objects was revealed and elaborated. What eventually became what I dub here the "3K" system was my personal search for order amidst this profusion of objects. At the same time, the "Three Domain" system of biologist Carl Woese was much in the news, uprooting previously cherished ideas in biological classification. Despite criticism of his system, biologists seemed to have a better handle on their objects of study than astronomers, and seemed to take comprehensive classification systems more seriously than their astronomical counterparts, who have historically had only piecemeal systems. The heated controversy over the planetary status of Pluto, in which I participated and voted on at the IAU in Prague in summer 2006, highlighted this deficiency in astronomy, as does the continual discovery of new exoplanets and other exotic objects.

I first broached the subject of a comprehensive classification system for astronomy publicly in broad outline in my Dibner Library Lecture, "Extraterrestrial Life and Our World View at the Turn of the Millennium," delivered at the Smithsonian Institution on May 2, 2000. It was elaborated in more detail in my LeRoy E. Doggett Prize Lecture at the winter meeting of the American Astronomical Society in Washington, D.C. in January 2006. And it was discussed in one chapter of my book *Discovery and Classifica-*

tion in Astronomy: Controversy and Consensus (Cambridge University Press, 2013). My duties as NASA Chief Historian prevented the further development and publication of the system until now, and I am thankful to Springer for publishing it with so many illustrations. Particular thanks to Maury Solomon, Hannah Kaufman, and Sharmila Sasikumar at Springer.

I am indebted to the fabulous and essential Smithsonian Astrophysical Observatory/NASA Astrophysics Data System (http://adswww.harvard.edu/) as the major source for much of the literature used and cited in the notes. The reader can access most of that literature by accessing that web site and entering author and date. I have also found very useful the following compilations and commentaries: Marcia Bartusiak, ed., *Archives of the Universe: A Treasury of Astronomy's Historic Works of Discovery* (New York: Pantheon Books, 2004); Kenneth Lang and Owen Gingerich, eds., *A Source Book in Astronomy and Astrophysics, 1900-1975,* (Cambridge, MA.: Harvard University Press, 1979); and Helmut Abt, ed., *American Astronomical Society Centennial Issue: Selected Fundamental Papers Published This Century in the Astronomical Journal and the Astrophysical Journal, Astrophysical Journal,* 525 (1999). The *Journal for the History of Astronomy,* edited for 40 years by Michael Hoskin at Churchill Cambridge, College, and now by James Evans, is a vast treasure trove of information that I have also used extensively, as is the *Journal of Astronomical History and Heritage,* edited by Wayne Orchiston. Both are available via ADS. For the latest astronomy news I have found very useful the Nova research highlights of the American Astronomical Society at https://aasnova.org, as well as the websites of particular observatories such as Hubble, Spitzer, Chandra and ground-based observatories such as ALMA and the European Southern Observatory. The Mikulski Archive for Space Telescopes (https://archive.stsci.edu/), located at the Space Telescope Science Institute, is another great resource.

In addition to primary and secondary literature, I have not hesitated to make cautious use of press releases, which often contain revealing quotations from discoverers not present in their technical publications. I have occasionally made use of newspaper reports, which often pinpoint the date the discovery was made public, although, as I am at pains to say in my previous book, "discovery" itself is an extended process. For illustrations, and much more, my debt to NASA and worldwide astronomical observatories will be evident. NASA's Planetary Data System (https://pds.nasa.gov/) is an almost inexhaustible source of imagery.

My thanks as well to the unparalleled astronomy treasures of the US Naval Observatory in Washington, D.C, and its ever-helpful Librarians over a period of many years: Brenda Corbin, Sally Bosken, and Gregory Shelton. My employment as an astronomer and historian there, only steps from its world-class library, kept me immersed in the astronomical literature for almost a quarter century. Last but not least, my thanks to those astronomers and historians who read all or part of the manuscript. Though inevitably opinions will vary on some of the classifications in this volume, any final faults are my own. My hope is that the book is useful to those trying to make sense of the ever-changing celestial landscape, and that it spurs provocative discussion about discovery, interpretation and classification.

This book may be considered a companion to my book *Discovery and Classification in Astronomy*, published by Cambridge University Press in 2013, where the Three Kingdom system was first published in full as an Appendix and briefly discussed in a section of Chapter 8. I have on occasion used passages from that volume in this book when appropriate for describing the history of discovery of particular objects, and have given the proper citation for readers who may wish to consult the earlier book for the broader context.

Ashburn, VA   Steven J. Dick
January, 2019

Contents

Introduction to Astronomy's Three Kingdoms

This book may be read on at least two levels. On one level, it is an illustrated guide to various classes of astronomical objects and their history of discovery. On another level, it is considerably more: a comprehensive classification system for those objects, the first in astronomy to cover all celestial objects in a consistent manner across astronomy's three realms of planets, stars, and galaxies. As such, it is intended to be useful to anyone interested in sorting out the bewildering array of celestial objects discovered over the last four centuries of telescopic astronomy. The system here is dubbed the *Three Kingdom*, or *3K System* (Fig. 1). It incorporates, to the maximum extent possible, specific classification systems already developed for specific objects such as stars and galaxies. Ironically, there is as yet no widely accepted classification system for planets, though that is changing with the rapid discovery of planets beyond our solar system. Those "exoplanets" now number in the thousands, with frequent new discoveries thanks to both ground-based observations and spacecraft such as Kepler, CoRoT, and the Transiting Exoplanet Survey Satellite (TESS). These new discoveries allow us to see classification systems in-the-making for the rapidly expanding class of exoplanets.

All classification systems are arbitrary to some extent, and although this is my own interpretation of a classification system, I believe most astronomers would agree with the majority of the classes I have delineated. Despite incorporating both existing classification schemes in astronomy and current precepts of the International Astronomical Union (as in the case of dwarf planets, for example), there is nothing "official" about the system. This is not unusual: classification systems for specific kinds of astro-

Astronomy's 82 Classes

Kingdom of the Planets

Family: Protoplanetary
Class P 1: Protoplanetary Disk

Family: Planet
Class P 2: Terrestrial (rocky)
Class P 3: Gas Giant
Class P 4: Ice Giant
Class P 5: Pulsar Planet

Family: Circumplanetary
Class P 6: Satellite
Class P 7: Ring
Class P 8: Radiation Belt

Family: Subplanetary
Class P 9: Dwarf Planet
Class P 10: Meteoroid
Subfamily: Small Bodies of Solar System
Class P 11: Minor Planet/Asteroid
Class P 12: Comet
Class P 13: Trans-Neptunian Objects

Family: Interplanetary Medium
Class P 14: Gas
Class P 15: Dust
Subfamily: Energetic Particles
Class P 16: Solar Wind
Class P 17: Anomalous Cosmic Ray

Kingdom of the Stars

Family: Protostellar
Class S 1: Protostar

Family: Star
Subfamily: Pre-Main Sequence
Class S 2: T Tauri
Class S 3: Herbig Ae/Be
Subfamily: Main Sequence (H burning -
            Luminosity Class V)
Class S 4: Dwarf
Class S 5: Subdwarf
Subfamily: Post-Main Sequence (He burning
            and higher elements)
Class S 6: Subgiant (Luminosity Class IV)
Class S 7: Giant (Luminosity III)
Class S 8: Bright Giant Class II)
Class S 9: Supergiant (Lumin. Class I)
Class S 10: Hypergiant (Lumin. Class 0)
Subfamily: Evolutionary Endpoints
Class S 11: Supernova
Class S 12: White Dwarf
Class S 13: Neutron Star/Pulsar
Class S 14: Black Hole

Family: Circumstellar
Class S 15: Debris disk
Class S 16: Shell (dying stars)

Kingdom of the Galaxies

Family: Protogalactic
Class G 1: Protogalaxy

Family: Galaxy
Subfamily: Normal
Class G 2: Elliptical
Class G 3: Lenticular
Class G 4: Spiral
Class G 5: Irregular
Subfamily: Active
Class G 6: Seyfert
Class G 7: Radio Galaxy
Class G 8: Quasar
Class G 9: Blazar

Family: Circumgalactic
Class G 10: Satellites and Stellar Streams
Class G 11: Galactic Jet
Class G 12: Galactic Halo

Family: Subgalactic
Class G 13: Subgalactic Object

Fig. 1 Astronomy's Three Kingdoms, 18 Family's, and 82 Classes, an umbrella system for classifying all classes of objects in the cosmos

Family: Systems
Class P 18: Planetary Systems/Exoplanets
Class P 19: Asteroid Groups
Class P 20: Meteoroid streams
Subfamily: Trans-Neptunian Systems
Class P 21: Kuiper Belt
Class P 22: Oort Cloud

Class S 17: Planetary Nebula
Class S 18: Nova Remnant
Class S 19: Core Collapse Supernova Remnant
Class S 20: Stellar Jet
Class S 21: Herbig-Haro Object
[See also Protoplanetary Disk (P 1); Planetary System, (P 18); Kuiper Belt (P 21); Oort Cloud (P 22)]

Family: Substellar
Class S 22: Brown dwarf

Family: Interstellar Medium
Subfamily: Gas (99%)
Class S 23: Cool Atomic Cloud (H I)
Class S 24: Hot Ionized Cloud (H II)
Class S 25: Molecular Cloud (H2)
Class S 26: White Dwarf Supernova Remnant
Subfamily: Dust (1%)
Class S 27: Dark Nebulae
Class S 28: Reflection Nebulae
Subfamily: Energetic Particles
Class S 29: Stellar Wind
Class S 30: Galactic Cosmic Rays

Family: Systems
Class S 31: Binary Star
Class S 32: Multiple Star
Class S 33: Association (OB)
Class S 34: Open Cluster
Class S 35: Globular Cluster
Class S 36: Population

Family: Intergalactic Medium
Subfamily: Gas
Class G 14: Warm Hot IGM
Class G 15: Lyman alpha blobs
Subfamily: Dust
Class G 16: Dust
Subfamily: Energetic Particles
Class G 17: Galactic Wind
Class G 18: Extragalactic Cosmic Rays

Family: Systems
Class G 19: Binary
Class G 20: Interacting
Class G 21: Group
Class G 22: Cluster
Class G 23: Supercluster
Class G 24: Filaments & Voids

Fig. 1 (continued)

nomical objects often gain favor simply because of their useful-
ness to practitioners in the field. In this case, I would be happy
if the Three Kingdom system proves useful for pedagogical reasons. The need for an overall "umbrella" classification system for
astronomy based on consistent principles has been growing. The
controversy in recent years over the status of Pluto as a planet,
the nature of numerous trans-Neptunian objects, and the debate
over whether certain objects beyond our Solar System are planets
or brown dwarfs form only the tip of the iceberg that is the more
general problem of classifying astronomical objects.

The Importance of Classification

Classification is a problem long known in science in general and
especially in the biological world, where "natural history," "taxonomy," and "systematics" form a significant part of the history
of biology. Although classification may seem to some a boring
subject, it stands in many ways at the foundation of science. In
his *Systema Naturae* (1735), Linnaeus stated that "The first step
in wisdom is to know the things themselves. This notion consists in having a true idea of the objects; objects are distinguished
and known by classifying them methodologically and giving them
appropriate names. Therefore, classification and name-giving will
be the foundation of our science." 250 years later, evolutionist Stephen Jay Gould reflected both the importance and the problematic
nature of classification when he wrote "taxonomies are reflections
of human thought; they express our most fundamental concepts
about the objects of our universe. Each taxonomy is a theory of
the creatures that it classifies."[1] This is certainly true of biology's "Five Kingdom" system, which Gould was addressing, and
its rival "Three Domain" system favored by molecular biologists.

The problems and promise of classification apply no less to
astronomy, where we replace living creatures with inanimate
objects. Allan Sandage, one of the great observers and classifiers
of galaxies, has noted that classification can have a real effect on
the science itself: "The first step of any new science has always

---

[1] Linnaeus, quoted in Paul L. Farber, *Finding Order in Nature: The Naturalist
Tradition from Linnaeus to E. O. Wilson* (Johns Hopkins University Press,
Baltimore and London: 2000), pp. 8-9; Stephen Jay Gould, in Foreword to Lynn
Margulis and Karlene V. Schwartz, *Five Kingdoms: An Illustrated Guide to
the Phyla of Life on Earth* (New York: W. H. Freeman, 2nd edition, 1988), p. x.

been to classify the objects being studied. If a proposed classification scheme is 'good,' progress is achieved. This leads to deeper understandings, and the subject advances. A 'bad' classification scheme, by contrast, usually impedes progress." The problem is not simple: "The initial classifier of any unknown subject necessarily begins with no idea of how to choose the key parameters. How then to produce a classification imbued with any fundamental physical significance or predictive power?"[2]

This is the problem that faced many astronomers who have worked on pieces of the classification puzzle involving planets, stars, and galaxies. Knowledge of key parameters is not so much the problem we now face in constructing a comprehensive classification system across all three Kingdoms of astronomy, thanks to the great advances in astronomy over the last century. But, as we outline below in describing astronomy's Three Kingdoms, problems of other kinds still abound for anyone trying to construct a classification system across such disparate objects. A few bold attempts have been made, ranging from Harlow Shapley's "Classification of Material Systems" based on size to Martin Harwit's original but perhaps too complex attempt to classify astronomical phenomena (rather than just objects) based on "phase space filters" that included wavelength, angular, spectral and time resolution, and polarization.[3] Neither of these attempts has caught on among astronomers. But the fact that a comprehensive system has been constructed for biology and other sciences provides cause for hope in astronomy.

Classification in Astronomy

Although astronomy is the oldest of the sciences, classification came to play an important role only in the late 19th and early 20th centuries. Only then, with the development of techniques like spectroscopy, was the true nature of the objects revealed. Astronomy's first classification efforts centered on stars and developed

[2] Allan Sandage, *The Mount Wilson Observatory: Breaking the Code of Cosmic Evolution*, vol. 1 of the Centennial History of the Carnegie Institution of Washington (Cambridge: Cambridge University Press, 2004), pp. 230-231.

[3] Harlow Shapley, *Flights From Chaos: A Survey of Material Systems from Atoms to Galaxies* (McGraw-Hill: New York, 1930); Martin Harwit, *Cosmic Discovery: The Search, Scope and Heritage of Astronomy* (Basic Books: New York, 1981), Chapter 4.

with advances in stellar spectroscopy, a "fingerprinting" technique that has proved surprisingly suited to unraveling a mystery that many thought might never be solved: the composition and nature of the stars. Each star has a unique fingerprint based on its composition and other factors, and these fingerprints can be classified. From the beginning of spectroscopy, astronomers could not resist. As historian David DeVorkin has shown, by the late 19th century there were eight competing systems of stellar classification, including five from the Italian astronomer Angelo Secchi and those of European astronomers Hermann Vogel and Norman Lockyer, as well as Edward C. Pickering's system at Harvard. It took 50 years before a negotiated stellar classification consensus was reached at the meeting of the International Solar Union in 1910.[4]

The Harvard classification system—the Henry Draper Catalogue of stellar spectra—triumphed. It is well known to astronomers, as is its association with Pickering and the women of "Pickering's harem" at Harvard (Fig. 2) who made it happen

Fig. 2 Part of the so-called "Pickering's harem," women who worked at the Harvard College Observatory and some of whom were pioneers in stellar classification. Pickering is at top center and Annie Jump Cannon is in the darker dress at top center. The image was taken May, 1913. Credit: Harvard College Observatory, courtesy AIP Emilio Segrè Visual Archives

[4]David H. DeVorkin, "Community and Spectral Classification in Astrophysics: The Acceptance of E. C. Pickering's System in 1910," *Isis*, 72 (1981), 29-49; DeVorkin, "Stellar Evolution and the Origin of the Hertzsprung-Russell Diagram," *Astrophysics and Twentieth-Century Astronomy to 1950*, Chapter A, Owen Gingerich, ed. (Cambridge, 1984), 90-108; and DeVorkin's synthetic masterpiece, *Henry Norris Russell: Dean of American Astronomers* (Princeton U. Press, 2000).

beginning in 1886: Williamina Fleming, Antonia Maury (Draper's niece), and Annie Jump Cannon, among others. This is a story that has been well told in Dava Sobel's book *The Glass Universe: How the Ladies of the Harvard Observatory Took the Measure of the Stars.* Piecemeal publications of Harvard spectral classifications began in 1890, but from 1918 to 1924 nine volumes of the *Harvard Annals* were issued with systematic spectral types for a quarter million stars—the basis for the spectral types now delineated as O, B, A, F, G, K, M. These types are recalled by thousands of students through the mnemonic "O Be A Fine Girl Kiss Me," to which was later added "Right, Now, Smack" (and just recently L and T classes, for which there is yet no mnemonic).

As Sobel recalls in her book, some astronomers were opposed to the system; aside from capital letters being used over a prototype nomenclature such as Procyonian, "they thought the progression O, B, A, F, G, K, M looked grotesque or random, as though signifying nothing."[5] In the end, however, it was utility and the sheer numbers of stars classified in the Harvard system that won the day. This is another lesson in classification: Fleming's original order had been strictly alphabetic: A, B, C, D, and so on. But with more spectral data (basically giving helium lines precedence over hydrogen) at a later date, Cannon placed the O and B types at the head of the line and consolidated others, giving rise to the system based on color and temperature that is still used today. In the end, the colors of the stars actually had great meaning for their true nature. And finding many ambiguities in classification, Cannon also introduced intermediate classes such as B 2 A, meaning the star was closer to B, but had some A characteristics. That, too, continues to this day, with ten gradations between the lettered spectral types.

Classification does not remain static as knowledge advances. In 1943, the Harvard system was elaborated with a second dimension characterizing a star's luminosity. This became known as the MKK (Morgan-Keenan-Kellman), or Yerkes system, known after revisions in 1953 simply as the MK system. In this case, Yerkes astronomers William Morgan (Fig. 3) and Philip Keenan and Yerkes "computer" Edith Kellman devised a system published as the *Atlas of Stellar Spectra, with an Outline of Spectral Classification*—a volume that became one of the canonical publications in the history of astronomy. As we shall see in Part 2, its defining feature

---

[5] Dava Sobel's, *The Glass Universe: How the Ladies of the Harvard Observatory Took the Measure of the Stars* (New York: Viking, 2016), pp. 142-143.

Fig. 3 William W. Morgan, the driving force behind the Morgan-Keenan-Kellman system for spectral classification. Unlike the phenomenological spectral line classifications of the Harvard system, Morgan sought to classify "the thing itself." Photograph by David DeVorkin, courtesy AIP Emilio Segrè Visual Archives.

is its luminosity classes labeled I to V, representing respectively supergiants, bright giants, giants, subgiants, and main sequence dwarfs, with each Class having specific standards defined across the Harvard spectral sequence (see Fig. 8.3). That history is not as well known among historians as it is among astronomers still living, who recall the controversial process of its adoption, even though (unlike the original Harvard system) it was never officially adopted, but gained currency only through widespread use.[6]

It is important to reiterate that classification systems evolve, and the MK system is a good example. The original five luminosity classes remain, but greatly elaborated and in many cases with new standard stars. To take only one example, because of the large range of luminosities among the supergiants, in the revised 1953 MK system, Luminosity class I was subdivided into Ia and Ib. In 1971 Philip Keenan suggested a new class, designated 0, to define hypergiants, and in the next revision of MK in 1973, Morgan and

---

[6] William Morgan, Philip Keenan, and Edith Kellman, *Atlas of Stellar Spectra, with an Outline of Spectral Classification* (Chicago: The University of Chicago Press, 1943), online at https://ned.ipac.caltech.edu/level5/ASS_Atlas/paper.pdf. On the personalities involved and the Yerkes context see Donald E. Osterbrock, "Fifty Years Ago: Astronomy; Yerkes Observatory; Morgan, Keenan and Kellman," in *The MK Process at 50 Years*, C. J. Corbally, R. O. Gray, and R.F. Garrison, eds. (San Francisco: ASP, 1994), http://adsbit.harvard.edu//full/1994ASPC...60..199O/0000199.000.html

Keenan adopted the new class. Similar changes have been made to other stellar classes as new discoveries have been made. On the other hand, some proposed changes such as the designation of a class VI for subdwarfs and a class VII for white dwarfs never came into widespread use; they are part of the MK system but not under those class designations. How and why such changes did or did not take place remains a subject ripe for historical study. In any event, the MK system has proved an essential classification system for astronomers, as seen in Richard Gray and Chris Corbally's magisterial volume *Stellar Spectral Classification* (2009).[7] Still, as we will see in the entries in Part 2, the MK system has its own limits when it comes to modern luminosity classes.

Even as early efforts at star classification were ongoing, a second effort at astronomical classification began in connection with the variety of galaxy shapes revealed with new telescopes and the new technique of photography. As Allan Sandage has pointed out in his 1975 review of galaxy classification—still the best review along with his 2005 article in *Annual Reviews of Astronomy and Astrophysics*—only the photographic surveys beginning about 1890 revealed the faint galactic structures that proved to be crucial to their classification. One such classification system extensively used in the early days was that of the German astronomer Max Wolf, published in 1908 before astronomers even reached consensus that the so-called "nebulae" were distant galaxies in their own right rather than gaseous objects within our Milky Way Galaxy. Wolf's system gradually fell by the wayside, and it was the classification system of Edwin Hubble (Fig. 4), presented in a 1926 paper and in his classic book *The Realm of the Nebulae* (1936), that won out in the natural selection of ideas. The *Hubble sequence* (Figs. 14.1 and 14.2) is the system still used in modified form today, though it does not represent the evolutionary sequence Hubble had hoped. For galaxies, as for stars, the classification system evolved: astronomers Knut Lundmark, Gerard DeVaucouleurs and Sidney van den Bergh, among others, proposed modifications, extensions, and variants of the Hubble sequence, some widely accepted and others not.[8]

---

[7] Richard O. Gray and Christopher J. Corbally, *Stellar Spectral Classification* (Princeton and Oxford: Princeton University Press, 2009).

[8] Alan Sandage, "Classification and Stellar Content of Galaxies Obtained from Direct Photography," in *Galaxies and the Universe*, edited by A. Sandage, M. Sandage and J. Kristian, 1975; online at http://nedwww.ipac.

Fig. 4 Edwin Hubble, classifier of "nebulae," who added a new realm or "Kingdom" to astronomy when he demonstrated some of these nebulae were galaxies outside our Milky Way. Courtesy AIP Segrè Visual Archives

Classification is no longer just for trained scientists. Over the last decade, as astronomers have become overwhelmed by huge galaxy surveys such as the Sloan Digital Sky Survey, a crowd-sourced astronomy project known as Galaxy Zoo has allowed citizen scientists to classify galaxies. Hundreds of thousands of people have participated in 15 versions of Galaxy Zoo since 2007, classifying millions of galaxies. What is lacking in expertise is gained in the sheer number of people using the basic pattern recognition abilities of the human brain. Based on the Galaxy Zoo experience,

---

caltech.edu/level5/Sandage/frames.html; Sandage, "The Classification of Galaxies: Early History and Ongoing Developments," *ARAA*, 43 (2005), 581-624; Ronald J. Buta, Harold G. Corwin, Jr., and Stephen C. Odewahn, *The de Vaucouleurs Atlas of Galaxies* (Cambridge: Cambridge University Press, 2007); Sidney van den Bergh, *Galaxy Morphology and Classification* (Cambridge: Cambridge University Press, 1998). For the personalities involved see Gale E. Christianson, *Edwin Hubble: Mariner of the Nebulae* (New York: Farrar, Straus and Giroux, 1995), and Dennis Overbye, *Lonely Hearts of the Universe* (New York: Harper Collins, 1991).

scientists have developed machine-learning codes, a form of artificial intelligence, that allow computers to classify galaxies in agreement with human classifications 90% of the time. [9]

It is important to understand that classification does not necessarily imply understanding. Even as they were classifying, astronomers were unaware of the true nature of the nebulous objects they viewed through their telescopes. Historian Robert Smith and others have shown the difficulties of the so-called "island universe theory," the many interpretations of spiral nebulae, and how even in the first decade of the 20th century many astronomers considered spiral nebulae solar systems in formation.[10] This highlights a remarkable fact: only a century ago, astronomers could not differentiate a structure 40 astronomical units in extent (our Solar System) from one millions of times bigger (the galaxies). They had as yet no idea of the scale of the universe, nor of the interrelatedness of its constituent parts through cosmic evolution.

Over the decades numerous smaller classification systems have been developed for particular classes of objects, ranging from asteroids and comets, to nebulae and star clusters, to binary stars and supernovae. We will discuss many of them in their separate entries in this volume. Each has been formulated in its own time in diverse ways that have proven useful to the communities that study these objects. But almost never did these classifiers have in mind consistent principles that applied across the diverse discipline of astronomy. This is what is attempted here.

Ironically, the most controversial area in astronomical classification in the last decade has been our own Solar System, as the controversy over Pluto's status as a planet demonstrates. This too is due to improved technologies (especially charge-coupled devices) that allow fainter and smaller objects to be seen more clearly than ever before. Pluto, Trans-Neptunian objects, exoplanets, and an entire zoo of objects challenge our attempts to create a comprehensive classification system. That attempt is made here in the very general way laid out in Fig. 1. While such an umbrella

---

[9] The Zooniverse project can be accessed at https://www.zooniverse.org/projects/zookeeper/galaxy-zoo/. Its history and astonishing success is described in Alison Klesman, "Zooniverse: A Citizen Science Success Story," *Astronomy*, 46, (October 2018), 26-35. On machine classification of galaxies see M. Banerji et al., "Galaxy Zoo: Reproducing Galaxy Morphologies Via Machine Learning," MNRAS, in press, preprint at https://arxiv.org/abs/0908.2033, reported at http://www.physorg.com/news194718001.html

[10] Robert Smith, *The Expanding Universe* (Cambridge: Cambridge University Press, 1982), Chapter 1, especially pp. 7-8 on spirals as solar systems.

system across all the domains of astronomy is no substitute for the meticulous classifications of objects based on their physical characteristics, my hope is that it will be a useful exercise for a broad audience wishing to make sense of the cosmos. The Great Pluto Debate also reminds us that classification can be an exciting process, one that involves not only scientists and scientific principles, but also the public and popular culture.

The Three Kingdom (3K) System[11]

Despite the clear importance of classifying planets, stars, galaxies and other objects of the universe, limited knowledge and historical circumstance have in the past combined to prevent a comprehensive classification system for astronomy. As in past classification systems in science, our attempt here makes no claim for uniqueness. Though the system is grounded in nature by the 82 classes of objects it distinguishes (Fig. 1), it is presented as only one way of ordering celestial objects. Mindful of Argentine novelist Luis Borges's famous classification of animals as "those that belong to the Emperor, embalmed ones, those that are trained, suckling pigs, mermaids, fabulous ones, stray dogs, those included in the present classification,"[12] the Three Kingdom system is offered here as a useful way of ordering those cosmic objects being unveiled in increasing detail around us. In short, it is only one way to make sense of the celestial landscape. Borges himself observed that "it is clear that there is no classification of the Universe not being arbitrary and full of conjectures."[13] And yet, classification systems still abound.

---

[11] Portions of this section are adapted from Steven J. Dick, *Discovery and Classification in Astronomy* (Cambridge University Press, 2013), pp. 263-270.

[12] In "The Analytical Language of John Wilkins," Borges describes "a certain Chinese Encyclopedia," the *Celestial Emporium of Benevolent Knowledge*, which contains these lines, which have influenced thinkers in disciplines ranging from anthropology to zoology.

[13] "John Wilkins' Analytical Language," in E. Weinberger *et al.*, ed. and trans, *The Total Library: Non-Fiction 1922-86* (London: Penguin Books, 2001), 229-232.

Definitions and Principles of Classification

Every classification system has a unit of classification. For physics, it is elementary particles. For chemistry, it is the elements, defined by atomic number in the periodic table. For biology, on the one hand it is species at the macro level, giving rise to biology's "Five Kingdoms" still favored by some macrobiologists. On the other hand it is genetic sequences of 16S ribosomal RNA at the molecular level, giving rise to Carl Woese's "Three Domains" of Archaea, Bacteria and Eucarya—favored by most molecular biologists.[14]

For astronomy, the unit of classification adopted here is the astronomical object itself. As strong and weak forces are dominant in particle physics, and as the electromagnetic force is dominant in chemistry (except for nuclear chemistry), so in astronomy it is the weakest but most far-reaching force of gravity that predominantly acts on and shapes these astronomical objects. Though other considerations such as hydrostatics, gas, and radiation pressure come into play, gravity is the determining factor for the structure and organization of planets, stars, and galaxies, their families, and classes of objects. To put it another way, the strong interaction holds protons and neutrons together and allows atoms to exist; the electromagnetic interaction holds atoms and molecules together and allows the Earth to exist; and the gravitational interaction holds astronomical bodies together and allows the Solar System, stellar systems, and galactic systems to exist.[15] Gravity is thus adopted here to serve as the chief organizing principle for a comprehensive classification system for all astronomical objects.

---

[14] On the three Kingdom versus five Kingdom controversy in biology see especially Jan Sapp, *The New Foundations of Evolution* (Oxford: Oxford University Press, 2009). On classification in physics and chemistry see Michael D. Gordin, *A Well-Ordered Thing: Dmitrii Mendeleev and the Shadow of the Periodic Table* (Basic Books, New York, 2004), Andrew Pickering, *Constructing Quarks: A Sociological History of Particle Physics* (Edinburgh: Edinburgh University Press, 1984), and Murray Gellman, *The Quark and the Jaguar: Adventures in the Simple and the Complex* (New York: W. H. Freeman, 1994).

[15] Paul Davies, *Cosmic Jackpot: Why Our Universe is Just Right for Life* (Houghton-Mifflin: Boston and New York, 2007), especially Chapter 4. Isaac Asimov has made the same point in his popular books; for example *Atom: Journey Across the Subatomic Cosmos* (New York: Penguin, 1992), p. 263.

In astronomy, as elsewhere in science, the names of the taxa themselves are an arbitrary function of language, even if sometimes (as in the case of Woese, who consciously chose "Domain" to avoid military language) there are specific reasons for a given name.[16] Even the name for the top-level grouping of organisms in biology is different for competing classification systems, in part to distinguish the different systems: "Kingdom" for the Five Kingdom system, "Domain" for the three-Domain system, and "Empire" for the now outdated two-Empire system of prokaryotes (non-nucleated) and eukaryotes (nucleated) cellular life. For biology's Five Kingdom system, the taxonomic levels in descending order of detail are Phylum/Division, Subphylum, Class, Order, Family, Genus, and Species. Thus humans are classified as Kingdom *Animalia*, Phylum/Division *Chordata*, Subphylum *Vertebrata*, Class *Mammalia*, Order *Primates*, Family *Hominoidea*, Genus *Homo*, and Species *Homo Sapiens.*

Astronomy need not slavishly follow taxon nomenclature for biology or any other science. Rather, for practical purposes the nomenclature should to the maximum extent possible make use of names already in use. In this volume we distinguish four taxonomic levels that incorporate at least some current astronomical usage: Kingdom, Family, Class and Type, though the Type taxon descends to a level of detail not comprehensively addressed in this volume. In particular, we distinguish 3 Kingdoms, 18 Families, and 82 Classes as the heart of the system. Throughout this volume, I am careful to use these terms only with their true taxonomic meanings defined below (and also to capitalize taxa to distinguish them from colloquial usage). Inevitably, such usage has not been systematic as astronomy has progressed over the centuries; for example, the terms "spectral classes" and "spectral types" are sometimes used interchangeably to describe the Harvard system of stellar classification. But more generally, astronomers have often referred to the discovery of new "classes" of objects, classes that later turn out to have multiple "types." Part of the problem is that there has been no superstructure connecting all classes of astronomical objects with a consistent foundation. The Kingdom and Family taxa serve that purpose in the Three Kingdom system proposed here.

---

[16] Sapp, *The New Foundations of Evolution*, pp. 262-263.

Delineating Astronomy's Three Kingdoms

Although several criteria might be used for astronomy's top taxonomic level, none is more straightforward than the natural distinction most often used in modern astronomy textbooks: planets, stars, and galaxies. These three domains we dub "Astronomy's Three Kingdoms." Textbooks often present the domains in this way for a reason: they constitute the central, even the entire, concerns of astronomy aside from the techniques used to study them and the cosmological conclusions they inform.

As one would expect, it has not always been so; the Kingdoms had to be discovered, distinguished, and understood at the most basic level. For most of history our own planetary system was the primary concern of astronomy, against which the fixed stars provided only the background. Beginning with ancient civilizations, skywatchers sought to determine the regular motions of the planets (literally "wanderers"), which they then employed for everything from astrology to mythology and religion. With the invention of the telescope a new dimension was added, as astronomers attempted to determine the physical characteristics of the planets—an endeavor that continues in the present day. Everything having to do with our Solar System—and by extension with other planetary systems now known to exist—constitutes the Kingdom of the Planets.

With the rise of large telescopes, especially those of William Herschel in the late 18th century, followed by those of John Herschel, Lord Rosse, and many others, the realm of the stars came to be a central concern of astronomy. In the mid-19th century, spectroscopy became a science; it was soon applied to astronomy, belying Auguste Comte's dictum that the composition of the stars would never be known. To the contrary, knowledge of stellar composition and stellar evolution became a mainstay of astronomy, and remains so today. During the 19th century, then, astronomy's second Kingdom was revealed in increasing detail, leading eventually to detailed classification systems.

As late as the 19th century, the Solar System and the "sidereal universe" were the two divisions most often found in canonical textbooks. Fully 80% of Princeton astronomer Charles Young's standard *Textbook of General Astronomy for Colleges and Scientific Schools* (1888 and 1898) dealt with Earth, Sun, Moon, and planets, with only three chapters dealing with stars, star clusters, and "nebulae." The same is true for Simon Newcomb and Edward

Holden's *Astronomy for High Schools and Colleges* at around the same time.[17] The story of how some of these nebulae (what we today term "galaxies") came to be recognized as conglomerations of stars largely takes place in the late-19th and early 20th century. It is the story of the rise, fall, and rise again of the "island universe theory," of "Astronomy's Great Debate," and of how Edwin Hubble's "realm of the nebulae" became the realm of the extragalactic universe, of which our galaxy was only a small part.[18] In the early 1920s, Young's successor at Princeton, Henry Norris Russell, had revised Young's textbook into two volumes, one on "the solar system," and one on "astrophysics and stellar astronomy." But Russell's chapter on "The Nebulae" now contained a section on "extragalactic nebulae," a sign that astronomy's third Kingdom had arrived.[19] Since the 1920s, planets, stars, and galaxies have been the canonical categories for textbook organization —and for astronomical research.

The Physical Basis for Astronomy's 18 Families

If astronomy's three Kingdoms have become fairly obvious over the last century, the taxonomic levels of classification beyond this first level are not so apparent. As mentioned above however, the gravitational force offers an organizing principle for the delineation of these Families, giving them an underlying physical basis and connecting astronomy's three Kingdoms with its well-recognized and familiar classes of objects.

Gravity is the dominant force at work in astronomical protosystems, as planets, stars, and galaxies form—by different mechanisms to be sure, but all subject to the laws of gravity as delineated

---

[17] Charles A. Young, *A Text-Book of General Astronomy* (Boston: Ginn and Company, 1900); Simon Newcomb and Edward Holden, *Astronomy for High Schools and Colleges* (New York: Henry Holt and Company, 1881).

[18] Richard Berendzen and Richard Hart and Daniel Seeley, *Man Discovers the Galaxies* (New York: Science History Publications, 1976). Robert W. Smith *The Expanding Universe: Astronomy's Great Debate, 1900-1931* (Cambridge: Cambridge University Press, 1982), and his two articles "Beyond the Galaxy: The Development of Extragalactic Astronomy 1885-1965, Part I," JHA, 39 (2008), 91-119; and Part II, JHA, 40 (2009), 71-107.Whitney, Charles A. *The Discovery of our Galaxy* (New York: Knopf, 1971).

[19] Henry Norris Russell, Raymond Smith Dugan, and John Quincy Stewart, *Astronomy: A Revision of Young's Manual of Astronomy* (Boston: Ginn and Company, 1926 and 1927).

by Newton in the late-17th century. Gravitational considerations (and secondarily hydrostatics for planets, and gas and radiation pressure in the case of stars,) determine the shape and size of planets, stars, and galaxies. Gravity shapes subplanetary, substellar, and subgalactic objects as well. The Newtonian laws of gravity determine the orbits of celestial bodies around each other, as described by Kepler's laws. As a universal force, gravity also acts on the interplanetary, interstellar, and intergalactic media, shaping their objects and controlling their dynamics. And it is gravity that is the force in building and maintaining higher systems of objects such as planetary systems, star clusters, and clusters and superclusters of galaxies. As astronomer James Kaler has remarked, gravity "is the driving and organizing force of the Universe, acting to assemble matter, and is responsible for the creation of stars and their embracing galaxies."[20] Because planets are an integral part of stellar evolution, gravity plays its central role there also.

Thus, each of astronomy's three Kingdoms may be divided into six Families: the eponymous object itself and five other Families denoted by prefixes representing their gravitational relationship to the central categories of planet, star or galaxy. For each Kingdom these astronomical Families are based on the object's origin (Proto-), location (Circum- and Inter-), subsidiary status (Sub-) and tendency to form systems (Systems). These considerations give rise to astronomy's 18 Families, and the symmetry of the six Families of each Kingdom belies their physical basis in gravity's action in all three Kingdoms.

As in biology, I have in some cases found it useful to distinguish an intermediate ranking, here labeled "Subfamily." For example, in the Subplanetary Family I have delineated the Subfamily "Small Bodies of the Solar System," an official category of the International Astronomical Union (IAU), as well as the "Trans-Neptunian Systems" Subfamily for the planetary systems Family. For the star Family I have delineated the Subfamilies known as Pre-Main Sequence, Main Sequence, Post-Main Sequence, and Evolutionary Endpoints. For the galaxy Family I have delineated the normal and active galaxy Subfamilies. In all cases these Subfamilies are useful in terms of the categories astronomers currently use. It is entirely possible that as more exoplanets are found, for example, more Subfamilies will be distinguished.

---

[20]James Kaler, *Extreme Stars at the Edge of Creation* (Cambridge: Cambridge University Press, 2001), p. 10.

Setting the organizing principle for defining astronomical Families still leaves the problem of defining the scope of the Families themselves. For example, for the purposes of this system an object is defined as circumplanetary, circumstellar, or circumgalactic if it orbits, surrounds, or emanates from a planet, star, or galaxy. Thus, in the planetary Kingdom a radiation belt is circumplanetary in the sense that it surrounds a planet, even while satellites and rings orbit a planet. In the stellar Kingdom, circumstellar objects include not only debris disks and shells but also planetary nebulae and supernova remnants from massive stars, because such remnants surround their much-diminished parent star. They also include jets and Herbig-Haro objects, because they emanate from young stars. In the galactic Kingdom, not only are satellite galaxies and star streams considered circumgalactic, but the jets and galactic halos typical of some galaxies are as well, because they surround the galaxy.

Defining Astronomy's 82 Classes: What Is a Class?

The Three Kingdom System contains 82 Classes of objects, as delineated in Fig. 1. But this begs the question: How does one define a class of astronomical objects? More specifically, how does one recognize a new class of objects? I have tackled these questions in my previous book *Discovery and Classification in Astronomy*. One can approach this question by looking at history, but only in part because (notable exceptions like stars notwithstanding) in astronomy classification has often been ad hoc, haphazard, and contingent on circumstances. History teaches us that the classification of astronomical objects has been based on many characteristics. For example, planets could be divided according to their physical nature (terrestrial, gas giant, and ice giants) or, as the recent discovery of planetary systems has taught us, by orbital characteristics (highly elliptical or circular), proximity to the parent star ("hot Jupiters"), size ("Super-Earths") and so on. Binary stars are often classified by the method of observation as visual, spectroscopic, eclipsing, and astrometric, or alternately by the configuration or contents of the system (a white dwarf binary), or even by its electromagnetic radiation (as in an X-ray binary).

History also shows that at the time of discovery, it is sometimes difficult to decide if a new class of object has been discovered. Perhaps by analogy with the Earth's Moon, Galileo decided relatively quickly that the four objects he first saw circling Jupiter were satellites. One might argue whether this constitutes the discovery of a new class, or whether one object—the Moon, discovered long before—represented the first discovery of a new class. The object Galileo first saw surrounding Saturn is quite a different case: it was not at all obviously a ring, and awaited the interpretation of Christian Huygens 45 years later. 350 years after that, it was not immediately evident that pulsars were neutron stars, or that quasars were active galactic nuclei. Nor was it obvious, even as the internal structures of the outer planets of our Solar System were being revealed, that a class of "ice giants" should be distinguished from the "gas giants," just as the gas giants were distinguished from terrestrial planets only in the 20th century after their gaseous nature was very gradually and painstakingly revealed.

Although history provides some insight, one can distinguish between how classes have historically been defined, and how they should be defined in a coherent and consistent system. In order to make the Three Kingdom system both intuitively useful and consistent, in distinguishing classes in all cases *I have opted for the classification that most closely describes the physical nature of the object, in particular its composition.* As it turns out, this is the criterion that astronomers often use in the astronomical literature, wherever possible. In the planetary Kingdom in our own Solar System, for example, rather than orbital characteristics, the definition of planetary classes has been based on their physical characteristics as rocky, gaseous, or icy in composition; pulsar planets have also been distinguished by being physically very different. It is tempting to define a class of "extrasolar" planets, but to do so would be to distinguish a new class based simply on location. We do not yet have enough evidence to declare new classes based on the physical nature of these objects; although thousands have been detected, their physical nature is often as yet obscure. New classes of planets will undoubtedly be uncovered as observations progress. So far many of the known extrasolar planets are likely gas giants, even though close to their stars, and thus are called "hot Jupiters." The first terrestrial extrasolar planets have also been claimed, and many more will be found with the TESS spacecraft. Whether or not objects such as the "styrofoam" planet (one of the first discovered by the Kepler spacecraft) constitute a new class of planet remains to be seen.

With the definition of "class" thus narrowed, the problem becomes having enough knowledge of the physical nature of an object to declare it a new class. Even then, there will be boundary questions, as Pluto showed, and the gas giant and ice giant planets before that. While such boundary questions may be considered problems by some, they are often also very interesting precisely because of their boundary status. They should in no way discourage the classifier but should rather be seen as an opportunity; whatever one may think of the IAU decision to "demote" Pluto, the discussion that has followed has been interesting and productive.

There will also be questions about which taxon level to place a particular kind of object, e.g., is it a Class or a Type? Such decisions depend on the entire superstructure of the classification system; they are in part subjective but also often commonsensical. In biology, different types of people may be distinguished based on skin color, but no one would classify them as a different species. Similarly one might wish to distinguish terrestrial planets with atmospheres from terrestrial planets lacking an atmosphere (or some other characteristic), but one would not want to say that Earth was a different Class of planet than Mercury, given the superstructure laid out in the Three Kingdom system and the definition of the Class of terrestrial planets as composed of rocky materials.

In the stellar Kingdom, as one might expect, such problems only proliferate. As noted above, stars were first classified on a temperature scale (the Harvard system with its familiar O and B stars and so on), and later on a luminosity scale, the Yerkes/MK system with its dwarfs, giants, and supergiants. Which to choose for our classes of stars? I have opted for the Yerkes/MK system, since it is a two-dimensional system based on spectral lines sensitive not only to temperature but also to surface gravity (g), and thus luminosity. This is in keeping with the theme throughout this book of the important role of the gravitational force in shaping astronomical classes of objects. As astronomers Gray and Corbally put it in their volume *Stellar Spectral Classification*, "Stars readily wanted to be grouped according to gravity as well as according to temperature, and this grouping could be done by criteria in their spectra."[21] Moreover, the MK system works in tandem with the Harvard system, as we will see throughout this volume.

The resulting luminosity classes (main sequence dwarfs and subdwarfs, subgiants, giants, bright giants, supergiants, and hypergiants, labeled from Roman numeral V to I and 0 respectively), together with the stellar endpoint classes (supernova, white dwarf,

---

[21] Richard O. Gray and Christopher J. Corbally, *Stellar Spectral Classification* (Princeton and Oxford: Princeton University Press, 2009), p. 10.

neutron star, and black hole) not only have significance in the evolutionary sequence, but also have a real history of discovery that can be uncovered. W. W. Morgan delineated these luminosity classes to begin with, because he realized each grouping of stars formed a sequence of near-constant surface gravity (known as log g), a characteristic revealed in their spectra.[22] This also confirms what should already be evident: gravity as a sculpting force for astronomical objects not only applies to the Family taxa discussed earlier, but also reaches down to the Class taxa, in the case of stars being recognized by the founders of the MK system as the dominating force for the luminosity classes.

In choosing luminosity for our stellar classes, we do not intend to subordinate the Harvard system of spectral types, which is still a central part of stellar spectral classification. As the originators of the Yerkes/MK system argued, it is simply the case that the MK system contains more information and better represents the physical nature of stars as astronomers gradually separated them over the 30 years from 1910 to 1940 into supergiants, bright giants, giants, subgiants, dwarfs, and subdwarfs (and eventually hypergiants). In other words, since 1943 with the Yerkes/MKK system, modern astronomy has had a formal two-dimensional temperature-luminosity system with distinct classes. This built on the Hertzsprung-Russell diagram (Fig. 8.2), which was literally a two-dimensional plot beginning around 1914. The O, B, A, F, G, K, M, R, N designations of the Harvard system may be seen as spectral Types of the luminosity classes, now extended to L and T for stars of even lower luminosity than M dwarfs, and for brown dwarfs. They too have their physical significance, based on color and temperature as one crosses from left to right in the HR diagram, and that is how we treat them in their respective luminosity class entries, as taxon Types below the level of Class in the 3K system. Thus the Sun may be fully described as a G 2 V star, where G 2 refers to the Harvard spectral type and V designates a main sequence dwarf. We will elaborate on and utilize these designations in Part 2. To some extent a third dimension has been added, taking into account compositional factors in the form of "metallicity," as one must, for example, in subdwarfs (S 5), as well as various other extensions for specific spectral types.

---

[22] Gray and Corbally, 9-10; Morgan, "On the Spectral Classification of the Stars of Types A to K, ApJ, 85 (1937), 380 ff.

The use of the Yerkes/MK luminosity classes for stellar classes here illustrates another important principle in the 3K classification system: wherever possible classes already in use are retained, usually in their most refined form. As noted earlier, however, the MK system has its own self-admitted limits, in that its classes of stars are defined only by the spectral lines. In the early 1990s the Hipparcos satellite determined distances ten times more accurate than ground-based parallaxes, and correspondingly more accurate luminosities. The data showed that many of the luminosities were in error, and in the post-Hipparcos era, the modern concept of a giant (shell H burning via CNO cycle) is by no means co-extensive with MK class III (defined by spectral line ratios). The process is again continued with the incoming Gaia data for more than a billion stars. Nevertheless, the general categories of stars remain, but with a broader definition than determined by the MK system. Science marches on.

Again in the stellar Kingdom, for the interstellar medium instead of "diffuse nebulae" (a morphological classification), classes are given in the 3K system according to physical constitution of the nebulae: gas (cool atomic neutral hydrogen, hot ionized hydrogen, and molecular), and dust (dark and reflection nebulae). These categories are used in astronomy and subsume classifications based on morphology that are historically contingent or based on the location of an object (emission or absorption nebulae). In the galactic Kingdom, galaxy morphology (elliptical, lenticular, spiral and irregular) also reflects compositional differences, so the principle still holds.

Having said all this, it is also important to state that in the interests of both accuracy and utility, I have followed the pronouncements and guidelines of the IAU where they exist. Thus, while one might think that a "dwarf planet" should be a class of planet along with the other four classes delineated in the planet Family, it is not so listed as a separate class in the 3K system because the IAU famously ruled in 2006 (to the chagrin of many scientists and much of the general public) that a dwarf planet is not a planet.[23] Neither is it one of the "small bodies of the Solar System" like asteroids, comets, and other trans-Neptunian objects. Rather, it is a type of Kuiper belt object, even though those are trans-Neptunian and this also stretches the official IAU definition.

---

[23] On this controversy see Steven J. Dick, *Discovery and Classification in Astronomy* (Cambridge University Press, 2013), pp. 9-30.

Moreover, I have where possible used terminology common in the field, except where it would cause confusion. Thus, while the term "asteroid families" is often used to denote the main asteroid belt, near-Earth objects, and the Trojans and Centaurs, I have used "asteroid groups" to avoid confusion with the Family taxon.

There is also the question of whether a proposed but unverified class of objects should be included in a classification system. I have refrained from doing so in the 3K system. In addition I have refrained from delineating a new class if an object's physical composition is unknown. "Floating planets" in interstellar space have been claimed, but it is not clear that their compositional nature requires a new class. A terrestrial planet ejected from its star system is still a terrestrial planet, though admittedly one on a very interesting journey!

The decisions do not end once the Classes are established. As in biology, there are ambiguities as to which Class (or even which Family or Kingdom) an object should be placed, even given the definitions of Family and Kingdom above. Is a globular cluster a circumgalactic object, or a hierarchical system of stars? It is both, of course, but for consistency I have adopted the following principle: *an object should always be placed in its most fundamental class*, where "fundamental" is defined as most closely related to its physical nature. While a globular cluster indeed orbits the galaxy, it does so no more and no less than all the other stars comprising the Galaxy; it is more fundamentally a very large association of stars that happens to be most readily visible on the outskirts of galaxies but is also present closer to the galactic nucleus. I have therefore classed it as under stellar systems as S 35.

To take another case, in the Kingdom of the planets, does the protoplanetary disk class belong in the Protoplanetary Family or the Circumstellar Family? Circumstellar disks may or may not form planets, thus protoplanetary disks are a separate class and are most fundamentally protoplanetary in nature. Distinguishing them observationally is another matter altogether, but astronomers have developed methods. Again, are classes under the "Systems" Family (asteroid groups, meteor streams, the Kuiper and Oort belts, and planetary systems) more properly classified in the planetary Kingdom or as circumstellar objects in the Kingdom of the stars? They are certainly the latter, but they are more fundamentally in the Kingdom of the Planets. At another level, a Herbig-Haro object (S 21) is a kind of emission nebula associated with young stars, but it is more fundamentally an object that emanates from a star, and is thus circumstellar until such time as it dissipates into the interstellar medium a few thousand years later.

In the Kingdom of the stars, are supernova remnants part of the interstellar medium or circumstellar objects? Here one runs directly into the vagaries of history. For most of history, supernovae appeared only to be large explosions whose exact cause was unknown. It turns out that remnants from so-called Type II supernovae (exploding stars greater than eight times the Sun's mass) and remnants from Type Ib and Ic (exploding stars greater than 15 times the Sun's mass) can be classified as circumstellar structures, because they surround the resulting neutron star or black hole until such time as they dissipate into the interstellar medium over millions of years. But so-called Type Ia supernovae that result from white dwarfs exploding as the result of accretion in a binary system cannot be classified as circumstellar, because there is most likely no star remaining after the explosion; only the heavy elements are dispersed immediately into the interstellar medium. And its binary companion has been unceremoniously ejected. Had astronomers known the true causes of these explosions before they gave them their Type names, their nomenclature would undoubtedly have been much different, perhaps "core collapse supernova remnant" and "white dwarf supernova remnant," here designated S 19 and S 26 and placed in the circumstellar and interstellar Families, respectively.

Occasionally, what may seem an inconsistency arises in delineating classes. For example, it may seem surprising that novae are not designated in the Class taxon like supernovae. Novae, after all, are stellar brightenings that seemed for a long time to be caused by explosions, if not as bright as supernovae. But novae are now known to be close binary stars (S 31) consisting of a white dwarf (S 12) and a normal star, in which hydrogen from the normal star is accreted onto the white dwarf. When sufficient material accumulates on the surface of the white dwarf, the temperature rises enough for the carbon cycle of nuclear fusion to ignite, causing the white dwarf to brighten by about ten magnitudes. A similar process happens for a white dwarf (Type Ia) supernovae (S 11), but in that case the accreted material accumulates to the point that the entire white dwarf explodes, releasing even more energy and generally obliterating the white dwarf and its companion star, leaving a supernova remnant (S 26) as part of the interstellar medium. The companion may also be ejected from the system, but in either case no star remains.[24]

---

[24] HST Release, October 27, 2004, "Stellar Survivor from 1572 A.D. Explosion

A nova remnant (S 18), on the other hand, like a core collapse supernova remnant, often still surrounds its parent star and is thus a circumstellar object. The nova itself is a type of binary star, and therefore one taxon below the class status. This leaves an unavoidable inconsistency: a Type Ia supernova is also a type of binary star, and deserves to be one taxon below the Class status. But because of historical contingency, it is placed in the Class with other supernovae. Whether core collapse supernovae and white dwarf supernovae should be given separate Class status or be placed one taxon down at the Type level is an open question. I have opted for the latter because the "Type" designation is common astronomical usage for supernovae, and because a white dwarf supernovae deserves to be in a separate class no more or less than a nova, which is also a Type of binary star.

Finally, in an entirely different category of placement difficulty, while a planet is a kind of circumstellar object, it is only cross-referenced in the circumstellar Family because it is the basis for an entire Kingdom of its own. A planetary system might also be considered in the circumstellar Family, but it is more fundamentally a system of planets in the planet Kingdom. The same applies for stars, which might be considered circumgalactic or subgalactic but are in fact the basis for their own Kingdom in this system.

Summary of Classification Principles in the Three Kingdom System

In summary, seven principles are employed in the 3K system, which is understood to include only "normal" baryonic matter:

(1) Kingdoms are delineated by the three central prototypes of objects in the universe—planets, stars, and galaxies—as enshrined in canonical textbooks since the 1950s.
(2) Families are delineated by the various manifestations of the gravitational force acting on astronomical objects, e.g. protoplanetary, planetary, circumplanetary, subplanetary, interplanetary, and systems.
(3) Classes are delineated based on the physical nature of the object, defined as physical composition wherever possible.
(4) An object should always be placed in its most specific Class.

Supports Supernova Theory," http://hubblesite.org/newscenter/archive/releases/2004/34/

(5) To the extent possible, Classes already in use are retained, as in the luminosity Classes of the MK system

(6) The recommendations of the International Astronomical Union are followed; e.g. a dwarf planet is not a Class of planet.

(7) Potential but unverified Classes are not included.

Relationships Among Classes: Cosmic Evolution

One of the hallmarks of biology is phylogeny. In a phylogenetic tree, classes of living creatures are evolutionarily related through the generation of variation and the Darwinian processes of natural selection and inheritance. Astronomy has plenty of variation, but no universal Darwinian process of selection and certainly no inheritance in a biological sense. Nevertheless, an evolutionary relationship undeniably exists among many of astronomy's classes of objects. It is a "transformational" rather than "variational" evolution. It has occasionally been suggested that cosmic evolution should more accurately be termed "cosmic development" or "cosmic transformation" to distinguish it from Darwinian evolution by natural selection. While this may be a good idea, the term "cosmic evolution" is the one that has caught on, and it is too late to change it. The important point is that astronomy has a phylogeny of its own, based on a variety of physical processes having to do largely with the laws of thermodynamics rather than natural selection.

In the broadest sense, all classes of astronomical objects are related, because all have developed from each other over the last 13.8 billion years since the Big Bang. But there are more particular relationships at many levels. Because planets form from stars and stars comprise galaxies, stars hold the central place in astronomical importance. The formation of the four classes of planets identified in the 3K system is dependent on the circumstances of their environment, which seems to provide limited possibilities. Great progress has been made in understanding planetary evolution in the century since Percival Lowell laid out some of the possibilities in *The Evolution of Worlds* (1909), as evidenced in the series of hefty *Protostars and Planets* volumes over the last several decades. But even with the discovery of hundreds of extrasolar planets, it is not clear that the number of planet classes will rise much above those already seen in our Solar System.

By comparison, the largest variety of evolutionary relationships exists in the Kingdom of the stars, where their gaseous composition, their tendency to exist in sometimes interacting multiple star systems with diverse components, and their spectacular evolutionary endpoints governed by the laws of physics, gives rise to astonishing variety. The stellar Subfamily names such as pre-main sequence, main sequence, post-main sequence, and evolutionary endpoints only begin to describe the variety in the Kingdom of the Stars. The stellar Classes elaborate on this variety, and numerous stellar Types exist below the Class taxon. Since the development of the Harvard system for stellar classification over a century ago, and since George Ellery Hale's *Study of Stellar Evolution* (1908), stellar evolution has become a central part of astronomy, still giving rise to new insights and discoveries.

Those congeries of stars known as galaxies have their own evolutionary relationships, still largely mysterious and widely debated since the time that Edwin Hubble recognized them in the 1920s as true "island universes" outside our own Galaxy. Since his book *The Realm of the Nebulae* was published in 1936, increasingly large telescopes with increasingly sophisticated detectors, both on the ground and in space, have unveiled a universe of billions of galaxies. Surprisingly, the number of classes of galaxies is somewhat limited, and their evolutionary relationships remain uncertain, even as the Hubble Deep Fields reveal galaxies in formation almost back to the beginning of time.

Uses of the System and Future Development

As Allan Sandage was quoted saying earlier, a good classification system must not only be useful, but should also lead to deeper understanding and advance its subject. The uses of the Three Kingdom system are at least threefold, all of which may potentially lead to deeper understanding for different audiences.

First, as a comprehensive system for all astronomical objects based on consistent physical principles, the 3K system brings a consistent set of classification principles to discussions such as the status of Pluto as a planet. It suggests that the definition of a planet should not be based primarily on hydrostatic equilibrium, or roundness, or dynamical considerations, but on physical constitution—just as stellar classification was based on consistent physical principles as determined by spectroscopy. Other criteria may indeed enter any classification decision, but they should be secondary. The 3K system thus brings consistency to astronomical classification, and more clarity in making classification decisions. In the process it might also, over the longer term, bring consistency to astronomical nomenclature as far as taxa such as Class and Type are concerned.

Secondly, the symmetric structure of the 3K system facilitates comparisons at three different scales. In the comparison of Families across Kingdoms, one can ask for example how the interplanetary, interstellar, and intergalactic media compare, and analyze what this tells us about the nature of the cosmos. Likewise for protoplanetary, protostellar, and protogalactic processes, and so on. Such comparisons are sometimes already made, but the 3K system cries out for such comparison in a systematic way. Comparisons of Classes across Kingdoms may also prove enlightening. Planetary rings, stellar rings, and galactic rings in the form of stellar streams have much in common as broken up remains, but at vastly different scales and energies. Similarly for planetary, stellar, and galactic jets, or subgalactic, substellar, and subplanetary objects. However, since the bedrock definition of a Class is that at least one representative object must have been observed, I have not included a Class of planetary jets, even though the discovery of brown dwarf jets in 2007 led to speculation that planetary jets might exist during the accretion phase of gas giants. Based on symmetry among Families in the Three Kingdoms, we might also predict the existence of such jets, as well as other objects. While some might argue that volcanic eruptions or water spouts from Europa or Enceladus might qualify as jets, this does not seem to me to be analogous to stellar and galactic jets formed by energetic processes. But one could argue.

Thirdly, there is an educational advantage for the teaching of astronomy. The 3K system allows students to perceive immediately where an object fits in the scheme of astronomical objects. In assessing a new discovery, for example, whether the object is a Type, Class, Family, or Kingdom should help a student to see its relative importance in the astronomical zoo. Definitive proof of a new Kingdom in astronomy would be vastly more important than, say, a new Type of subgiant star. Moreover, the decision as to whether a particular Class should be placed in a particular Family can lead to fruitful discussion among students, and also even scientists. For example, the question of whether a globular cluster is circumgalactic or not will lead students to realize that these objects are not found just surrounding the Galaxy, but also within the Galaxy, and so on.

Finally, as new discoveries are made in astronomy, the Three Kingdom system may well be elaborated. For the most part the additions and revisions will be made at the Class and Type level as new classes of planets are discovered, new classes of baryonic dark matter objects are revealed, or newly detected objects are analyzed, such as the mysterious "G objects" at the center of our Galaxy that look like gas clouds but behave like stars.[25] It is not out of the question that a new Family could be added, though this seems unlikely given our definition of Family. At the Kingdom level, surprisingly, one can already glimpse a possible new entry: the universe itself may be one of a class of objects in what has been called the multiverse. Because this is a Kingdom that so far we have not seen but only inferred from concepts like the anthropic principle, it has not been included in the 3K system at present. Only time will tell. More fundamentally, we must always remember we are classifying baryonic objects composed of protons, electrons, and neutrons, and that baryonic matter constitutes only 4.6% of the matter and energy content of the universe. Non-baryonic dark matter is 23%, and dark energy (believed to be responsible for the accelerating universe) is 72%. But we have no idea what that dark matter and dark energy may be. Classification of the known objects notwithstanding, plenty of work remains for future astronomers.

For readers who are looking for their favorite variable stars and do not find them among the Classes of stars defined here, I should point out that in the Three Kingdom system they are usually found at a taxon below the Class level, which I designate as Type. Thus, variable stars are spread among the many classes of stars, and I discuss some of these Types of variables in their specific Class designation. For cxample, RR Lyrae variables are Types of the giant and bright giant classes, while Cepheid variables are found mainly among the supergiants, as are the R Corona Borealis variables that are the Herbig Ae/Be stars. A few stars, such as T Tauri and Herbig Ae/Be stars (S 2 and 3) are themselves always variable at the class level, though they have their own Types. Types of other objects are also discussed under their appropriate Class. An elaborated version of the Three Kingdom system would list all of these Types. Variable stars have their own classification system, most generally those based on the physical properties of the stars (intrinsic variables that pulsate, erupt, or explode) and those based on external properties such as eclipsing binaries (extrinsic variables). We will run into both under their appropriate Classes.

---

[25] W. M. Keck Observatory, "More Mystery Objects Detected Near Milky Way's Supermassive Black Hole," News Release, June 6, 2018, http://www.keckobservatory.org/recent/entry/G-Objects

In the Three Kingdom system, Classes are designated "P" for the Planetary Kingdom (P 1–P 22), S for Stellar Kingdom (S 1–S 36), and G for Galactic Kingdom (G 1–G 24), for a total of 82 Classes. Again, I emphasize that words such as "Classes" and "Types" are always capitalized when referring to a taxon level, in contrast to their more generic use. It is interesting that the Kingdom of the Stars has 50% more Classes than the other two Kingdoms, and many more Types at the next descending taxon, belying the rich variety of this Kingdom and highlighting the fact that neither planets nor galaxies could exist without stars (free-floating planets and starless galaxies notwithstanding)! This numeration system leaves room for the inevitable revisions that are common in any classification system and that also serve as a tracer of progress in astronomy.

Some may bridle at the effort to "shoe-horn things into boxes," as some people characterize the process of classification. But ordering Nature has been and remains an honorable and productive endeavor in the history of science. At a minimum, this exercise should be of pedagogical use, helping students and astronomy enthusiasts understand the place of these objects in the celestial landscape. As I stated at the outset, I will be happy if that is its only use. In any case, I offer Astronomy's Three Kingdoms as an innovative way—I would argue one of considerable power—to view the celestial landscape, and to highlight the amazing nature of the objects that populate our universe, as well their relationships in the epic of cosmic evolution of which we are all a part.

Abbreviations

A & A	Astronomy and Astrophysics
Abt	AAS Centennial Issue
	American Astronomical Society Centennial Issue: Selected Fundamental Papers Published This Century in the Astronomical Journal and the Astrophysical Journal, Helmut Abt, ed., *Astrophysical Journal*, 525 (1999).
ADS	Astrophysical Data System of NASA and the Smithsonian Astrophysical Observatory
AJ	Astronomical Journal
AJSA	American Journal of Science and Arts
AN	Astronomische Nachrichten
ApJ	Astrophysical Journal
ARAA	Annual Reviews of Astronomy and Astrophysics
AREPS	Annual Reviews of Earth and Planetary Science
ASP	Astronomical Society of the Pacific
BAAS	Bulletin of the American Astronomical Society
BAIN	Bulletin of the Astronomical Institute of the Netherlands
Bartusiak	*Archives of the Universe: A Treasury of Astronomy's Historic Works of Discovery*, Marcia Bartusiak, ed. (New York: Pantheon Books, 2004)
BEA	Biographical Encyclopedia of Astronomers

BJHS	British Journal for the History of Science
CMWO	Contributions of Mt. Wilson Observatory
Dick	*Discovery and Classification in Astronomy: Controversy and Consensus* (Cambridge University Press, 2013)
DSB	Dictionary of Scientific Biography
HCO	Harvard College Observatory
HR Diagram	Hertzsprung-Russell Diagram
HSTPR	Hubble Space Telescope Press Release
JAHH	Journal of Astronomical History and Heritage
JBAA	Journal of the British Astronomical Association
JHA	Journal for the History of Astronomy
JHB	Journal for the History of Biology
JRASC	Journal of the Royal Astronomical Society of Canada
Lang and Gingerich	*A Source Book in Astronomy and Astrophysics, 1900-1975*, Kenneth R. Lang and Owen Gingerich, eds., (Cambridge, MA.L Harvard University Press, 1979).
MK System	Morgan-Keenan System (post-1953)
MKK System	Morgan-Kennan Kellman System (pre-1953)
MNRAS	Monthly Notices of the Royal Astronomical Society
NDSB	New Dictionary of Scientific Biography
OHI	Oral History Interview
PA	Popular Astronomy
PASP	Publications of the Astronomical Society of the Pacific
PNAS	Proceedings of the National Academy of Sciences
PTRSL	Philosophical Transactions of the Royal Society of London
QJRAS	Quarterly Journal of the Royal Astronomical Society
S&T	Sky and Telescope
SciAm	Scientific American
SSR	Space Science Reviews

Part I
The Kingdom of the Planets

Montage of planetary images taken by NASA spacecraft, ranging from Mercury at top to Neptune at bottom. Credit: NASA/JPL

The Kingdom of the Planets is the realm of astronomy most familiar in popular culture. This largely has to do with the fact that its objects are closest to our home planet and thus are most easily observable both from the ground and—uniquely among the Three Kingdoms—from *in situ* spacecraft. If "Earth" is our first celestial address, "Solar System" is the second, and we have been surveying it since the invention of the telescope more than 400 years ago. We now know that it harbors not only classes of large objects such as planets, satellites, and rings, but also smaller objects such as asteroids, meteoroids, and comets, as well as the gas and dust of the interplanetary medium—not to mention energetic particles that whiz through the system from a variety of sources including its central Sun. And only in the last quarter century have we empirically discovered what was long conjectured to be the case: that the Solar System is not alone, that almost all stars have planets, many of them in planetary systems both similar and diverse compared to ours. The close-up scrutiny of our Solar System therefore tells us something about systems that exist around stars throughout the universe—but only to a point.

The discovery of our Solar System is a continuous process extending even until today. Five planets beyond Earth were known since ancient times, wandering as they were among the seemingly fixed stars. William Herschel added Uranus in 1781, Johann Galle clinched Neptune in 1846, and Clyde Tombaugh picked out the slow motion of Pluto in 1930 amidst thousands of other faint background objects. Gas giant planets like Jupiter were distinguished from terrestrial planets like Earth only in the 19th century, and ice giants only in the late 20th. And in an indication that discovery never ends as new details are revealed, in 2006 the International Astronomical Union controversially declared Pluto a "dwarf planet," a new class that was officially different from a planet, leaving the Solar System with only eight, much to the chagrin of the public and some astronomers. Beyond our Solar System, the true nature of the thousands of "exoplanets" in other stellar systems remains for the most part unknown, while the pulsar planets first detected in 1992 are so different that they deserve a class of their own. One thing was for sure: the discovery and classification of exoplanets (P 18) would remain a major activity for astronomers in the 21st century. Not only that, but new techniques in the 21st century, particularly at infrared, millimeter, and submillimeter wavelengths, are unveiling the birth of planets in the form of protoplanetary disks around other stars.

Meanwhile in the circumplanetary Family (Chapter 3), satellites, rings, and radiation belts had their own stories of discovery. Our Moon was a class of one until Galileo's discovery of the four moons of Jupiter in 1610. By 1950, only 30 satellites were known, but today they number at least 187, including 12 discovered around Jupiter in 2018 alone. The incredible diversity of all these satellites has been a major source of study, and astronomers now appreciate satellites and possible "exomoons" as objects of study in themselves, and even as possible locations for life. Galileo discovered the rings of Saturn in 1610, a new class of objects that we now know to be common around gas giant planets. The discovery of the Van Allen radiation belts at the beginning of the Space Age completed the repertoire of known circumplanetary objects, a class soon discovered around some of the other planets as well.

In the subplanetary Family (Chapter 4), dwarf planets and meteoroids (in the form of meteor showers and storms) are often in the news. Those objects now officially designated "small bodies of the Solar System" have also drawn attention throughout history, especially in the form of those spectacular comets that visit the Earth, if only briefly. The first asteroid, Ceres, was discovered on New Year's Day 1801, sparking the beginning of a new class of objects that now number in the tens of thousands. As we will see, in another story of classification, Ceres has now been declared a dwarf planet. Why? You will find out in entries P 9 and P 11.

Asteroids are not just distant objects of little concern to the citizens of Earth: near-Earth objects (NEOs) are the subject of close scrutiny for their potential to strike the Earth and cause catastrophic damage. Finally, beyond the terrestrial and giant planets is what has been called the "third zone" of the Solar System, a zone we now know is filled with objects generally classed as Trans-Neptunian, some of which gather into Trans-Neptunian systems such as the Kuiper Belt and the Oort Cloud.

Not to be forgotten are objects even smaller than these subplanetary objects, composing the interplanetary medium Family (Chapter 5) of gas, dust, and energetic particles emanating from the solar "wind" and from the more distant regions of interstellar space. Though invisible to the human eye, they play an important part in the planetary Kingdom and the ecology of the Solar System.

Finally, as we have hinted in the case of the Kuiper Belt and the Oort Cloud, and as our planetary system itself attests, astronomical objects tend to gather together into systems under the force of gravity. These systems include not only well-known asteroid groups such as the main belt between Mars and Jupiter, but also many other asteroid conglomerations such as the Trojans, Centaurs, and near-Earth asteroids (NEAs). Similarly, meteoroids exist not only at random, but also tend to gather in meteoroid streams, most often because they are the remnants of comets that intersect the Earth's orbit, causing sometimes spectacular meteor showers or meteor storms.

The Kuiper Belt and the Oort Cloud, the third (and perhaps the fourth) zones of the Solar System, extend the Sun's influence far beyond the planets we usually think of as the outer boundaries of the system. They contain object believed to be leftovers from the Solar System's formation, which are occasionally propelled toward Earth in the form of comets.

Altogether, the 22 classes of objects we designate in the Kingdom of the Planets make it a busy and interesting place, an object of continuous study and a source of amazement to those who take the time to contemplate Earth's immediate surroundings.

1. The Protoplanetary Family

Class P 1: Protoplanetary Disk

All things must be born, and for planets birth begins with a protoplanetary disk, a circumstellar disk of gas and dust that originates during star formation, when a protostar (S 1) condenses out of a molecular cloud (S 25). The central star is believed to be a T Tauri or higher mass object known as a Herbig Ae/Be star (S 2 and S 3) where planets have not yet had time to accrete. Recent research indicates that in massive starburst clusters such as those in the Carina nebula, the ages of these pre-main sequence stars varies from half a million to 20 million years. Protoplanetary disks are distinguished from debris disks (S 15) left over after planet formation, which may be equivalent to (or younger versions of) the Kuiper belt (P 21) in our Solar System. One form of protoplanetary disk is known as a proplyd, an externally illuminated photoevaporating disk first observed in the Orion Nebula. Proplyds represent what our Solar System must have looked like at an age of about one million years. The lifetime of protoplanetary disks is not expected to exceed several tens of millions of years before the formation of kilometer-sized planetesimals and then plancts. Not all disks will result in planets; simulations have shown that rings may arise around stars that do not form planets.

It had long been theorized since Laplace's nebular hypothesis in the 18th century that a circumstellar cloud of material was a necessary part of star formation. Prior to the 1980s, the existence of protoplanetary disks was inferred from such theories, as well as models of the protosolar nebula. The challenge was empirical confirmation. Observations reported in 1979 indicated the presence of compact ionized sources around some stars, and by 1987 one of the possible models of these sources was a circumstellar shell around young stars. Meanwhile in the mid-1980s, the Infrared Astronomical Satellite (IRAS) found infrared excesses around some T Tauri and main sequence stars, indicating the presence of cold circumstellar dust. Many of these disks were erroneously interpreted as protoplanetary. Only later were the class of protoplanetary disks

© Springer Nature Switzerland AG 2019

S. J. Dick, *Classifying the Cosmos*, Astronomers' Universe,

https://doi.org/10.1007/978-3-030-10380-4_1

separated from debris disks, the former containing more dust and still accreting onto the central young star, the latter containing less dust as well as planets and planetesimals. Different types of protoplanetary disks may exist depending on the environment in the star-forming cluster.[1]

Hubble Space Telescope observations first proved the existence of protoplanetary objects in the form of proplyds (Fig. 1.1). In 1992 the astronomer C.R. O'Dell and his colleagues at Rice University were making observations to study the fine structure of the Orion nebula when they unexpectedly found the circumstellar disks, previously thought to be stars, but now rendered visible to Hubble by being near a hot ionized cloud of hydrogen known as an H II emission nebula (S 24). "We knew that such circumstellar clouds are a necessary part of the formation of new stars," O'Dell recalled. "We also knew that the Orion Nebula Cluster was very

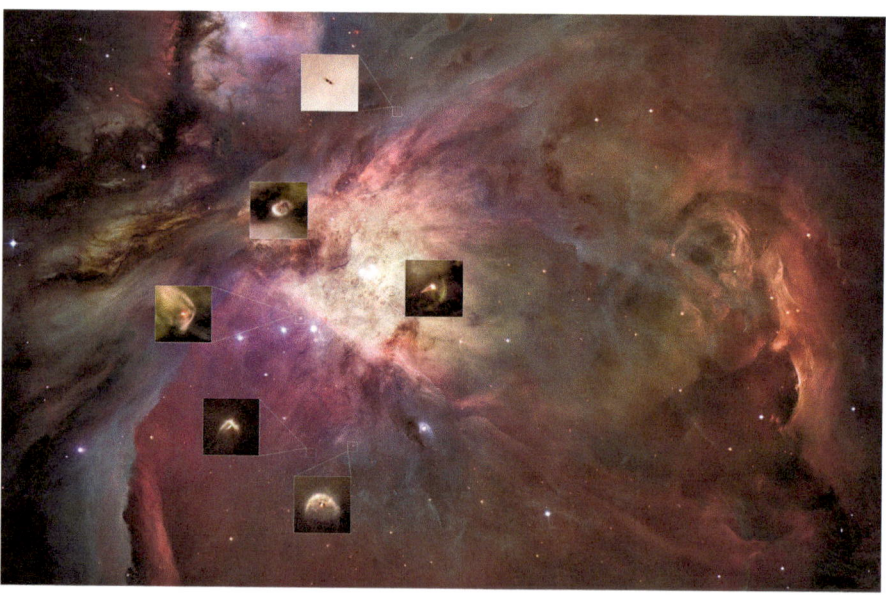

Fig. 1.1. Protoplanetary disks in the Orion Nebula, captured by the Hubble Space Telescope, are believed to be fledgling planetary systems. Only six of many are highlighted here. Credit: NASA/ESA/Hubble

---

[1] Two early reviews of this subject with history are Aki Roberge and Inga Kamp, "Protoplanetary and Debris Disks," in Sara Seager, ed., *Exoplanets* (Tucson: Univ. Arizona Press, 2010), and Jonathan P. Williams and Lucas Cieza, "Protoplanetary Disks and their Evolution," ARAA, 49 (2011), 67–117. A more recent review is Quentin Kral et al., "Circumstellar Discs: What Will be Next?," in Hans Deeg and Juan Antonio Belmonte, eds., *Handbook of Exoplanets* (New York: Springer, 2017).

young. What we had not expected was that these circumstellar clouds would be rendered easily visible because of the Orion Nebula. With all the wisdom of hindsight, we now understand that this must be the case, and we should have been planning special observations just to detect the proplyds." Instead, this was "a good example of serendipity in science."[2]

O'Dell coined the term "proplyd" to describe these extended disks of dust around 15 newly formed stars observed with the Hubble Space Telescope in Orion. The name (from PROtoPLanetarY DiskS) was created following the suggestion of O'Dell's wife, Gail Sabanosh, who joked that for him the phrase "protoplanetary disks" was too much of a tongue twister. The Hubble press release, which touted the objects as "the strongest evidence yet that many stars form planetary systems," quoted O'Dell as saying "these young disks signify an entirely new class of object uncovered in the universe."[3]

Not everyone agreed that these were indeed protoplanetary disks. But by 1994, O'Dell and his colleagues had found 56 proplyds around 110 stars in Orion and demonstrated that they are likely pancake-shaped disks of dust rather than shells of dust—further evidence of their protoplanetary nature. By 1996 they announced "145 compact sources that can be classified as proplyds."[4] Depending on the perspective and the illumination conditions, the circumstellar material is sometimes seen as a bright-rimmed object, and other times as a dark object silhouetted against a bright background. An Atlas of protoplanetary disks in the Orion Nebula published in 2008 contained 178 proplyds (defined as "externally ionized protoplanetary disks) and 28 "silhouette disks," defined

---

[2] C. R. O'Dell, *The Orion Nebula: Where Stars are Born* (Cambridge, MA: Harvard University Press, 2003), pp. 131–132. The discovery paper is C. R. O'Dell, Zheng Wen and X. Hu, "Discovery of New Objects in the Orion Nebula on HST Images – Shocks, Compact Sources, and Protoplanetary Disks," ApJ, 410 (1993), 696–700.

[3] HST Release, "NASA's Hubble Space Telescope Discovers Protoplanetary Disks Around Newly Formed Stars," Dec. 16, 1992, http://hubblesite.org/newscenter/archive/releases/1992/29/text/

[4] HST Release, "Hubble Confirms Abundance of Protoplanetary Disks around Newborn Stars," June 13, 1994, http://hubblesite.org/newscenter/archive/releases/1994/24/text/. C. R. O'Dell and Zheng Wen, "Postrefurbishment Mission Hubble Space Telescope Images of the Core of the Orion Nebula: Proplyds, Herbig-Haro Objects, and Measurements of a Circumstellar Disk," ApJ, 436 (1994), 194–202; O'Dell, C. R., & Wong, S. K., "Hubble Space Telescope Mapping of the Orion Nebula. I. A Survey of Stars and Compact Objects," AJ, 111 (1996), 846–855.

as "disks seen only in absorption against the bright nebular background." Several other types, also distinguished by physical appearance, are evidence of a new field still in great flux.[5]

The Orion star-forming region was not alone. In 2003, astronomers Nathan Smith, John Bally, and colleagues announced the discovery of numerous "candidate" proplyds in the Carina nebula, an H II region like Orion, some 7,300 light years distant. Their discovery comes as a surprise because Carina is powered by much hotter and more massive stars, including eta Carina, one of the most massive stars known. It was thought such stars would have blown away the proplyds, or at least given them much shorter lifetimes. The Carina proplyds were considerably large than those in Orion and were labeled "candidate proplyds" because of the possibility that they were remnants of the molecular cloud and not protoplanetary in nature, in which case they would be interesting examples of ongoing star formation. Since their initial discovery, protoplanetary disks have also been imaged in the Carina Nebula by the Atacama Large Millimeter/submillimeter Array (ALMA).[6] The observations were made using the 4-meter telescope of the Cerro Tololo Inter-American Observatory (CTIO) in Chile.

Proplyds observed thus far in the Orion Nebula vary from 100 to 1000 astronomical units across; by comparison, our Solar System out to Pluto is 40 astronomical units. Those in Carina are up to ten times larger—100 times the diameter of the Solar System. They are associated with the stellar jets and Herbig-Haro Objects (S 20 and S 21) that come with starbirth. Spectra taken with the Spitzer Space Telescope have indicated the first building blocks of planets and life embedded in protoplanetary disks, including silicates and water, methanol, and carbon dioxide ice.

The next great step in imaging protoplanetary disks came with the construction of ALMA at an altitude of 16,500 feet in the Chilean Atacama desert. This amazingly innovative instrument

---

[5] L. Ricci, M. Robberto and D. R. Soderblom, "The Hubble Space Telescope/ Advanced Camera for Surveys Atlas of Protoplanetary Disks in the Great Orion Nebula," AJ, 136 (2008), 2136–2151.

[6] NOAO Release, "Substantial Population of Stellar Cocoons Found in Surprisingly Harsh Environment," January 8, 2003, http://www.noao.edu/ outreach/press/pr03/pr0301.html. Nathan Smith , John Bally , and Jon A. Morse, "Numerous Proplyd Candidates in the Harsh Environment of the Carina Nebula," ApJ Letters, 587 (2003), L105–L108. The ALMA imagery is reported in A. Mesa-Delgado et al., "Protoplanetary Disks in the Hostile Environment of Carina," ApJ Letters, 825 (2016), L 16. A few isolated proplyd-like objects have also been seen elsewhere, as in the Lagoon and Trifid Nebulae and in NGC 3603.

Fig. 1.2. Protoplanetary disk surrounding the young star HL Tau. Credit: ALMA (ESO/NAOJ/NRAO)

began imaging these disks even in the midst of its testing and verification process in 2014. The 66 high-precision dish antennas of ALMA are ideal for directly imaging the formation of planets at millimeter and submillimeter wavelengths. The image released in 2014 shows a young star known as HL Tau (Fig. 1.2) surrounded by a disk comprised of multiple rings and gaps as the emerging planets sweep their orbits clear of dust and gas—a striking confirmation of the nebular hypothesis. The system, located about 450 light years from Earth, is less than one million years old. ALMA has since imaged many such systems of various ages, including the Orion proplyds and in 2018 the object known as AS 209, located 410 light years away in the Ophiuchus star forming region. In the same year, astronomers using ALMA detected three Jupiter-mass planets still forming—actual protoplanets rather than fully formed ones. They did so by detecting unusual patterns in the flow of carbon monoxide gas in the protoplanetary disk surrounding HD 163296. That system is about 4 million years old and is located about 330 light years from Earth in the direction of the constellation Sagittarius.[7]

---

[7] ALMA release, "ALMA Discovers Trio of Infant Planets around Newborn Star," June 12, 2018; http://www.almaobservatory.org/en/press-release/alma-discovers-trio-of-infant-planets-around-newborn-star/. A recent paper that gives some of the history and science of protoplanet detection is C. Pinte et al., "Kinematic evidence for an embedded protoplanet in a circumstellar disc," http://www.eso.org/public/archives/releases/sciencepapers/eso1818/eso1818a.pdf

Other ground-based telescopes have also captured protoplanetary disks, including the SPHERE instrument at optical wavelengths on the Very Large Telescope in Chile's Atacama desert. The project released images of three planet-forming disks in 2016. The instrument, an acronym for Spectro-Polarimetric High-contrast Exoplanet Research, uses a coronagraph to block light from the young stars to detect the surrounding disks. In another first, in 2018 astronomers using SPHERE directly imaged a planet still forming around its star. The parent star was a five to ten-million-year-old T Tauri or dwarf star known as PDS 70, 370 light years from Earth in the constellation Centaurus. The planet, named PDS 70b following the usual nomenclature, has an upper mass limit of 5–14 Jupiter masses, a temperature around 1,000° C, and is at a distance of 20 astronomical units from its parent star, giving it an orbital period of around 120 years.[8]

The discovery of circumstellar disks in the 1980s and proplyds in the 1990s still begged the question of whether planets would eventually form to comprise planetary systems (P 18). The answer was not long in coming. The first unambiguous planets around Sun-like stars were discovered beginning in 1995, and thousands are now known. ALMA and other telescopes are now advancing our knowledge of the stages before these planets were formed. Surprisingly, studies of interplanetary dust (P 15) are also adding to our knowledge of the formation of planetary systems.

More images and information on protoplanetary disks is available at the Hubble Space Telescope, Spitzer Telescope, and ALMA sites. For a Hubble Atlas of 30 protoplanetary disks in Orion see https://www.spacetelescope.org/news/heic0917/; for Spitzer http://spitzer.caltech.edu/search/image_set/20?search=protoplan etary&x=11&y=10; and for ALMA http://www.almaobservatory. org/en/home/. A European Southern Observatory video zooming in on the planet forming around PDS 70 can be found at: https:// www.youtube.com/watch?v=y8nRsiAK92Y

---

[8] On PDS 70b see ALMA release, "First confirmed image of newborn planet caught with ESO's VLT, June 2, 2018, https://www.eurekalert.org/pub_ releases/2018-07/e-fci062918.php

2. The Planet Family

Class P 2: Terrestrial (Rocky)

Once formed, planets come in a variety of classes that astronomers have distinguished over a period of several centuries. We are partial to the class of terrestrial planets because one of them is our home. In general, a terrestrial planet is one whose constituents are rocky and therefore similar to the Earth in composition, by comparison with the gas giant or ice giant planets (P 3 and P 4). By anyone's definition, Mercury, Venus, Earth, and Mars are classified as the terrestrial planets in our Solar System. It is notable that these are the four inner planets of the system, certainly a clue to their formation mechanism.

Following the protoplanetary stage (P 1), terrestrial planets are believed to form by accretion of particles, which eventually produce 10-kilometer "planetesimals," the building blocks of planets. Because solar heating vaporizes ices and prevents lighter elements like hydrogen and helium from condensing near the Sun, mostly rocky material was left for the terrestrial planets, while further from the Sun ices and the lighter elements prevailed and were captured by the cores of the gas giants and ice giants. Pluto likely formed by a different mechanism, and for this and other reasons, in 2006 the International Astronomical Union famously downgraded it to a "dwarf planet" (P 9), which in turn was—inexplicably to many—ruled not to be a class of planet at all. The classification issues here are far more than academic, driven by the discovery of objects in the outer Solar System either larger than or approaching the size of Pluto. That is a problem of definition at the Family level; here we consider the problem at the class level, i.e. distinguishing different kinds of planets.

Although planets (Greek for wanderers) had been distinguished from stars by their motions since antiquity, it was not until Copernicus's heliocentric theory that the Earth became a planet, and the planets potential Earths. Just how much the planets are like the Earth is a problem that has occupied astronomers ever since. The essential realization that some planets were more

© Springer Nature Switzerland AG 2019
S. J. Dick, *Classifying the Cosmos*, Astronomers' Universe,
https://doi.org/10.1007/978-3-030-10380-4_2

13

Earth-like than others was made only gradually between the 17th and 19th centuries with advances in telescopic observation, and the separation of the class of gas giants came only in the late 19th century (see P 4).[1]

It is notable that long before much thought had been given to classification systems in astronomy, planets were distinguished by their physical composition. For terrestrial planets the essential parameter was their rockiness, or to put it more scientifically, the presence of silicates. Although that is the essential characteristic of terrestrial planets, it was gradually realized that they differ widely in other ways. A brief survey of the four terrestrial planets of our Solar System highlights the similarities and differences in this class of objects, and how much the physical characteristics of an object may vary and still be considered to belong to the same class.

Mercury, the innermost planet of our Solar System, and since the downgrading of Pluto also the smallest one, held its secrets the longest. Precisely because it is so close to the Sun, telescopic observations were difficult, and only with the flyby mission of the Mariner 10 spacecraft in 1974 were its secrets revealed. Heavily cratered, almost totally lacking an atmosphere, and with temperatures ranging from 800° F in direct sunlight to -300° F at the bottom of polar craters, it is inhospitable to any indigenous life or future human spaceflight. Its large, partially molten iron core, 80% of the planet's radius, generates a magnetic field about 1% the strength of the Earth's.

The surface of Mercury exhibits a wide range of crater sizes and some 15 impact basins—testimony to a violent past. Beginning in 2008, the MESSENGER spacecraft began to study the planet in more detail, with three flybys and an orbital mission from 2011 to 2015. With camera resolution almost 100 times greater than Mariner 10, it resolved features down to 59 feet. MESSENGER found evidence of past volcanic activity, small and geologically young fault scarps indicating the planet might still be geologically active, and confirmed an extensive metal-rich planetary core. Inexplicably, it also found Mercury's crust is 5% sulfur by mass, 100 times its abundance on Earth. And it confirmed an almost unbelievable suspicion dating back to ground-based radar observations in 1991: that some of the permanently shadowed deep impact craters at the poles contain water ice tens of meters thick.

---

[1] For more context see Dick (2013), pp. 53–57.

Mercury is not only the smallest known terrestrial planet but also the densest and exhibits the oldest surface.

In 2018, the joint European-Japanese BepiColombo mission began a seven-year journey to Mercury to continue studying this enigmatic example of a terrestrial planet. The mission will deliver two spacecraft to polar orbit around Mercury.[2]

Because of its proximity and similarity in size to Earth, Venus (Fig. 2.1) was long known as the Earth's "sister planet." Spacecraft observations, however, radically changed that notion in the 1960s.

Fig. 2.1. Venus revealed. The different colors represent different elevations in terms of the planet's radius, ranging from 6,048 km to 6,062 km, a difference of 14 km over the entire surface. Blue and green represent lower topographies, brown and green higher ones. This hemispheric view, as revealed by more than a decade of radar investigations culminating in the 1990–1994 Magellan mission, is centered at 0° east longitude. The effective resolution of this image is about 3 km. Credit: NASA, JPL, USGS

---

[2] Emily Lakdawalla, "Return to the Iron Planet," S&T, 136 (November, 2018), 22–28.

The planet is now known to be surrounded by a dense carbon dioxide atmosphere, causing a greenhouse effect with temperatures rising to 900° F, hotter than Mercury. The atmospheric pressure at the planet's surface is 92 times that at the Earth's surface. Moreover, Venus has clouds of sulfuric acid and sulfuric acid rain in its upper atmosphere.

The surface of Venus remained largely a mystery until in the early 1990s, when the Magellan spacecraft radar revealed evidence of volcanism in the form hundreds of volcanoes and smooth plains thought to originate from volcanic flow. About 70% of the planet's surface is covered with these lava plains. Evidence from the European Space Agency's Venus Express spacecraft indicates some volcanoes may still be active. Magellan also revealed mountains, plains, canyons, and about 1,000 impact craters—much fewer than on either Mercury or Mars. Most of the craters are relatively pristine, indicating a global resurfacing event about a half-billion years ago. Unlike Earth, Venus has no active plate tectonics and no internally generated magnetic field.

Venus illustrates how different a planet can be and still be classified as terrestrial. But because of its similar size and density, Venus's internal structure is believed to be similar to Earth's, with a core, mantle, and crust. There is however no evidence for plate tectonics. Although Europe and Japan have had successful orbital missions, it may be some time before NASA spacecraft return to the planet to answer some of the mysteries posed by them and Magellan. To planetary astronomers this is a shame, since more observations might reveal much about how the climate diverged so far from that of Earth.[3]

Earth, the planet we all know and love, is 71% covered with water—its most distinguishing characteristic compared to the other planets (Fig. 2.2). Its atmosphere is 78% oxygen and 21% nitrogen, with other trace elements. Unlike the other terrestrial planets, Earth's surface has been constantly modified by weather, so that any impact craters it once possessed have been largely obliterated. Mountains, deserts, plains, and a variety of other landforms dominate the 29% of the surface not covered by water. The internal structure of the Earth consists of a thin silicate crust some 35 kilometers in diameter, a viscous solid mantle, and an inner and outer core. Unlike Venus, Earth possesses a strong magnetic field caused by its iron core, protecting the surface from solar wind (P 16) and other particles trapped in its radiation belts (P 8).

---

[3] Shannon Hall, "Destination Venus?," S&T, 136 (Sept. 2018), 14–21.

Fig. 2.2. Earth is the prototype terrestrial planet, but unlike the other planets of this class in our Solar System, its surface is 71% water. Credit: NASA/ NOAA/GSFC/Suomi NPP/VIIRS/Norman Kuring

Beyond Earth is Mars, for centuries an object of fascination particularly because of its possibility of hosting life. Unlike Earth, Mars harbors no oceans, has a very thin atmosphere averaging less than 1/100th of the Earth's (6 millibars), and is geologically dead. Spacecraft imaging of Mars, beginning with Mariner IV in 1965 and continuing through a variety of orbiters and landers, has revealed a variety of landforms. Its northern plains were shaped by volcanic flow, in stark contrast to its southern highlands, and the planet exhibits more than 40,000 craters. Mars is also home to extinct volcanoes, including the enormous Olympus Mons, three times the height of Mt. Everest and the largest in the Solar System. Vallis Marineris, its largest canyon, is almost ten times the length of the Earth's Grand Canyon.

One of the planet's most striking features is its polar ice caps, composed of water and carbon dioxide ice. While Mars at present has no magnetic field, paleomagnetic studies of rocks indicate it had one in the past. Although water cannot exist in liquid form at current Martian atmospheric pressure, observations of dry river channels and other features show that Mars once harbored large quantities of water, and gullies and other observations indicate water may still be present in large quantities below the Martian surface. In fact, in 2018 astronomers using the Mars Express spacecraft announced the discovery of a 12-mile-wide body of briny water beneath the South polar cap, kept in liquid form only by the salty brine.[4] Such discoveries drive the search for extraterrestrial life, which has been carried out by spacecraft ranging from the Viking Landers in 1976 to the Phoenix and Mars Exploration Rovers, Curiosity, MAVEN, and the Mars Reconnaissance Orbiter.

Thus, although all terrestrial planets are rocky, they vary greatly in other features. Some (Mercury) have practically no atmosphere, while others (Venus) exceed Earth's atmospheric pressure by almost a hundred times. Some have large planetary cores and active magnetic fields (Mercury and Earth), others (Venus and Mars) have no magnetic field at all. Some are geologically active (Venus and Earth), others geologically dead (Mars and Mercury). Temperatures range from 900° F to -300° F. All of these features of terrestrial planets are in stark contrast to the other three classes of planets. Terrestrial planet densities, which range from 3.9 grams/cc for Mars to 5.4 grams/cc for Mercury, are in stark contrast to the gas and ice giants, which range from .7 grams/cc for Saturn to 1.64 grams/cc for Neptune. Pulsar planets likely have much higher densities. All of these factors need to be taken into consideration in the classification of exoplanets now being found by the thousands.

Types of terrestrial planets have not yet been formally distinguished because so few are known, but searches to reveal terrestrial planets beyond the Solar System will likely change that. Numerous extrasolar gas giant planets have been discovered over the last two decades, but the first extrasolar planet confirmed as Earth-like in composition was announced in early 2009.[5] It was given the unwieldy name CoRoT-7b because it is the first planet discovered around the star CoRoT 7, which is in turn named after the French-lead CoRoT spacecraft (Convection, Rotation, and planetary Transits) active between 2006 and 2012. CoRoT 7b is a rocky body with the density of Earth, 1.7 times Earth's diameter, and

---

[4] R. Orosei et al., "Radar evidence of subglacial liquid water on Mars," *Science*, 361 (2018), 490–493.

[5] A. Leger et al., "Transiting exoplanets from the CoRoT space mission. VIII. CoRoT-7b: the first super-Earth with measured radius," A&A, 506 (2009), 287–302; D. Queloz et al., "The CoRoT-7 planetary system: two orbiting super-Earths," *A&A*, 506 (2009), 303–319.

five times its mass. But it has an estimated surface temperature of 3300° F to 4700° F, giving it the nickname of "lava planet" and "planet from Hell." The planet was discovered photometrically by the transit method, the diminution of light crossed in front of its 12th magnitude yellow G9 or K0 dwarf star, located 400 light years away in the constellation Monoceros. The planet circles its Sun in only 20.5 hours, so close that the star would be 30 degrees wide in the planet's sky. Since 2009 many more terrestrial planets have been discovered, including Kepler 10-b with a diameter about 1.6 times that of Earth. It is estimated that one in five Sun-like stars has an Earth-sized planet in their habitable zones. These discoveries may well allow distinctions of Types of terrestrial planets, one taxon down in the classification system.

Some have suggested planets intermediate in mass between Earth and Jupiter or Neptune be called "super-Earths," hinting at a new class of planets based on size or mass. As MIT astronomer Sara Seager wrote already in 2010, "Super-Earths are unofficially defined as planets with masses between about 1 and 10 Earth masses. The term is largely reserved for planets that are rocky in nature rather than for planets that have icy interiors or significant gas envelopes." Two years later, Harvard astronomer and theorist Dimitar Sasselov largely agreed, arguing that super-Earths are "a crucial class of planets" composed mostly of rock but perhaps some ice.[6] Size and mass alone, however, would not be a good way to distinguish a new class based on criteria of physical constitution. The bottom line is that it is not yet known whether a new class could be proclaimed based on the composition of newly discovered exoplanets, or whether these will eventually be seen as Types of terrestrial planets. Adding to the confusion, super-Earths are also referred to as mini-Neptunes or (for those of higher mass) as gas dwarfs. Whether a new class or not, by this definition hundreds of super-Earths or super-Earth candidates have been discovered beginning with Gliese 876 d in 2005 and Gliese 581c in 2007. The latter may be in the habitable zone of its parent star Gliese 581, located only 20 light years from Earth. And in 2017 astronomers detected an atmosphere around super-Earth 55 Cancri e, eight times the Earth's mass and almost twice its radius. The scorchingly hot planet orbits very close to its parent star, and its retention of an atmosphere is a mystery.

---

[6] Sara Seager, "The Hunt for Super-Earths," S/T (Oct 2010), 30–36; Dimitar Sasselov, *The Life of Super-Earths* (New York: Basic Books, 2012), chapter 5.

Astronomers have also suggested a hypothetical class of planet between rocky and gaseous planets, known as "ocean planets" with a density less than rocky planets.[7] Ocean planets in fiction include C. S. Lewis's Perelandra (Venus), Stanislaw Lem's Solaris (a sentient ocean) and Kamino in episode II of Star Wars. Whether an ocean planet should be considered a new Class or a Type of terrestrial planet will depend on actual observation. The delineation of the terrestrial Class of planets into types remains for the future, when a larger sample of terrestrial exoplanets have been discovered by observatories such as NASA's TESS (Transiting Exoplanet Survey Satellite) and the European CHEOPS (Characterising Exoplanets Satellite), both launched in 2018. With TESS observing 85% of the sky out to 300 light years—an estimated two million stars (primarily main sequence red dwarfs, part of class S 4)—astronomers hope to discover at least 50 rocky exoplanets.

NASA's Planetary Data System at https://pds.nasa.gov is an immense archive of data from its planetary missions, including imagery.[8] Supplementary information and images from the MESSENGER mission to Mercury are available at https://www.nasa.gov/mission_pages/messenger/main/index.html. Examples of Venus missions can be found at https://www2.jpl.nasa.gov/magellan/, https://nssdc.gsfc.nasa.gov/planetary/magellan.html, and http://sci.esa.int/venus-express/. NASA's Earth Science programs are described at https://science.nasa.gov/earth-science/earth-science-data. The latest news from missions to Mars is available at https://solarsystem.nasa.gov/planets/mars/overview/ and at individual mission websites.

---

[7] A. Leger et al., "A New Family of Planets? Ocean Planets," *Icarus*, 169 (2004), 499–504.

[8] NASA's Planetary Data System is so massive it can be intimidating. Instructions for navigating the system can be found at its website. Another good entrée is Emily Lakdawalla, "Dig into NASA's Planetary Data System," S&T (May 2018), 28–35.

Class P 3: Gas Giant

It has been said that Earth and the other small bodies of the Solar System are the leftovers remaining after the formation of the giant planets. Though we now take them for granted, the existence of giant planets composed largely of gas was one of the great surprises in the history of astronomy. We now know that a gas giant planet is one whose main constituents are hydrogen and helium, and whose gaseous/liquid component is larger than the core, which may be rocky or metallic. In the deeper interior above the core, the majority of the hydrogen and helium exists in liquid or metallic form due to the effects of high pressure. This is followed by a layer of molecular hydrogen, with ammonia and water clouds at the surface. Jupiter and Saturn are the two gas giants in our Solar System by this definition, which not only distinguishes gas giants from terrestrial planets (P 2), but also from ice giants (P 4). Jupiter has long been called the "giant planet" because of its mass (318 Earth masses) and size (1,300 Earths would fit inside), and it gradually became the prototype for the outer "Jovian" planets of our Solar System. Only with spacecraft observations did astronomers realize at a more refined level of analysis that Uranus and Neptune may be different enough from Jupiter and Saturn in structure and composition to constitute a possible new class of "ice giants." But even today not everyone accepts ice giants as a separate class of planet.

The seemingly simple question of when Jupiter came to be considered a class separate from the terrestrial planets has no simple answer. The title "gas giant" implies a much more sophisticated understanding than the simple title "giant planet," because size is much easier to determine than structure and composition. While Jupiter was known to be a giant planet for centuries, the realization that it was composed primarily of gas came about only gradually.[9] To be sure, Jupiter's density was known with accuracy by the mid-19th century, when Bessel calculated it had 388 times the mass of the Earth but only 1.35 times the density of water. This was one quarter of the Earth's density, and enough to hint that it was quite different than terrestrial planets. But it was an average density and thus yielded no guarantee of the structure of the planet. The differential rotation of Jupiter's atmospheric features

---

[9] Thomas Hockey, *Galileo's Planet* (Institute of Physics Publishing: Bristol and Philadelphia, 1999), pp. 173 ff. For more see Dick (2013), p. 54, from which this paragraph is drawn.

could be measured at a remarkably rapid ten hours—the fastest in the Solar System. But these might be only atmospheric features covering a solid surface, masking what by analogy with the terrestrial planets would be its underlying rocky composition. Lacking sufficient knowledge, astronomers sometimes assumed to a first order approximation that Jupiter's composition was similar to the Sun, since the Sun and planets arose out of the same protosolar nebula. Indeed, Jupiter was most often compared to the Sun rather than the other planets in terms of composition. But the Sun itself was not believed to be totally gaseous either at this time.

According to historian Thomas Hockey, only after the mid-19th century did research on the planet turn from measurement of physical parameters such as its position, diameter, and rotation rate, to more qualitative descriptions of its form and substance, descriptions that would eventually lead to knowledge of its true nature.[10] Only then could astronomers apply the new arsenal of spectroscopy, thermodynamics, and kinetics to the mystery of Jupiter's structure and composition. Surprisingly, the same held true of the Sun, which until 1859 was believed to be a solid object with a luminous atmosphere. By the 1860s, laboratory evidence and physical theory indicated that even the Sun's interior could be gaseous.[11] By the 1870s, some astronomers and popularizers such as R.A. Proctor were concluding the Jupiter was far from Earth-like, though they still believed there was a "surface" somewhere below the clouds. The question was how far below. By 1876 some astronomers were speculating the surface could be thousands of miles below the visible cloud features. As the British astronomer George H. Darwin (second son of Charles Darwin) wrote, "That planet must be very much denser in the centre than at the surface. Is it not possible that Jupiter may still be in a semi-nebulous condition, and may consist of dense central part with no well-defined bounding surface? Does not this view accord with the remarkable cloudy appearance of the disk, and the remarkable belts?"[12]

---

[10] Thomas Hockey, *Galileo's Planet*, pp. 64 ff.

[11] DeVorkin, "Stellar Evolution and the Origin of the Hertzsprung-Russell Diagram," *Astrophysics and 20th-Century Astronomy to 1950*, Owen Gingerich, ed., (Cambridge: Cambridge University Press, 1984), pp. 90–108: 90–91.

[12] George H. Darwin, "On internal densities of Planets, and on an oversight in the *Mécanique Céleste*," MNRAS, 37 (1876), 77–88: 83; Hockey, pp. 168–174. For more see Dick (2013), p. 55, from which this paragraph is drawn.

Building on the work of Harold Jeffreys in the early 1920s, in 1932 Rupert Wildt identified ammonia and methane in the spectrum of Jupiter, and he and others went on to note that its low density was consistent with a dominant composition of hydrogen similar to the Sun. Even then, Jeffreys suggested in 1934 that Jupiter had a rock core, and Wildt calculated Jupiter's inner core comprised 50% of the diameter of the planet, with an atmosphere perhaps 10,000 km deep.[13] By the 1940s observations of the masses, radii, and rotation periods of the giant planets, together with new knowledge of the high-pressure behavior of matter, constrained the composition of Jupiter and Saturn to be predominantly hydrogen. Hydrogen however is difficult to detect at planetary temperatures, and it was not actually detected until 1960 by Carl Kiess and his colleagues. The presence of helium awaited the Pioneer 10 encounter in 1973.[14] Even in 1940, astronomers talked in terms of the atmospheres of the giant planets, arguing only that "the giant planets Jupiter, Saturn, Uranus and Neptune possess atmospheres of considerable extent and density." It is revealing that the term "gas giant" appears to have first been used by science fiction writer James Blish in his story "Solar Plexus," as anthologized in 1952.[15] It was not present in the original 1941 version of the story. Thus, in Blish's mind, undoubtedly a reflection of the new scientific realities, the outer planets as gas giants became a common concept at mid-20th century, even if their exact composition was still unknown.

Over the first 60 years of the Space Age, the Pioneer, Voyager, Galileo, Cassini-Huygens, and Juno spacecraft have revealed astonishing facts about Jupiter and Saturn, and just how much they differ from the class of terrestrial planets. Jupiter (Fig. 2.3) is 11 times the radius of the Earth and has an upper atmosphere composed of 90% hydrogen and 10% helium by volume

---

[13] Mark Marley, *Encyclopedia of the Solar System* (Academic Press, 1999), p. 340; David Leveringon, *A History of Astronomy from 1890 to the Present* (Springer: London, 1995), p. 65.

[14] C. C. Kiess, C. H. Corliss and H. K. Kiess, "High-Dispersion Spectra of Jupiter," Ap. J, 132 (1960), 221; Fran Bagenal, Timothy Dowling and William McKinnon, *Jupiter: The Planet, Satellites and Magnetosphere* (Cambridge: Cambridge University Press, 2004), p. 59; Mark S. Marley, "Interiors of the Giant Planets," in Paul R. Weissman, Lucy-Ann McFadden and Torrence V. Johnson, *Encyclopedia of the Solar System* (Academic Press: San Diego, 1999), pp. 339–355: 340; Dick (2013), p. 56.

[15] "Solar Plexus," in *Beyond Human Ken*, Judith Merril, ed. (1952); R. M. Petrie, "The Atmospheres of the Planets," JRASC, 34 (1940), 137–145:143; Dick (2013), p. 56.

Fig. 2.3. This full-disc image of Jupiter was taken on April 21, 2014 with Hubble's Wide Field Camera 3 (WFC3). Credit: NASA, ESA, and A. Simon (Goddard Space Flight Center)

Fig. 2.4. Saturn's northern hemisphere seen from Cassini in 2016, from a distance of 1.9 million miles (3 million km). The giant hexagonal jet stream is seen at top. Credit: NASA/JPL-Caltech/Space Science Institute

(75 and 24% respectively by mass), with traces of ammonia, methane, and even water. This is indeed similar to the composition of the Sun. The composition of Saturn (Fig. 2.4) is similar,

but with 96% hydrogen and only 3% helium. Jupiter's gaseous cloud layer extends downward about 1,000 km. Below that, as the gas becomes more and more compressed, a liquid and metallic hydrogen layer extends downward about 78% of the radius of the planet. Thus, most of Jupiter consists of "metallic hydrogen," a form of degenerate matter consisting of protons embedded in a sea of electrons. Below that, Jupiter was believed to have a solid core whose composition remained unknown, although even in the Space Age some still scientists thought it had no core at all. The Juno spacecraft, which entered Jupiter's orbit July 4, 2016 and passed 2,600 miles (4,200 km) above the equatorial cloud tops, found that Jupiter has a partially dissolved core much bigger than expected, out to perhaps half the radius of Jupiter, though this result is still open to interpretation. Saturn's interior structure is believed to be similar to Jupiter's.[16]

Prior to Juno, Jupiter's magnetic field was known to be stronger than Earth's. But Juno found Jupiter's maximum magnetic field to be 7.76 gauss, twice as strong as expected and ten times stronger than the strongest magnetic field at the Earth's surface (.66 gauss). This most intense planetary magnetic field in the Solar System is unevenly distributed, hinting at the nature of the dynamo below.[17] This magnetic field also gives rise to occasional bursts of radio emission, first detected by Kenneth Franklin and Bernard Burke in 1955. By contrast, Saturn's magnetic field is weaker than Earth's. Jupiter's magnetosphere—the "sphere of influence" of its magnetic field above the ionosphere, is so extensive that if it were visible from the Earth, it would be the size of the full Moon. The magnetospheres of Jupiter and Saturn give rise to the most powerful auroras in the Solar System. The two ice giants, Uranus and Neptune, exhibit them as well.

Both Jupiter and Saturn are known for their spectacular clouds, believed to be composed of ammonia crystals, and their colorful latitudinal bands, believed to be caused by winds. Beginning in 1979, the Voyager spacecraft found complex vortices

---

[16] S. J. Bolton et al., "Jupiter's Interior and Deep Atmosphere: The Initial Pole-to-pole passes with the Juno Spacecraft," *Science*, 356 (26 May, 2017), pp. 821–825. For a popular overview of discoveries related to Jupiter and Saturn see Francis Reddy, "Unveiling a Giant," *Astronomy*, 45 (2018), 20–27, and Liz Kruesi, "Close Encounters with the Ringed Planet," ibid., 28–33.

[17] J. E. P. Connerney et al., "Jupiter's magnetosphere and aurorae observed by the June spacecraft during its first polar orbits," *Science* 356 (2017), 826–832.

within these cloud bands. The changing colors are most likely due to upwelling materials. Both Saturn and Jupiter have ferocious storms; whereas 300 mph winds on Jupiter are not uncommon, Voyager clocked Saturnian winds at up to 1,200 miles per hour. Jupiter's Great Red Spot is a storm system larger than the Earth that has persisted spinning in its counterclockwise direction for more than three centuries, though it has shrunk by about one-third since the Voyager flybys of 1979. For the first time, the Juno spacecraft found that its depth was roughly 200 miles. Juno's images of both of Jupiter's poles showed a chaotic scene of Earth-sized cyclonic storms (Fig. 2.5). In contrast to Jupiter, both the

Fig. 2.5. Jupiter's south pole as seen from the Juno spacecraft from an altitude of 32,000 miles (52,000 km). The oval features in this composite image are cyclones up to 600 miles (1,000 km) in diameter. Credit: NASA/JPL-Caltech/SwRI/MSSS/Betsy Asher Hall/Gervasio Robles

Voyager and Cassini spacecraft imaged a hexagonal jet stream around Saturn's north pole, indicating polar dynamics quite different from that of Jupiter.

While Saturn has long been famous as the ringed planet, Jupiter (and the two ice giants as well) also has a less extensive system of rings (P 7). Nevertheless, Saturn remains "Lord of the Rings." Both planets have the most extensive system of satellites (P 6) in the Solar System.

Most of the extrasolar planets (P 18) found thus far are believed to be gas giants, though most are "hot Jupiters" orbiting very close to their parent star. On the basis of composition, these are likely to be gas giant or ice giant planets rather than new classes of planetary objects. However, one attempt to classify extrasolar planets distinguishes five classes of planets based on planetary mass, semimajor axis, eccentricity, stellar mass, and stellar metallicity.[18] Whether these are truly new classes of objects or are declared types of gas giants remains to be seen. Likewise, whether the "super-Earths" now being discovered by the Kepler spacecraft are new classes or terrestrial in nature is the subject of much current debate.

As with the terrestrial planets, NASA's Planetary Data System at https://pds.nasa.gov is an immense archive of data from its planetary missions. The Voyager missions are still being monitored at https://voyager.jpl.nasa.gov/mission/. After 13 years of spectacular exploration and a "Grand Finale" tour of Titan and Saturn's rings, the Cassini spacecraft dove into Saturn's atmosphere and disintegrated on September 15, 2017. Cassini-Huygens results are at https://saturn.jpl.nasa.gov. More Juno images of Jupiter can be seen at https://photojournal.jpl.nasa.gov/mission/Juno and https://www.theatlantic.com/photo/2018/01/gorgeous-images-of-the-planet-jupiter/550595/.

---

[18] S. Marchi, "Extrasolar Planet Taxonomy: A New Statistical Approach," ApJ, 666 (2007), 475–485.

Class P 4: Ice Giant

An ice giant is a planet whose outer constituents are made of elements heavier than hydrogen and helium—probably water, ammonia, and methane—and whose dense cores may dominate the gaseous component. By this definition Uranus and Neptune are the two ice giants in our Solar System, as distinguished from the gas giants Jupiter and Saturn (P 3). The ice giants are also distinguished by their smaller masses, 14.5 and 17 Earth masses for Uranus and Neptune respectively, compared to 318 and 95 Earth masses for Jupiter and Saturn. This class derives its name because water, ammonia, and methane are considered "ices" in the conditions under which they exist on Uranus and Neptune. Ice giants did not emerge as a new class of planet until the Voyager 2 spacecraft observations of 1986 and 1989 allowed close study of their structure and constituents. Even now, controversy exists as to whether they should be distinguished as a separate class of objects, with some astronomers still calling them gas giants or Jovian planets. As such, they raise the question of when and why a new class of object should be distinguished.

Unlike the gas giants, which were known to the ancients as naked-eye objects in the night sky, Uranus and Neptune had to be "discovered." The road leading to their true nature was long and tortuous. Uranus had been observed as a star-like object long before the German/British astronomer William Herschel claimed it as a planet in 1781. British Astronomer Royal John Flamsteed observed the object six times in 1690, and the French astronomer Pierre Lemonnier observed it twelve times between 1750 and 1769. Neither recognized it as a planet. On March 17, 1781, Herschel, in his private residence in Bath, England and using a telescope with a 6.5-inch mirror of his own construction, spotted what he described as a comet while undertaking observations for stellar parallax. Indeed, the article reporting the discovery is titled "Account of a Comet."[19] Based on its estimated distance, however, over the next several years several astronomers opined that it could be a planet. In 1783, Herschel himself wrote that "By the observation of the most eminent Astronomers in Europe it appears that the new star, which I had the honour of pointing out to them in March 1781,

---

[19] William Herschel, "Account of a Comet," *PTRSL*, 71 (1781); reprinted in part in Bartusiak, 130–131.

is a Primary Planet of our Solar System."[20] Although Herschel suggested it be named "Georgius Sidus" (George's star) after his patron George III, this seemed too nationalistic to some, and by 1850 even the British had officially accepted the name "Uranus" based on a suggestion of the German astronomer Johann Bode.

Like Uranus, Neptune had also been observed as a star before it was recognized as a planet. In fact, its first observation predates the first observations of what turned out to be Uranus, and was recorded by none other than Galileo Galilei himself, in two drawings on December 28, 1612 and January 27, 1613, made while the object was near Jupiter. It was not realized that Galileo had observed Neptune until 1980, when astronomer Charles T. Kowal and historian Stillman Drake made the discovery in Galileo's notebooks.[21] In 2009, Australian physicist David Jamieson claimed that Galileo had not only observed Neptune but had also detected its motion, an important step toward realizing it was a planet. Galileo, however, did not make that claim. It was only 234 years after Galileo's first accidental Neptunian observation, on September 13, 1846, that the German astronomer Johann Gottfried Galle, acting on predictions made by the French astronomer Urbain Jean Joseph Le Verrier, spotted Neptune and claimed it as a "new planet."[22]

The initial discovery of Uranus and Neptune as planets, however, hardly constituted the discovery of a new class of objects. To the contrary, it only ensured their place in the Family of planets, not yet even in the class of gas giants. By the mid-20th century, based on the gross similarity of their features, they had taken their place as gas giants, presumed to have similar structures as the gas giants. Their true physical nature as a possible separate class remained unknown until the Voyager 2 spacecraft observations provided data that constrained models of their interiors (Figs. 2.6 and 2.7).

A hydrogen-rich atmosphere is believed to extend from the observed cloud tops to 85% of Neptune's radius, 80% in the case of Uranus. Models show the density and pressures of the deep interior mimic laboratory experiments of an artificial icy mixture known

---

[20] *The Scientific Papers of Sir William Herschel* (London: Royal Society and Royal Astronomical Society, 1912), J. L. E. Dreyer, ed., pp. 100, reprinted in Bartusiak, 132–133. See also Dick (2013), pp. 43–44.

[21] Charles T. Kowal and Stillman Drake, "Galileo's Observations of Neptune," *Nature* 287 (1980), 311–313.

[22] J. G. Galle, "Account of the discovery of the planet of Le Verrier at Berlin," *MNRAS*, 7 (November 13, 1846), 153. See Dick (2013), pp. 49–51.

Fig. 2.6. This infrared composite image of the two hemispheres of Uranus obtained with Keck Telescope adaptive optics. Credit: Lawrence Sromovsky, University of Wisconsin-Madison/W.W. Keck Observatory

as "synthetic Uranus," and so is likely icy. The total amount of hydrogen and helium in Uranus and in Neptune is about two Earth masses, compared to 300 Earth masses for Jupiter. Although other models exist, in 1999 astronomer Mark Marley declared, "Given the relatively small amounts of gas compared to ices in Uranus and Neptune, these planets are aptly termed 'ice giants,' whereas Jupiter and Saturn are indeed 'gas giants.'" By 2006, Heidi Hammel declared, "Planetary scientists now appreciate that the planets Uranus and Neptune differ from Jupiter and Saturn in more than just size: their interiors differ in composition and phase; their zonal winds differ in magnitude and direction; and their magnetic fields differ in structure. Hence we refer to these mid-sized planets as 'Ice Giants' to distinguish them from their larger 'Gas Giant' cousins."[23] It is notable that in both cases, the declaration of a new

---

[23] Mark S. Marley, "Interiors of the Giant Planets," in Paul R. Weissman, Lucy-Ann McFadden and Torrence V. Johnson, *Encyclopedia of the Solar System* (Academic Press: San Diego, 1999), pp. 339–355: 352; also Heidi Hammel, "The Ice Giant Systems of Uranus and Neptune," in *Solar System Update*, Phillippe Blonde and John Mason, eds. (Springer, 2006), p. 251–265: 251. Dick (2013), pp. 56–57.

Fig. 2.7. Voyager 2 view of Neptune taken in 1989 at a range of 4.4 million miles from the planet, 4 days and 20 hours before closest approach. The picture shows the Great Dark Spot and its companion bright smudge. On the west limb the fast moving bright feature called Scooter and the little dark spot are visible. Credit NASA/JPL

class was made on the basis of physical characteristics, including composition. The solid core of Neptune is believed to be about one Earth mass, above which is a layer of water and other melted ices that constitute about 80% of the planet's mass.

Uranus and Neptune are sometimes considered failed gas giant planets, in the sense that (because of the lower density of the solar nebula at their distances compared to Jupiter and Saturn) they took longer to accrete their icy/rocky cores, and by the time they had accreted to the size where they could capture the nebular gases, those gases may have been blown away as part of the stellar evolutionary process.

NASA's gallery of Uranus and Neptune-related photos can be found at:

https://solarsystem.nasa.gov/planets/uranus/overview/
https://solarsystem.nasa.gov/planets/neptune/galleries

Class P 5: Pulsar Planet

Even more surprising than the existence of gas giants and ice giants was the discovery of planets around those extremely dense neutron stars (S 13) known as pulsars, one of the exotic evolutionary endpoints of high-mass stars. Indeed, the discovery was so astonishing that it pushed the boundaries of what it meant to be a planet. But once accepted as a planet, the status of these objects as a distinct class is richly justified. Their composition is likely very different from other types of planets, since they must be built from stellar rubble (iron, nitrogen, and other heavy elements) that survived a supernova explosion ending in a neutron star. In 1992, astronomers Aleksander Wolszczan and Dale Frail announced the first two pulsar planets orbiting the object with the unwieldly name PSR B1257+12.[24] (Pulsar numeration simply refers to the pulsar's coordinates, in this case right ascension 12 hours 57 minutes and declination 12 degrees above the celestial equator). The pulsar is located some 980 light years from the Sun in the constellation Virgo. Later a third planet was confirmed, and their masses were refined to .02, 4.3 and 3.9 Earth masses, with orbital periods of 25, 66, and 98 days respectively. This means all three planets would fit within the orbit of Mercury in our Solar System. A fourth planet with a mass of .0004 Earth masses and a period of 1,250 days has also been claimed.

The existence of pulsar planets was conjectured as far back as 1970, as was the idea that such planets would be observable as a change in the frequency of the rotation period of the pulsar. Unsubstantiated claims of such observed variation attributed to pulsar planets were also published about the same time.[25] The first confirmed discovery of pulsar planets followed a long period of speculation and spurious claims, including one by Matthew Bailes, Andrew Lyne, and colleagues in *Nature* in 1991, and a famous retraction at a meeting of the American Astronomical Society at

---

[24] A. Wolszczan, and D. A. Frail, "A Planetary System around the Millisecond Pulsar PSR 1257+12" 1992, *Nature*, 355, 145–147; A. Wolszczan, "Confirmation of Earth--Mass Planets Orbiting the Millisecond Pulsar PSR B1257+12" 1994, *Science*, 264, 538.

[25] F. Curtis Michel, "Pulsar Planetary Systems," ApJ, 159 (Jan, 1970), L25–L28; J. G. Hills, "Planetary Companions to Pulsars," *Nature*, 226 (1970), 730–31; M. J. Rees, V. L. Trimble, and J. M. Cohen, "Planet, Pulsar, 'Glitch' and Wisp," *Nature*, 229 (1971), 395–396. Dick (2013), p. 299.

which Lyne was applauded for revealing an honest mistake in the data analysis—the failure to remove the effect of the eccentricity in the Earth's orbit around the Sun. Lyne was part of the Jodrell Bank pulsar group and a Director of Jodrell Bank radio telescope Observatory.

The first confirmed pulsar planets were detected during a search for pulsars that began in 1990 with the giant 305-meter radio telescope at Arecibo, Puerto Rico. Since their discovery in 1967, pulsars were known to emit pulsing radio emission with such regularity that they constituted the most precise natural clocks in the universe. Their precise emission is tied to their rotation, an astounding 6.22 milliseconds in the case of the planet pulsar, which emits its radio-wave beams like a rotating lighthouse. If a pulsar has one or more orbiting planets, the timing will be off by tiny amounts, ranging from milliseconds for Earth-masses to microseconds for asteroid masses, as the pulsar is tugged one way and then another by the planets. Multiple planets can also be separated by this radio-timing technique because they are seen as higher order effects. The pulsar timing method is so powerful that it has been estimated that pulsar planet moons may one day be detected.[26]

Although this was believed to be the leading edge of numerous pulsar planets, in the decades since the discovery of the first pulsar planetary system, only one other pulsar planet has been confirmed. Located in the M4 globular cluster 12,400 light years away in the constellation Scorpius, the planet is about 2.5 Jupiter masses in a 100-year orbit around a binary system consisting of the 11-millisecond pulsar PSR B1620-26 and a companion white dwarf (S 12). The planet was discovered in 1993 and confirmed by the Hubble Space Telescope in 2003. In confirming the white dwarf companion and accepting the planetary companion previously discovered, the authors of the Hubble study concluded, "This implies that planets may be relatively common in low-metallicity globular clusters and that planet formation is more widespread and has happened earlier than previously believed."[27] This planet may be

---

[26] Dick (2013), 299–300.

[27] D. C. Backer et al., "A Second Companion of the Millisecond Pulsar 1620-26," *Nature*, 365 (1993), 817; S. E. Thorsett et al., "PSR B1620-26: A Binary Radio Pulsar with a Planetary Companion?," *ApJ Letters*, 412 (1993), L33; the Hubble confirmation of the white dwarf is S. Sigurdsson et al., "A Young White Dwarf Companion to Pulsar B1620-26: Evidence for Early Planet Formation," *Science*, 301 (2003), 193–196.

12.7 billion years old—the age of the cluster—and some astrono-mers have dubbed it "Methuselah" after the oldest living person in the Bible. Several other candidate planets have been claimed around pulsars, most notably PSR J1719-1438, but they have not yet been confirmed.

The burning question in the wake of these discoveries is how could planets form in such a harsh environment? In 2006, NASA's infrared Spitzer Space Telescope shed light on the formation of pulsar planets by detecting a disk surrounding a pulsar, believed to be a remnant of the supernova explosion that had not yet formed into planets. The pulsar, dubbed 4U 0142+61 and located 13,000 light years away in Cassiopeia, was once a large, bright star with a mass between 10 and 20 times that of our Sun. The star likely survived for about 10 million years before collapsing under its own weight about 100,000 years ago and blasting apart in a super-nova explosion. Any planets the star may have originally have had would have been incinerated, and the debris disk may form a second generation of planets such as discovered by Wolszczan in 1992. "We're amazed that the planet-formation process seems to be so universal," Dr. Deepto Chakrabarty, principal investigator of the new research, said at the announcement of the discovery. "Pulsars emit a tremendous amount of high energy radiation, yet within this harsh environment we have a disk that looks a lot like those around young stars where planets are formed."[28] The disk, which Spitzer detected from its infrared glow, orbits its parent pul-sar at a distance of about one million miles and contains about ten Earth masses.

Despite this early optimism, pulsar planets are now believed to be rare because they form only under very specific circum-stances, including the destruction of a companion star with a low mass, and the necessity of a "dead zone" region of low turbulence that is probably rare in disks around pulsars due to irradiation from the pulsar. Given these constraints, astronomers estimate that only 1% of neutron star progenitors even have the potential to form pulsar planets, and not all of them actually do. That pulsar planets were the first planets to be discovered, when they are now known to exist around almost all other stars, is one of the amazing

---

[28]Z. Wang, D. Chakrabarty et al., "A debris disk around an isolated young neutron star," *Nature*, 440 (2006), 772; the quotation is from Spitzer Press Release, April 5, 2006, "NASA's Spitzer Finds Hints of Planet Birth around Dead Star," http://spitzer.caltech.edu/news/235-ssc2006-10-NASA-s-Spitzer-Finds-Hints-of-Planet-Birth-Around-Dead-Star

facts of astronomical history. It is explainable only by the fantastically accurate timing techniques used in their discovery.[29]

An interesting planet classification question arose in 2007 when NASA's Swift and X-ray Timing Explorer (RXTE) satellites detected a planet-mass object around pulsar SWIFT J1756.9, located at a distance of 25,000 light years. The object's minimum mass was determined to be about seven times that of Jupiter, orbiting the pulsar every 54.7 minutes at a distance of only 230,000 miles, about the same as the Earth-Moon distance. But it was also determined that this object was the helium-rich core of a former star whose gas was accreting onto the pulsar. Astronomers conjecture that the original system billions of years ago consisted of a very massive star and a smaller star. The massive star exploded as a supernova, leaving a neutron star behind, which continues to pull material from the smaller star. Models suggest that the low-mass star is dominated by helium, and Christopher Deloye of Northwestern University noted that "Despite its extremely low mass, the companion isn't considered a planet because of its formation [mechanism]. It's essentially a white dwarf that has been whittled down to planetary mass."[30] This episode shows that for astronomers, a planetary mass orbiting a star is not enough to classify it as a planet—its history and formation also need to be taken into account.

Because pulsar planets are so distant, and because radio techniques are used for their discovery, no optical imagery exists. Artists, however, have imagined these planets and their surroundings. See for example http://spitzer.caltech.edu/images/1610-ssc2006-10b-Stellar-Rubble-May-Be-Planetary-Building-Blocks and http://spitzer.caltech.edu/images/1612-ssc2006-10c-Extreme-Planets.

---

[29] Rebecca G. Martin, Mario Livio and Divya Palaniswamy, "Why are Pulsar Planets Rare?", ApJ, 832:122 (2016), pp. 1–9.

[30] H. A. Krimm et al., "Discovery of the Accretion-powered Millisecond Pulsar SWIFT J1756.9-2508 with a Low-Mass Companion," ApJ, 668 (2007), 147–150; quotation from NASA Goddard Press release, "NASA Astronomers find Bizarre Planet-Mass Object Orbiting Neutron Star," September 12, 2007, http://www.nasa.gov/centers/goddard/news/topstory/2007/millisecond_pulsar.html

3. The Circumplanetary Family

Class P 6: Satellite

The term "natural satellite" most commonly refers to a body such as the Moon orbiting a planet, but it also applies to bodies orbiting minor planets (P 11), dwarf planets (P 9), and Kuiper belt objects (P 21). As of 2018, 187 satellites are known to orbit the planets of our Solar System, as well as over 300 that orbit minor planets and four that orbit dwarf planets. 12 new satellites of Jupiter were announced in 2018 alone, giving it a grand total of 79 (for now!). 58 satellites orbiting Kuiper belt objects have been discovered thus far, including the six largest of those distant denizens. Planetary satellites in our Solar System range in size from several thousand miles in diameter (including our Moon, the four Galilean moons of Jupiter, Saturn's Titan, and Neptune's Triton) to 10 miles for the two moons of Mars, and even smaller for some satellites embedded in planetary rings (P 7).

Just as planets are a normal byproduct of stellar evolution, so too are planetary satellites a normal byproduct of planetary formation. This makes it almost certain that satellites exist around the planets of other stars. The Hunt for Exomoons with the Kepler spacecraft collaboration has identified several candidates, including one in 2018 around the planet Kepler 1625b, a gas giant some 4,000 light years away in the constellation Cygnus. The exomoon candidate appears to also be a gas giant about the mass of Neptune, so perhaps it can be more appropriately classified a double planet. The close-up study of planetary satellites in our Solar System is a golden opportunity to understand what must exist in numerous other planetary systems.

Confronting the bewildering array of satellites in our Solar System, it is important to remember that, judged by size alone, not all satellites are created equal, and that only a few objects comprise by far the most mass of planetary satellites in our system. Nevertheless, there is more to satellites than size, and they have a very wide range of intriguing physical characteristics that reveal information about their formation history. In 1957, only 30

© Springer Nature Switzerland AG 2019
S. J. Dick, *Classifying the Cosmos*, Astronomers' Universe,
https://doi.org/10.1007/978-3-030-10380-4_3

planetary satellites were known; it was not until the Space Age (when the term "satellite" was also applied to artificial bodies) did spacecraft reveal the full panoply of natural satellites. Those same planetary missions also revealed for the first time the great physical variety of those worlds.

Because compositional differences were unknown until the 1980s, astronomers have most often classified planetary satellites by a combination of their shapes, sizes, and orbital characteristics, data that was easier to determine. In 1986, when astronomer Joseph Burns and Mildred Matthews issued what was at the time the definitive volume on satellites with 45 collaborating authors, Burns divided planetary satellites into three types based on their orbits and sizes. The first type was *regular satellites*, the classical major satellites that are large and spherical and whose orbits are prograde, nearly circular, close to the plane of their planets' equators, and close to their parent planet, likely formed at the time of planetary formation. The second type was *collisional shards*, much smaller, irregularly-shaped satellites often extremely close to the planet or co-orbital with a regular satellite, with almost zero eccentricity and inclination, possibly the remnants after the planet and larger satellites had formed, and mostly discovered since 1979 by the Voyager spacecraft. The third type was *irregular satellites*, characterized by highly inclined and sometimes retrograde orbits. They were believed to be small and probably captured objects usually located in the outer parts of their planet's gravitational field.[1]

Under this "gross classification scheme," as Burns put it, the *regular satellites* include the four Galilean satellites of Jupiter, the eight classical satellites of Saturn, and the five classically known satellites of Uranus (Fig. 3.1). The *collisional shards* include the two moons of Mars (Fig. 3.2); satellites of the gas giants sometimes embedded in ring systems such as Saturn's F-ring "shepherds" Janus and Epimetheus, or Amalthea, Adrastea, Thebe, and Metis in Jupiter's much thinner ring system; as well as a raft of the satellites of Uranus and Neptune. The *irregular satellites* of Jupiter fall into several clusters based largely on their orbital inclination

---

[1] Joseph A. Burns and Mildred Shapley Matthews, eds., "Some Background About Satellites," in *Satellites* (Tucson: University of Arizona Press, 1986), pp. 1–38, especially pp. 16–19 for satellite classification. This classification has been adopted by other astronomers, e.g., Nicolas Thomas, "Physical Processes Associated with Planetary Satellites," in I. P. William and N. Thomas, eds., *Solar and Extrasolar Planetary Systems* (Berlin and Heidelberg: Springer, 2001), pp. 173–190.

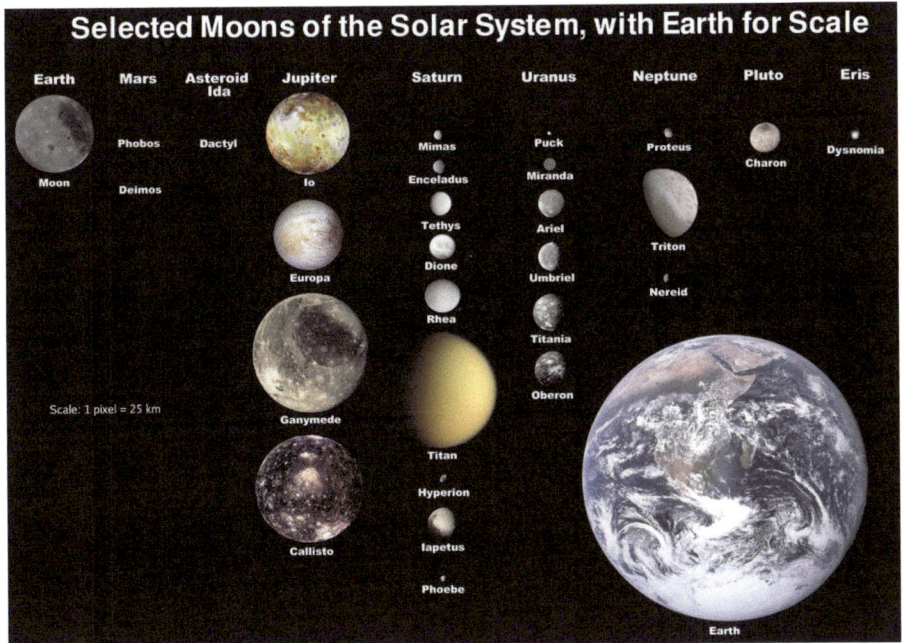

Fig. 3.1. Selected, mostly "regular" satellites of the Solar System, showing their diversity in size and physical nature. Credit: NASA

Fig. 3.2. The two moons of Mars are classified as collisional shards. This color image of Phobos was imaged by the Mars Reconnaissance Orbiter on March 23, 2008, from a distance of about 4,200 miles (6,800 km). The illuminated part of Phobos seen in the image is about 13 miles (21 km) across. The most prominent feature in the image is the large crater Stickney in the lower right. With a diameter of 5.6 miles (9 km), it is the largest feature on Phobos and was named after the wife of the moon's discoverer, Asaph Hall. Credit: NASA/JPL-Caltech/University of Arizona

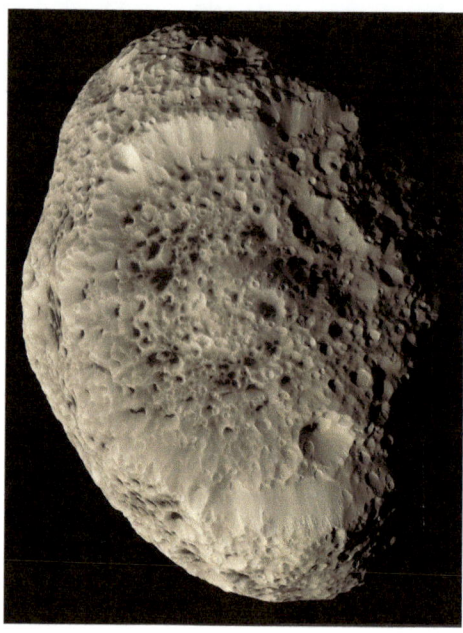

Fig. 3.3. Hyperion, one example of an irregular satellite. Its sponge-like appearance remains mysterious. Taken during the Cassini flyby on September 26, 2005. Credit: NASA/JPL/SSI/Gordan Ugarkovic

with respect to Jupiter's equator, some of them prograde and some retrograde. Some of the satellites of Saturn (Phoebe and Hyperion, Fig. 3.3), Uranus (Caliban and Sycorax), and Neptune (Nereid) are also irregular satellites.

This leaves a few objects classified as *unusual satellites*. The Earth's Moon and Pluto's Charon are in this category because they are very large compared to the size of their parent planet, and they likely did not form as part of the protoplanetary process, as other regular satellites are presumed to have originated. The Moon is believed to have originated as the result of a collision of another Mars-sized object with Earth, and Charon's origin remains speculative but may have been similar. Neptune's moon Triton, another satellite in this category, resembles a regular satellite but has a retrograde orbit, leading to theories that it is a captured Trans-Neptunian Object (P 13), probably from the Kuiper Belt (P 21) or the associated scattered disk. Altogether, only 19 planetary satellites are large enough to be round. Despite our greatly increased knowledge of satellites, no refined classification system has been widely accepted based on physical characteristics, although to some extent the Burns system from the 1980 reflects compositional differences. By analogy with planets, their satellites could be classified as rocky or icy, but they are too small to be composed of gas, Titan's atmosphere notwithstanding.

Each satellite has its own discovery story, but credit for discovering this new class of objects other than the Moon goes to Galileo and can be attributed to technological advance. Using his newly invented telescope, Galileo observed Jupiter from January 7 to March 2, 1610. Upon spying three objects in the vicinity of Jupiter, he at first thought they were fixed stars, but the fact that they were aligned made him suspicious. Returning to them over the next two months, he could see they were circling the planet. Even after this realization, he referred to the system as "Jupiter and his adjacent planets." Galileo called them the "Medicean planets" after his patron. Even though he did not call them moons or satellites, there is no doubt he knew they circled Jupiter; today we would say he had discovered a new class of object whose prototype was our Moon.[2] These were the only moons known in the Solar System until Christian Huygens discovered Titan around Saturn on March 25, 1655. Only four other moons were discovered in the 17th century: Iapteus and Rhea in 1671 and 1672, and Tethys and Dione in 1684, all by Giovanni Domenico Cassini and all around Saturn. Thus, nine satellites were known at the end of the 17th century.

It took over a century for the next set of satellite discoveries in the 1780s: two more around Saturn (Enceladus and Mimas), and two around the newly discovered planet Uranus (Titania and Oberon), all requiring for their detection the large telescopes built by William Herschel, the discoverer of Uranus. These were the only satellites discovered during the century, so at the end of the 18th century, the total stood at only 13 satellites.

Another 50 years elapsed before William Lassell detected a moon around the newly discovered planet Neptune in 1846, as well as Ariel and Umbriel around Uranus in 1851. Lassell and Bond also independently detected Hyperion around Saturn in 1848. It was another quarter century before astronomer Asaph Hall, working with the 26-inch refractor of the US Naval Observatory in Washington, D.C. (then the largest in the world), discovered the two moons of Mars in 1877, the only other moons ever discov-

---

[2] Galileo Galilei, *Sidereus Nuncius, or The Sidereal Messenger*, translated with introduction, conclusion and notes by Albert van Helden (Chicago: University of Chicago, 1989), pp. 64–86. See also Dick (2013), pp. 34–39.

ered in the inner Solar System. The 19th century was rounded out with E. E. Barnard's discovery of Amalthea around Jupiter, and William H. Pickering's discovery of Phoebe around Saturn—the first to be detected by photography, another new technology that would enhance the discovery of future satellites. By the end of the 19th century, 21 satellites other than our own Moon were known to exist.

In the 20th century, the floodgates of satellite discovery opened. Nine more satellites were discovered prior to the Space Age, seven of them around Jupiter. But then—again an indication of the march of new technology—numerous satellites of the outer planets were found by the Voyager spacecraft, the Cassini spacecraft, and ground-based telescopes employing new detectors such as charge-coupled devices (CCDs), bringing the total to 175. With the 12 new Jovian satellites announced in 2018, the total stands at 187. Meanwhile, the first satellite of an asteroid (Dactyl around Ida) was found by the Galileo during its flyby of Ida in 1993, and the first satellite of a Kuiper belt object (other than the 1978 discovery of Charon around Pluto, now considered a KBO), was found around the object known only as 1998 WW31 in 2002.[3]

The fact that the terrestrial planets of the inner Solar System contain only three moons compared to more than 184 for the gas and ice giants of the outer Solar System surely has to do with their different formation mechanisms. But even for our Moon and the moons of Mars, the mechanisms must have been very different. The Moon, after all, is one of the largest satellites of the Solar System (almost to the point of making the Earth-Moon system a "double planet"), while Phobos and Deimos circling Mars are among its smallest, perhaps captured, satellites.

The wide variety of satellite characteristics has defied classification based on their physical natures. The Galilean satellites of Jupiter seen in Fig. 3.1 are a case in point. Among the most curious and unique is Io. Unlike most of the other moons of the Solar System, it has a liquid or solid iron core and is composed of silicate rock, with a mantle similar to the terrestrial planets. Moreover, with over 400 active volcanoes, it is the most volcanically active object in the Solar System. It derives its heat energy from gravitational (tidal) interactions with Jupiter.

---

[3] Michael E. Brown et al, "Satellites of the Largest Kuiper Belt Objects," ApJ, vo. 639 (March 1, 2006), L43–L46.

Europa is also primarily rocky but has an outer layer of water and a surface layer of mostly ice. Its linear surface features indicate a tectonic deformation of ice, and the paucity of impact craters indicates a repeated and rapid resurfacing by melted ice. The thickness of the surface ice remains unclear, and there is intense interest in whether life might be supported in the oceans below the surface.

Because Ganymede has a magnetic field, this means it has a liquid iron core. It has a surface layer of ice perhaps 600 miles thick above a silicate mantle. Like Europa, it has rapid resurfacing, but also preserves impact craters. Callisto is dominated by impact craters but also has an ocean of liquid water. From Io to Callisto, with increasing distance from Jupiter, there is less and less heating from tidal energy, accounting for the differences in geologic activity.

While Saturn has a diverse retinue of at least 62 satellites, 96% of the mass of those satellites is in the aptly named Titan (Fig. 3.4)—larger by volume than the planet Mercury.

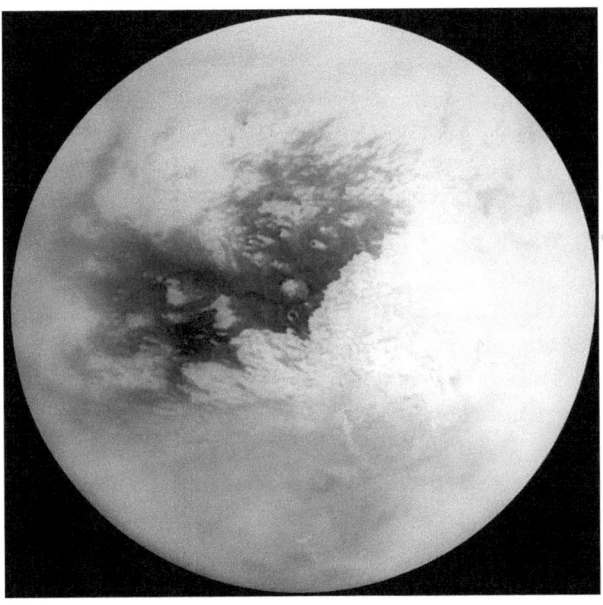

Fig. 3.4. Mosaic of Titan's surface processed to remove the effects of Titan's atmosphere. The images were taken in 2005 by the Cassini spacecraft from a distance of about 150,000 miles. The large dark region is known as Shangri-la. Credit: NASA/JPL/Space Science Institute

Titan, composed primarily of water ice and rocky material, has an atmosphere 60% denser than Earth's, dominated by nitrogen and hydrocarbon elements. Titan was investigated in detail by the Cassini-Huygens mission and found to have lakes of methane at its north and south poles—the only other body in the Solar System aside from Earth known to have large bodies of standing liquid on the surface.

But size is not the only indicator of interest. Enceladus is the only geologically active Saturnian moon, emitting jets of gas, dust, and ice, and harboring a subsurface ocean between its icy crust and rocky core. In 2018, planetary scientists announced the Cassini spacecraft had detected complex carbon-rich organic molecules in an ice jet spewing from cracks in the icy crust in the Southern Hemisphere, an indication that there could be life on the satellite.[4] Mimas is sometimes called the "Death Star," because the huge 90-mile impact crater makes the satellite superficially resemble the Death Star in the movie Star Wars. Hyperion is irregularly shaped and has a surface resembling a sponge. Although only seven of Saturn's moons are spherical, some astronomers count 24 of its satellites as *regular* and 38 as *irregular*, indicating the fluid nature of the satellite classification system.

As for the outer planets, among the moons of Uranus, five stand out for their size—ten to 30 times larger than its remaining moons. The last was discovered by Gerard Kuiper in 1948. At Neptune, Triton and Nereid rule in terms of size. The dwarf planet Pluto's moon Charon was discovered by James W. Christy at the U S Naval Observatory in 1978 and is tidally locked with Pluto, so it always faces the same side to that dwarf planet. The Hubble Space Telescope discovered two new moons of Pluto in 2005 (Nix and Hydra), another (Kerberos) in 2011, and a fifth (Styx) in 2012. Surprisingly, the New Horizons spacecraft discovered no new moons when it observed the Pluto system up close in 2015, but it greatly enhanced our knowledge of Charon as well as Pluto. Because of their great distance and small size, little is known of the many satellites surrounding the other dwarf planets, asteroids, and Kuiper belt objects.

---

[4] Frank Postberg et al., "Macromolecular organic compounds from the depths of Enceladus," *Nature*, 558 (2018), 564–568.

It is remarkable that by standards of size and composition alone the Galilean moons of Jupiter, Saturn's Titan and our own Moon would be classified as planets. However, their circumplanetary nature trumps their size and composition and renders them satellites rather than planets. Even Galileo, who called the four largest moons of Jupiter the "Medicean planets," realized they were secondary to their gigantic parent body. In the case of the Moon, however, this is not so clear, making plausible the term "double planet," analogous to "double star." And the announcement in 2018 of an "exomoon" the size of Neptune around the planet Kepler 1625b, which itself is several times the mass of Jupiter, gives more credence to the double planet concept.[5]

Images of planetary satellites can be accessed by clicking on the planet images at https://photojournal.jpl.nasa.gov/index.html and selecting specific satellites as the targets. The International Astronomical Union's "Gazetteer of Planetary Nomenclature," including all natural satellites and their discoverers, may be found at http://planetarynames.wr.usgs.gov/append7.html, http://en.wikipedia.org/wiki/List_of_moons. A complementary list with imagery is at http://en.wikipedia.org/wiki/Natural_satellite and a useful timeline of discovery can be found at http://en.wikipedia.org/wiki/Timeline_of_discovery_of_Solar_System_planets_and_their_moons. A database of asteroids with satellites, and much more information, is accessible at http://www.johnstonsarchive.net/astro/asteroidmoons.html.

---

[5] HST Release, "Astronomers Find First Evidence of Possible Moon Outside Our Solar System," http://hubblesite.org/news_release/news/2018-45

Class P 7: Ring

Planetary rings consist of particle- to house-sized objects orbiting a planet, forming one of the great spectacles of the Solar System. Like circumstellar protoplantary disks (P 1), rings are extremely flattened structures, but even more so due to the action of gravity on small scales. Four of the eight planets in the Solar System (all the gas and ice giants) have rings, but the ring system of Saturn—detected more than 350 years before the others—is by far the most complex and spectacular, giving it the popular moniker "Lord of the Rings." Saturn's rings are composed of almost pure water ice, with only slight contamination by rocky silicates. Other planetary rings have varying compositions that depend on their diverse origins. Planetary rings and satellites are in some ways connected, including small satellites that are embedded in rings, "shepherd" satellites that confine and sculpt rings, and past satellites that may have disintegrated into rings.

Despite a long history of both theory and observation prior to the 20th century, only spacecraft observations revealed the complex nature of Saturn's ring system, with its braids, waves, spokes, "propellers," and numerous divisions. Comparison with other planetary rings has also helped astronomers understand the evolution of ring systems and their differences. But many mysteries remain, especially regarding their origin, age, and dynamical complexity. They provide an excellent local laboratory for studying other flattened systems ranging from protoplanetary disks to spiral galaxies. Rings have not yet been detected around extrasolar planets but will likely be found around at least some of the many gas giants that have been detected over the last several decades.

Planetary rings illustrate well the differences between detection, interpretation, and physical understanding of an object. Galileo first detected what we now know as planetary rings when he turned his telescope to Saturn on July 30, 1610. But he was perplexed, noting that "[T]he star of Saturn is not a single one, but an arrangement of three that almost touch each other and never move or change with respect to each other; and they are placed in a line along the zodiac, the one in the middle being about three times larger than the other two on the sides..." Although Galileo observed Saturn many more times and noted different configurations of his first observations, it was not until 1655 that Christiaan Huygens gave the correct interpretation of these bodies as rings. Historian Albert van Helden has shown how difficult the interpretation of the new phenomenon was, puzzling astronomers for decades and beyond. Only in the mid-19th century did James Clerk Maxwell demonstrate that the rings could not be a single solid body. The actual composition of the rings remained unknown until spectroscopy revealed they were composed of icy particles.[6] Another great surprise came in 2009 when the Spitzer Space Telescope discovered a huge, nearly invisible dust and ice ring extending from 3.7 and 7.4 million miles from Saturn, much further than its classical rings. The ring glows only in the infrared and is likely caused by dust originating with collisions of comets or other matter with the satellite Phoebe, which orbits within the ring.

Like other classes of astronomical objects, gravity is the dominant force sculpting the Saturnian ring system (Fig. 3.5) and all

Fig. 3.5. Saturn's rings as seen by the Cassini spacecraft from a distance of 1.27 million miles on August 22, 2009. Particle sizes range from a grain of sand to mountains; the cause of the different ring colors is still unknown. Credit: NASA/JPL-Caltech/Space Science Institute

---

[6] Galileo Galilei to Grand Duke Cosimo II, in *Sidereus Nuncius, or The Sidereal Messenger*, translated by Albert van Helden (Chicago: University of Chicago Press, 1989), p. 102. Albert van Helden, "Annulo Cingitur: the Solution of the Problem of Saturn," *JHA*, 5 (1974), 155–174, and "Saturn and his Anses," *JHA*, (1974)5, 105–121. Stephen G. Brush, Elizabeth Garber and C. W. F. Everitt, eds., *Maxwell on Saturn's Rings* (Cambridge, MA: MIT Press, 1983). See Dick (2013), pp. 39–42 for more details on Galileo's discovery and its aftermath.

other ring systems. Orbiting the planet at 20,000 to 40,000 miles per hour, and with a mass that approximates the moon Enceladus, Saturn's rings and their underlying dynamics only gradually came to be understood. Astronomer Frank Shu was the pioneer in modern ring physics, and understanding was furthered when Scott Tremaine and Peter Goldreich adapted spiral galaxy dynamics to Saturn's rings in a series of papers between 1978 and 1982.

Although Saturn's classical rings span 280,000 km in diameter (only slightly less than the distance from the Earth to the Moon), they are at most places only 10 to 20 meters thick. This is likely due to particle collisions, which keep most of the particles in a very thin plane. There are, however, areas where moonlets squeezed out material that may rise vertically to 1 or 2 km. Ring pioneer Carolyn Porco likes to imagine skimming the surface of the rings in a spaceship and coming upon one of these kilometer-sized walls. Horizontal ring gaps are also believed to be caused when particles are forced out of a region, due either to external "resonance" influences of moons or to small moons actually embedded in the rings that clear their orbit. Two "shepherd" moons can also force particles into a very thin ring. The largest gap, named the Cassini Division after its 17th century discoverer Giovanni Cassini, is believed to exist because of a resonance effect: the distant moon Mimas continually imparts a gravitational nudge to this region, which is exactly half of Mimas's orbital period, clearing it of most particles.

For almost four centuries, Saturn remained the only planet in the Solar System known to have rings. Then, on March 10, 1977, James L. Elliot, Edward W. Dunham and Douglas J. Mink discovered the rings of Uranus. The discovery was made by from the Kuiper Airborne Observatory by observing occultations of a star by Uranus. Instead of observing a single occultation by the planet, the star blinked five times, leading to the interpretation that Uranus too was surrounded rings, though much less robust than those of Saturn. They have since been imaged even from the ground (Fig. 2.6).[7] Aside from Saturn, this remains the only other ring system discovered from the ground (an airborne Observatory being close enough to qualify for "ground" by comparison with Earth orbit).

---

[7]J. L. Elliot, E. Dunham, E. and D. Mink, D. (1977), The rings of Uranus," *Nature*, 267 (1977), 328–330; Elliott, Dunham and R. L. Millis, "Discovering the rings of Uranus," S&T, 53 (1977), 412–416.

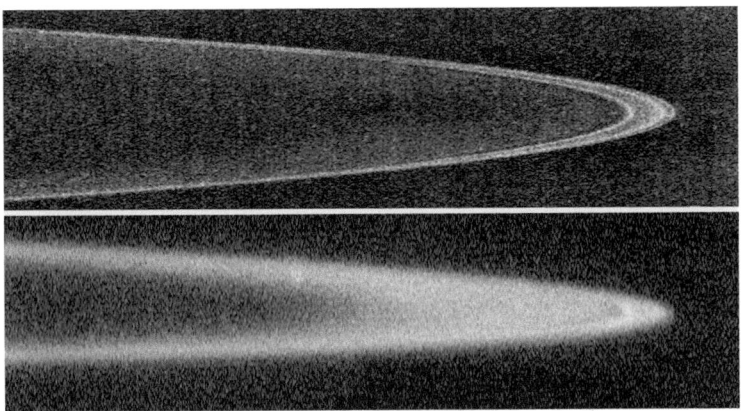

Fig. 3.6. Images of Jupiter's main ring obtained by the New Horizons during its gravity assist on the way to Pluto. The top image was taken on approach in backscattered light, and the lower one was taken after the spacecraft had passed Jupiter on February 28, 2007. The top image shows three well-defined rings confined in their orbits by three shepherding moons, plus material between the rings. Credit: NASA/Johns Hopkins University Applied Physics Laboratory/Southwest Research Institute

As with so many other aspects of the outer Solar System, it was the Pioneer and Voyager spacecraft that really lifted the veil on the rings of the outer planets. In 1979 Voyager 1 detected the Solar System's third set of planetary rings, a thin and extremely faint ring system around Jupiter (Fig. 3.6), as reported by Brad Smith and his colleagues in the journal *Science*. These rings are now known to be about 3,700 miles (6,000 km) wide and consist of four components: the main ring, a toroidal halo ring interior to the main ring, and two giant outer gossamer rings coinciding with the orbits of Amalthea and Thebe.[8] Additional rings of Uranus were found in 1986 by Voyager 2, and two outer rings were discovered in 2003–2005 by the Hubble Space Telescope, for a total of 13 known rings. The discovery of the rings of Neptune by Voyager 2 in 1989

---

[8] The discovery is announced in B. A. Smith, L. A. Soderblom, T. V. Johnson, *et al.*, "The Jupiter System Through the Eyes of Voyager 1," *Science*, 204 (1979): 951–957, 960–972. The classic volume on planetary ring studies in the wake of Pioneer 10 and 11 and Voyager 1 and 2 was Rick Greenberg and Andre Brahic, *Planetary Rings* (Tucson: Univ. of Arizona Press, 1984). That has now been updated and to some extent superseded by M. S. Tiscareno and C. D. Murray, eds., *Planetary Ring Systems* (Cambridge: Cambridge Univ. Press, 2017), which also takes into account Galileo, Cassini and other spacecraft observations. A very useful diagram comparing planetary ring systems is found on page 7.

completed the discovery of ring systems among the gas giants of the Solar System.[9] Altogether, the Voyager cameras acquired over 27,000 images of the planetary ring systems of the outer planets.

The Saturnian system of rings remains by far the most complex, followed by Uranus, and then the simpler systems of Jupiter and Neptune.[10] The ring systems of Jupiter and Saturn have also been studied in detail by the Galileo and Cassini spacecraft and by the Hubble Space Telescope. Cassini revealed hundreds of spiral density and bending waves, as well as fine-scale structure of gravitational instabilities in the rings, new ringlets, new moons near the rings, a moon stealing particles from the narrow F ring, moonlets causing propeller-shaped features, and features resembling straw and rope. Up to and including Cassini's Grand Finale dive into Saturn between April and September 2018, the spacecraft obtained some of the best images of the rings, indicating their dynamic changes even on time scales as short as hours, months, and years, and certainly the three decades between the Voyager and Cassini observations.[11]

Spacecraft observations have shown that composition can vary with the ring system. Jupiter's translucent gossamer rings consist largely of dust, believed to have been ejected from the Jovian moons Metis and Adrastea. Saturn's largely opaque rings are mostly composed of water ice. The New Horizons spacecraft observed Jupiter's rings in 2007 during its critical gravity assist en route to its successful rendezvous with Pluto (Fig. 3.6). This was the first opportunity to observe Jupiter up close (from four million miles) since Cassini had passed by in 2000 on its way to Saturn. Ring astronomer Mark Showalter designed an observing program to make high-resolution maps of the ring as Io occulted it, revealing its fine structure. Not only was this accomplished, but as a bonus, one of Io's volcanoes was also erupting at the time, resulting in the first time-lapse sequence of an erupting extraterrestrial

---

[9] B. A. Smith, L. A. Soderblom, D. Banfield, *et al.*, "Voyager 2 at Neputne: Imaging Science Results," *Science* 246 (1989), 1422

[10] Carolyn Porco, "Planetary Rings," in Paul R. Weissman, Lucy-Ann McFadden and Torrence V. Johnson, *Encyclopedia of the Solar System* (Academic Press: San Diego, 1999), pp. 457–475. For another pre-Cassini view of planetary rings see Larry W. Esposito, "Planetary Rings," *Reports on Progress in Physics*, 65 (2002), 1741–1783.

[11] On the Grand Finale results, including the closest imagery of the rings, see Keith T. Smith et al., "Diving Within Saturn's Rings," *Science*, 362 (October 5, 2018), 44–51.

volcano.[12] Uranus has narrow dark rings with broad bands of dust. And Neptune's rings include incomplete ones that appear as arcs.

In 2007, Cassini spacecraft scientists reported that Saturn's ring system, once thought to be several tens of millions of years old, may date to the birth of the Solar System itself. "The evidence is consistent with the picture that Saturn has had rings all through its history," said astronomer Larry Esposito. "We see extensive, rapid recycling of ring material, in which moons are continually shattered into ring particles, which then gather together and re-form moons." But ten years later, some of the same Cassini scientists concluded the rings were only a few hundred million years old.[13] It is possible that parts of the Saturnian ring system may have different ages, ranging from millions to billions of years, but the latest evidence is in favor of the younger age. The Uranian system also appears to be young, perhaps only 600 million years old. Even the youngest systems imply a continual source of debris such as a small embedded satellite, otherwise the debris would dissipate or spiral inward toward the parent planet. The question of age is related to the question of the origin of the rings around the gas giants. One possibility is tidal disruption of a large satellite as it migrated inward toward Saturn. Disruption of a passing large Kuiper belt object (P 21) is also possible. The question of the origin of planetary ring systems remains a subject of intense debate.

Planetary ring nomenclature has not been consistent among ring systems, but rather has depended on circumstances of the discovery. Thus, the major rings of Saturn have been named A through G, with the A ring at about 130,000 km; B, C, and D successively closer, with D at about 70,000 km; and E, F and G being farther out than A (see Fig. 3.7). The E ring is the most distant and by far the broadest at 180,000 to 480,000 km. The Cassini Division is the gap that separates the A from the B rings, the B ring carrying most of the mass of the rings. The C ring is diaphanous and transparent. With the resolution of spacecraft imagery, structures within the rings and gaps have also been named, although individual objects cannot themselves yet be resolved. By contrast, the rings of Jupiter are named 1979 J1R, 1979 J2R, and 1979 J3R,

---

[12] The observations at Jupiter are described in Alan Stern and David Grinspoon, *Chasing New Horizons: Inside the Epic First Mission to Pluto* (New York, Picador, 2018), pp. 175–181.

[13] NASA/Cassini Press Release, Dec. 12, 2007, "Saturn's Rings May be Old Timers," December 12, 2007, http://www.nasa.gov/mission_pages/cassini/media/cassini20071212.html. The young age for Saturn's rings is reported in *Science*, 358 (2017), 14113–1514.

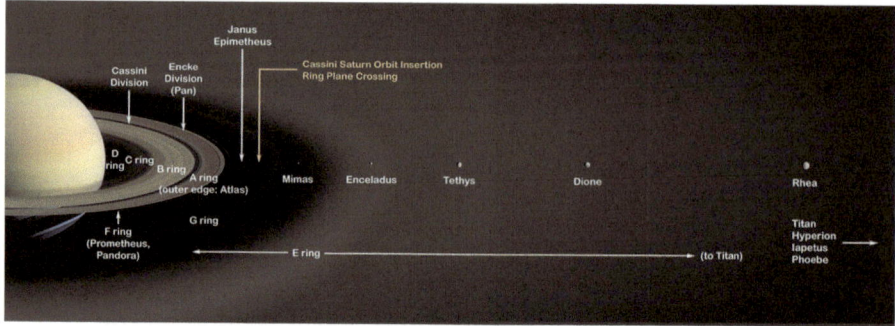

Fig. 3.7. Saturn's ring system is a complex structure larger than the distance from the Earth to the Moon. The seven main rings are named in their order of discovery. Credit: NASA/JPL

or Halo, Main, and Gossamer respectively as one moves further from the planet. The first five rings of Uranus were named alpha through epsilon and retain those names, while the five main rings of Neptune are named after various observers of Neptune including (in order of increasing distance from the planet) Galle, LeVerrier, Lassell, Arago, and Adams. Its lesser rings or arcs are given more contingent names like Liberte, Egalite, Fraternite, and Courage.

On March 6, 2008, Geraint Jones and his colleagues reported in *Science* that the Saturnian moon Rhea may have a faint ring system. If true, this would indicate a new type of object. New observations by Cassini, however, have largely ruled out the Rhea ring hypothesis.[14] In 2017, astronomers reported a ring around the dwarf planet Haumea, and at least two centaur asteroids are also believed to have rings.

A robust planetary rings research community now exists. A recent volume on *Planetary Ring Systems*, cited above, includes 58 authors and covers all known ring systems by location, type, and topic. The transition from the term "planetary rings" to "planetary ring systems" is one indication of the advances in the field.

Images of Saturn and its rings can be accessed by clicking on the Saturn image at https://photojournal.jpl.nasa.gov/index.html and selecting "S rings" as the target. The Voyager and Cassini websites also show ring imagery, and much else.

---

[14] Geraint H. Jones *et al.*, "The Dust Halo of Saturn's Largest Icy Moon, Rhea," *Science*, 319 (2008): 1380–1384; Matthew S. Tiscareno, Joseph A. Burns, Jeffrey N. Cuzzi, and Matthew M. Hedman, "Cassini imaging search rules out rings around Rhea," *Geophysical Research Letters*, 37 (2010), L14205.

Class P 8: Radiation Belt

A planetary radiation belt is a reservoir of energetic charged particles surrounding a planet and trapped in the magnetosphere created by the planet's internally generated magnetic field. The Earth's belt is known as the Van Allen radiation belt, named after James Van Allen, who discovered it in 1958 with instruments aboard Explorer 1, the first American satellite. The Earth has an inner radiation belt extending 1,000 to 8,000 miles above Earth composed primarily of protons, and an outer belt composed primarily of electrons extending 12,000 to 25,000 miles from Earth (Fig. 3.8). The size of these belts can vary depending on solar activity and other factors. For example, the inner belt can sometimes move to within 300 miles of the Earth (enough to be a danger to some satellites), and the outer belt can extend to 36,000 miles. In other words, the belts are very dynamic depending on conditions. A transient, third radiation belt was discovered in 2012. The Earth's inner radiation belt is believed to originate from the collisions of cosmic rays (S 30) in the upper atmosphere, while the outer belt is fed by the solar wind (P 16) and contains particles with much lower energy.

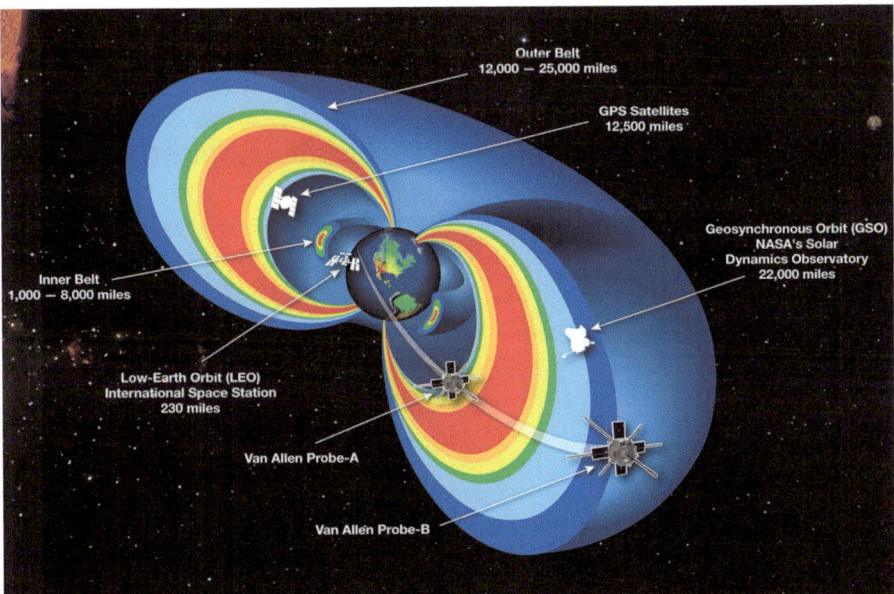

Fig. 3.8. A cutaway illustration of the donut-shaped Van Allen radiation belts with two Van Allen probes flying through them. The belts can affect the performance of nearby satellites. Credit: NASA

Extensive radiation belts also surround the gas giant planets in our Solar System. Evidence exists that Mars may have had a magnetic field in the past, but the terrestrial planets aside from Earth have extremely weak magnetic fields, too weak to maintain a belt of trapped particles. Mercury, for example, has a magnetic field 100 times weaker than Earth's. Despite its iron core extending out to 80% of its radius, most of it has solidified, leaving only a small liquid component to generate the magnetic field. The radiation belts of Earth and the giant planets are a concern both for the electronics of robotic spacecraft and for the health of humans in spaceflight. As on Earth, when particles from the radiation belts enter planetary atmospheres, they may cause spectacular displays of light known as aurorae.

Jupiter's magnetic field is by far the strongest of any planet in the Solar System, about ten times that of Earth (7.76 gauss compared to a maximum of .66 gauss at Earth's surface). It generates a magnetosphere—its sphere of influence above the ionosphere, where it controls charged particles and excludes most of the solar wind—with a "magnetic moment" some 20,000 times that of Earth's. The Jovian magnetosphere is about 1,200 times the volume of the Earth's, extending 100 Jupiter radii; if visible from Earth, it would appear several times the size of the full Moon. Its magnetotail extends 650 million km, almost to the orbit of Saturn.[15] By comparison, the Earth's magnetosphere extends to only about ten Earth radii. Jupiter's magnetic field is generated by a thick layer of metallic hydrogen in its outer core.

Saturn's magnetic field is slightly weaker than Earth's (about .2 gauss), but it's magnetosphere is 540 times stronger and generates the second largest magnetosphere in the Solar System, extending to about 20 Saturn radii, by the same mechanism. Accordingly, both planets have radiation belts, Jupiter's extending to about 100 Jupiter radii and Saturn's concentrated at two to three Saturn radii. Saturn's largest satellite Titan moves through the outer edges of its magnetosphere. Uranus and Neptune also have radiation belts, stronger than Earth's but weaker than Jupiter and Saturn's, each extending about 25 radii from the planet. For unknown reasons, Neptune's magnetic axis is tilted 47°, and Uranus's 59° to their axes of rotation. By comparison, Earth's is tilted 11°.

---

[15] For more on these concepts and the comparison of planetary surface magnetic fields and magnetospheres see Margaret Kivelson and Fran Bagenal, "Planetary Magnetospheres," *Encyclopedia of the Solar System*, at http://www.igpp.ucla.edu/public/mkivelso/Publications/299-Encyclopedia%20519-540.pdf

The discovery of both terrestrial and planetary radiation belts awaited the Space Age. When in 1958 physicist James Van Allen placed a Geiger counter on board Explorer 1, it was designed to observe cosmic rays, not a radiation belt. Much to everyone's surprise, the instrument demonstrated the existence of a large population of energetic particles trapped by the Earth's magnetic field. Three months later, Explorer 3 showed that the radiation belt was permanent, widespread, and intense, and spacecraft have mapped the nature and extent of the Van Allen belts in more detail ever since.[16] Explorer 4, Pioneer 3, and Luna 1 were the pioneer mappers of the terrestrial radiation belt in the late 1950s. Numerous spacecraft have studied it since then, notably the current Van Allen Probes.

There was some reason to expect Jovian radiation belts before they were discovered. Bernard Burke and K.L. Franklin, both of the Department of Terrestrial Magnetism at the Carnegie of Washington, had discovered powerful radio emissions from Jupiter in 1955 at a frequency of 22.2 megahertz, corresponding to decametric wavelengths. In 1959, Frank Drake and Hein Hvatum suggested these were caused by relativistic electrons trapped in a Jovian magnetic field "similar to the terrestrial Van Allen belts."[17] But only with the observations of the Pioneer spacecraft in 1973–1974 did it become clear that Jupiter really had particles trapped in a magnetosphere, just as Drake and Hvatum had suggested. Not only was the discovery of the Van Allen belts a surprise, but also the term "magnetosphere" was not even coined until 1959 by Cornell physicist Thomas Gold. As Van Allen himself has

---

[16] Early reports were James Van Allen et al., "Observation of High Intensity Radiation by Satellites 1958 Alpha and Gamma," *Jet Propulsion* (Sept. 1958), 588–592, reprinted in part in Bartusiak, 477–483; Van Allen et al., "Radiation Observations with Satellite 1958," *J. Geophysical Research*, 64 (1959), 271–286; and Van Allen and Louis Frank, "Radiation Around the Earth to a Radial Distance of 107,400 km," *Nature*, 183 (1959), 430–434. For the context and more on Van Allen see Abigail Foerstner, *James Van Allen: The First Eight Billion Miles*, (Iowa City: Univ. of Iowa Press, 2007).

[17] B. F. Burke and K. L. Franklin, "Observations of a Variable Radio Source Associated with the Planet Jupiter," JGR, 60 (1955), 213–217; Frank Drake and S. Hvatum, "Non-thermal microwave radiation from Jupiter," AJ, 64 (1959), 329–330; James A. Van Allen and Louis A. Frank, "Radiation Around the Earth to a Radial Distance of 107,400 km," *Nature*, 183 (1959), 430–434. Bagenal et al, 2.

pointed out, both the term "magnetosphere" and "radiation belt" are misnomers in the sense that particles rather than electromagnetic radiation are involved, and the particles are not trapped in a sphere, but rather exist in toroidal-shaped areas within the "sphere of influence" of the Earth's magnetic field. Nor, he pointed out, do radiation belts involve radioactivity: "nearly all of the particles actually present in natural radiation belts derive from hydrogen and other common, stable atoms."[18] On the other hand, electromagnetic radiation does originate in the radiation belts due to plasma instabilities, and the Earth radiates about 100 million watts into interplanetary space.

Unlike Jupiter, the magnetospheres of the other outer planets were almost entirely conjectural before their discovery. Saturn's magnetosphere was discovered in 1979 by the Pioneer 11 spacecraft. It was further studied in 1980–81 by the Voyager spacecraft, and was studied by Cassini since 2004. In 2008, Cassini detected a transient radiation belt around Saturn near the orbit of its satellite Dione, 337,000 km from the planet's core, making it the third planet for which variations in the magnetic field could be detected. Cassini's Grand Finale dive into Saturn revealed an asymmetry in the magnetic equator shifted about 2,800 km north of the planetary equator.[19] The Uranian magnetosphere was first observed by Voyager 2 in 1986, and Neptune's in 1989. The Uranian configuration is particularly interesting because the magnetic axis of the planet is offset 59° from its rotational axis, yielding radiation belts very different from the usual toroids.

The origin of the particles trapped in planetary magnetospheres is distinctive for each planet. For the Earth, the main source seems to be the neutrons produced when galactic cosmic rays (S 30) and the solar wind particles bombard gases in Earth's atmospheres. A fraction of these are ejected back into space, where they decay into energetic protons and electrons trapped in

---

[18] James A. Van Allen, "Magnetospheres, Cosmic Rays and the Interplanetary Medium," in *The New Solar System*, J. Kelly Beatty and Andrew Chaikin, eds. (Cambridge: Cambridge University Press, 3d edition, (1990), pp. 29–40: 32–33. For an online history of magnetospheric physics see physicist David Stern's pages at http://www-istp.gsfc.nasa.gov/Education/Intro.html

[19] E. Roussos et al, "Discovery of a transient radiation belt at Saturn," *Geophysical Research Letters*, 35 (2008); Michele K. Dougherty et al., "Saturn's magnetic field revealed by the Cassini Grand Finale," *Science*, 362 (October 5, 2018), 46.

the toroidal belts shaped by the terrestrial magnetic field. For the outer gas giants, the magnetospheres are affected both by their satellites and their rings. Jupiter's magnetosphere is dominated by particles vented from volcanic eruptions of its satellite Io, and its inner moons absorb or scatter some of its particles. At Saturn, the inner satellites Rhea, Dione, Tethys, Enceladus, and Mimas, along with some of the rings, absorb and diffuse the energetic electrons and protons. And at Uranus the satellites Miranda, Ariel, and Umbriel, as well as particulate matter in the rings, absorb energetic particles.

The class of radiation belts is likely repeated for many of the exoplanets now being discovered, and based on our knowledge of the Solar System their magnetospheres will undoubtedly be shaped by the planetary environments in ways both expected and unexpected.

More information, imagery, videos, and the latest data from the Van Allen probes can be found at: https://www.nasa.gov/van-allen-probes.

4. The Subplanetary Family

Class P 9: Dwarf Planet

With the class of "dwarf planet," we come face-to-face with the heated public and scientific controversy that followed the 2006 decision by the International Astronomical Union (IAU) to "demote" Pluto and create this new class of objects. It was one of the most notorious decisions in the history of the IAU and in astronomical classification. The decision might not have been so controversial if a dwarf planet were ruled to be a type of planet; however, it was ruled instead to be a subplanetary object. We classify it here following those decisions.

A dwarf planet as defined by the IAU orbits the Sun, is massive enough to be rounded by its own gravity, is not a satellite, and has not cleared its orbit of other materials. As the New Horizons mission has shown, Pluto is definitely rounded and has many planetary features, but it has not "cleared its orbit," in the sense that it both crosses the orbit of Neptune and shares the Kuiper Belt (P 21) with numerous other objects. It is that criterion, pushed by dynamical rather than planetary astronomers during the IAU deliberations, that places Pluto in the dwarf category by fiat.[1] The IAU currently recognizes five dwarf planets: Ceres, Pluto, Haumea, Makemake, and Eris, all but Ceres in the Kuiper Belt (Fig. 6.5). As of 2018, astronomer Michael Brown listed ten objects that were nearly certainly dwarf planets, and hundreds of others ranging from "possibly" to highly likely to be dwarf planets. More are constantly being discovered; in 2018 astronomers announced another candidate, nicknamed "the Goblin," in an elongated orbit outside of Pluto's and thus at the outer fringes of the Solar System.

---

[1] International Astronomical Union, 2006 resolutions on Pluto at http://www.iau.org/public_press/news/detail/iau0603/; IAU, "The IAU draft definition of "planet" and "plutons"", Press release, August 16, 2006; http://www.iau.org/public_press/themes/pluto/; Owen Gingerich, "The Inside Story of Pluto's Demotion," *S&T*, 112 (November, 2006), 34 ff.; Steven Soter, "What is a Planet?," *SciAm* (January, 2007), and Soter, "What is a Planet?," *AJ*, 132 (2006), 2513–2519; Tyson, Neil deGrasse, *The Pluto Files: The Rise and Fall of America's Favorite Planet* (New York: W.W. Norton, 2009).

© Springer Nature Switzerland AG 2019
S. J. Dick, *Classifying the Cosmos*, Astronomers' Universe,
https://doi.org/10.1007/978-3-030-10380-4_4

If Pluto is not a planet (classes P 2 through 5), neither is it a "small body of the Solar System" like asteroids, comets, and Trans-Neptunian objects (classes P 11–13), because it is rounded. As a sub-planetary object it falls between the official definitions of a planet and a small body of the Solar System. A dwarf planet is not necessarily a Trans-Neptunian object; at least one of them (Ceres) is within the orbit of Jupiter, and another (Pluto) is inside Neptune's orbit part of the time. Hence its designation here as a separate class. In honor of Pluto, dwarf planets beyond Neptune's orbit are sometimes termed "plutoids," though the term is not in common usage.

Because the case of Pluto (Fig. 4.1) became so controversial among both scientists and the public, it is important to see how the new class of "dwarf planet" came to be created. When Clyde Tombaugh discovered Pluto on February 18, 1930 by "blinking" photographic plates from the 13-inch telescope at Lowell Observatory in Flagstaff, Arizona, it was only a point of light, detected among the background stars by its extremely slow

Fig. 4.1. The dwarf planet Pluto, as seen by the New Horizons spacecraft July 14, 2015. Four images were combined to create this global view. The images, taken from a distance of 280,000 miles (450,000 km), show features as small as 1.4 miles (2.2 km. The north polar region is at top, with bright Tombaugh Regio to the lower right of center and part of the dark Cthulhu Regio at lower left. Part of the dark Krun Regio is also visible at extreme lower right. Credit: NASA/Johns Hopkins University Applied Physics Laboratory/Southwest Research Institute

motion.[2] That motion translated to a 248-year orbital period, plac-ing it at the edge of the Solar System. It was a fantastic discovery, but Pluto at that time was not recognized as a new class of object, nor could it be, without knowing its mass. The mass, size, and density of Pluto were for decades considered to be similar to the planet Mars. But this was very uncertain until astronomer James W. Christy, using photographic plates taken with the 61-inch astrometric reflector of the US Naval Observatory (located four miles from Lowell Observatory), discovered its satellite Charon in 1978. Using Kepler's laws of motion, it was then determined that Pluto had a mass only 1/400th of the Earth and a diameter of less than 1,500 miles, considerably smaller than our Moon. A low-grade debate began even then about the status of Pluto.[3]

The situation ramped up considerably in the early 1990s, when astronomers began to discover a variety of objects beyond Pluto in what is known as the Kuiper belt. Objects given names like Sedna, Orcus, and Quaoar began popping up, all slightly smaller than Pluto. The situation was brought to a head in 2003, when Caltech astronomer Michael Brown (previously mentioned) discovered an object believed to be larger than Pluto, designated 2003 UB313 (now known as Eris).[4] Was it the 10th planet, or was Pluto not a planet? And since many more similar objects are now known to exist in our Solar System beyond Pluto, with more surely to be discovered, what is their status?

These questions were brought up for a vote at the trien-nial meeting of the IAU General Assembly in Prague in 2006. Dispassionately put, the status of Pluto and the Kuiper belt objects were questions of classification, of moving an object from one class to another with new knowledge in the case of the former, or of classifying newly discovered objects in the case of the latter. As we will see in the case of minor planets (P 11), their status was changed from planets to minor planets in the course of the 19th century as more were discovered. The case of the class of dwarf

---

[2] William G. Hoyt, *Planets X and Pluto* (University of Arizona Press, 1980); Tombaugh's own account is Clyde Tombaugh and Patrick Moore, *Out of the Darkness: The Planet Pluto* (New American Library: New York, 1980).

[3] Stern, A. "The Pluto-Charon System" ARAA, 30 (1992), 185–233; Alan Stern and Jacqueline Mitton, *Pluto and Charon: Ice Worlds on the Ragged Edge of the Solar System* (New York: John Wiley, 1999), R. M. Marcialis, "The Discovery of Charon: Happy Accident or timely Find?", JBAA, 99 (1989), 27–29. For more details see Dick (2013), pp. 10–17.

[4] Govert Schilling, *The Hunt for Planet X: New Worlds and the Fate of Pluto* (New York: Copernicus Books, 2009).

planets is a close parallel to that series of events. Currently, only the largest members of the asteroid belt and the Kuiper belt are designated as dwarf planets, including Ceres.

The criteria for dwarf planets have been criticized from a number of perspectives. For example, of the five designated dwarf planets, only Ceres and Pluto have actually been observed to be round. Theory indicates that a rocky object must be about 900 km (the diameter of Ceres) in order to be round, while an icy body would have to be greater than 400 km (the diameter of Mimas, the smallest icy satellite). The roundness criterion thus highlights the fact that the physical composition of the dwarf planets may vary considerably, from rocky (Ceres) to icy (those in the Kuiper belt). To this day, Alan Stern, the Principal Investigator of the New Horizons mission to explore Pluto, and many other planetary scientists do not accept the designation of Pluto as a dwarf planet, arguing that matters of classification should not be decided by vote.[5]

In any case, with the official IAU definition in mind, we can see how fraught the classification problem can be based on the history of discovery of those objects now designated dwarf planets. Ceres, the only dwarf planet that is not trans-Neptunian, was discovered on January 1, 1801 by Giuseppe Piazzi. Official government Almanacs designated it for years as a planet, then a minor planet/asteroid, and since 2006 as a dwarf planet. Pluto, discovered in 1930, was designated a planet until 2006, even though after 1978 it was known to be smaller than our Moon. Then came those discoveries that precipitated the classification crisis: Haumea, discovered by Michael Brown of Caltech in 2004 and by J.L. Ortiz in 2005 in Spain (yet another controversy), now known to have two moons; Eris, at first thought to be the largest known dwarf planet but in reality about the same size (1,476 miles in diameter) as Pluto, discovered in January 2005 by Brown; and Makemake, discovered March 31, 2005, also by Brown. Many other Trans-Neptunian objects are candidates for dwarf planet status, but it is not yet known if they are in hydrostatic equilibrium.[6]

The end of the dwarf planet story has not yet been written; indeed, it is likely just beginning, as more becomes known about the physical nature of dwarf planets and new objects are discovered. In 2015, NASA's Dawn mission visited Ceres (Fig. 4.2) and revealed a combined water ice and rocky surface, a differentiated rocky core and icy mantle, and perhaps even an internal ocean of liquid water.

---

[5] Alan Stern and David Grinspoon, *Chasing New Horizons: Inside the Epic First Mission to Pluto* (New York, Picador, 2018), pp. 170–174.

[6] Michael Brown, *How I Killed Pluto and Why it Had it Coming* (2010); http://en.wikipedia.org/wiki/List_of_dwarf_planet_candidates

Fig. 4.2. The dwarf planet Ceres seen by the Dawn spacecraft on February 19, 2015, from a distance of about 29,000 miles (46,000 km). The heavily cratered surface ranges from shallow to peaked craters. Credit: NASA/JPL-Caltech/UCLA/MPS/DLR/IDA

Several mysterious bright spots were seen inside a crater, believed to be highly reflective ice or salts, or a cyrovolcanic feature.

In the same year, NASA's New Horizons mission visited Pluto, making it and Ceres the first dwarf planets (and in the case of Pluto, the first trans-Neptunian object and the first Kuiper belt object) to be observed in detail by spacecraft. Figure 4.1 reveals how Pluto was much more diverse than anyone expected. As Alan Stern and David Grinspoon summarized these phenomena in their riveting description of the New Horizons mission, Pluto showed "Ground fogs, high-altitude hazes, possible clouds, canyons, towering mountains, faults, polar caps, apparent dune fields, suspected ice volcanoes, glaciers, evidence of flowing (and even standing) liquids in the past, and more." Surprisingly, Pluto was found to be geologically active, with both ancient cratered and fresh-looking landscapes, and a 1,000-km-wide churning nitrogen glacier dubbed Sputnik Planitia. Even more surprising, there are indications of an interior liquid water ocean, analogous to those thought to exist inside Europa and Enceladus. "Could there be Plutonian life forms swimming deep underneath the planet's icy surface?" Stern and Grinspoon provocatively ask. Only further exploration will tell.[7]

The latest spectacular images of Pluto are found at the New Horizons mission site http://pluto.jhuapl.edu/Multimedia/Science-Photos/byTopic.php. Michael Brown's dwarf planet website is accessible at http://web.gps.caltech.edu/~mbrown/dps.html.

---

[7] Alan Stern and David Grinspoon, *Chasing New Horizons*, especially pp. 275 ff.

Class P 10: Meteoroid

Meteoroids are debris particles ranging in size from grains of sand to boulders traveling in solar orbits that may occasionally intersect the orbit of the Earth. Or, as the International Astronomical Union says more formally, a meteoroid is "a solid object moving in interplanetary space, of a size considerably smaller than an asteroid and considerably larger than an atom." At the upper end of meteoroid mass, this leaves considerable room for ambiguity in distinguishing between a meteoroid and an asteroid; some organizations have defined objects as large as 50 meters to be meteoroids. Moreover, the lower end of meteoroid mass—the so-called micrometeoroids—overlaps the range of interplanetary dust particles (P 15). The distinction at both boundaries is semantic, yet real enough that astronomers speak of a separate class of meteoroids, due not only to their composition but also their origin.

When entering the Earth's atmosphere a meteoroid is termed a meteor, and if it survives impact it is a meteorite. Because meteoroids are believed to originate as a trail of debris shed from a larger body like a comet, they commonly travel in collections called meteoroid streams (P 20). If the Earth's path and a meteoroid stream intersect, a meteor shower or meteor storm may result. Meteors meeting the Earth head-on may enter the atmosphere at a velocity of 44 miles per second, of which 18 miles per second is the Earth's own velocity through space; those not entering head-on have a correspondingly slower speed depending on their direction. It has been estimated that one trillion meteoroids pepper the Earth each day, amounting to 110 tons per day, or 40,000 tons annually. From their characteristic spectra visible as they enter the atmosphere, they have been determined to consist of sodium, iron, copper, and various silicates.

Prior to the Space Age, knowledge of the existence of meteoroids in outer space depended entirely on the observation and interpretation of meteor showers in the Earth's atmosphere, the only way they became visible. The phenomenon of "shooting stars" had been widely observed throughout history; cuneiform records of meteors from the ancient Near East date back at least to 1200 BC, but they must have been uncomprehendingly observed long before that. The Roman poet Virgil wrote, "Oft you shall see the stars, when wind is near, shoot headlong from the sky and through the night leave in their wake long whitening seas of flame."[8] The question was: What

---

[8] Virgil, *Georgics*, Book I, lines 365–367; Judith Kingston Bjorkman, "Meteors and Meteorites in the Ancient Near East," *Meteoritics*, 8, no. 2 p. 91, online at ADS.

was their origin? From the time of Aristotle's *Meteorology* in about 340 BC, meteors were believed to be atmospheric phenomena for more than two millennia (thus their name). Author Mark Littmann called this "perhaps the longest run of any scientific error," because unlike comets, which Tycho Brahe proved to be astronomical phenomena in 1577, the celestial explanation of meteors took much longer. [9] Aristotle was actually half correct: meteors are an atmospheric phenomenon, but with a celestial cause.

Although the suspicion that meteors were of cosmic origin dates at least to Edmond Halley in the 18th century, it was not until 1863 that they were definitively proven to be astronomical. The proof was based largely on observations of what we now call the Leonid meteors, so called because their radiant is in the constellation Leo. The Perseid "August meteors" also played a role, but the Leonid "November meteors" tended to storm every 33 years and thus demanded a more immediate explanation. The Leonid storm of 1833 peaked in the Eastern part of North America and gave birth to modern meteor studies. Yale astronomer Denison Olmstead wrote:

> Probably no celestial phenomenon has ever occurred in this country, since its first settlement, which was viewed with so much admiration and delight by one class of spectators, or with so much astonishment and fear by another class. For some time after the occurrence, the "Meteoric Phenomenon" was the principal topic of conversation in every circle, and the descriptions that were published by different observers were rapidly circulated by the newspapers, through all parts of the United States.[10]

From these many observations Olmsted collated data relating to weather, time and duration, number, variety, sound, and apparent origin, and concluded that the meteors of 1833 originated beyond the Earth's atmosphere. However, the clinching argument for the cosmic origin of meteors came only in 1863, when Yale astronomer Hubert A. Newton determined from past cycles that the November meteors repeated in intervals of sidereal years, not tropical years—a sure sign of a celestial phenomenon.[11]

---

[9] Mark Littmann, *The Heavens on Fire: The Great Leonid Meteor Storms* (Cambridge: Cambridge University Press, 1998), p. 35. Dick (2013), 208–211.

[10] D. Olmsted, "Observations on the meteors of 13 Nov. 1833," AJSA, 25 (1834), 363–411 and AJSA, 26 (1834), 132–174; Steven J. Dick, "Observation and interpretation of the Leonid meteors over the last millennium," JAHH, 1, (1998), 1–20.

[11] H. A. Newton, "Evidence of the cosmical origin of shooting stars derived from the dates of early star-showers," AJSA, 36 (1863), 145–149; Dick (2013), pp. 208–211.

The proof that meteoroids actually land on the Earth was met with similar skepticism. Though the idea was accepted by many in Europe, responding to Yale chemist Benjamin Silliman's investigation of a meteorite fall in Connecticut in 1807, President Thomas Jefferson famously (though possibly apocryphally) remarked, "I would more easily believe that [a] Yankee professor would lie than that stones would fall from heaven."[12] Since then, numerous meteorites have been identified and analyzed. Meteorites landing on Earth range in size from those resulting in catastrophic impacts, caused by asteroid or comet impacts, to smaller objects weighing less than a pound, caused by meteoroids. A leading candidate for the famous Tunguska event, which took place in central Siberia on June 30, 1908, is a meteoroid about five miles in diameter. Numerous eyewitness reports exist for this event, including some only 40 miles away.

Occasionally, meteoroids interact with humans in a more direct way. On January 18, 2010, a half-pound meteorite traveling 200 miles per hour crashed into a doctor's office in Lorton, Maryland near Washington, D.C. The "Lorton meteorite" splintered into several pieces and became the object of a dispute over ownership rights. It was estimated to be worth $50,000 in the meteorite market, and although the doctors wanted to donate it, the landlords weren't so sure. Pending a decision on these rights, it was placed in the custody of the Smithsonian's National Museum of Natural History, which holds about 15,000 meteorites—about half of the world's collections. A year after the fall, the landlords dropped their claim, the Smithsonian paid the doctors $10,000, the doctors donated it to Doctors Without Borders, and the Lorton meteorite took its permanent place in the museum.

The International Meteorite Organization maintains an extensive and regularly updated website on meteoroids, meteors, fireballs, and related phenomena at http://www.imo.net/, including photos and videos, observing methods, fireball reports, and upcoming meteor showers. The American Meteor Society also hosts similar information at https://www.amsmeteors.org/home.html. On meteor showers and storms, see entry P 20 (Meteoroid streams).

------

[12] On the circumstances and conjectural nature of this remarks see John G. Burke, *Cosmic Debris: Meteorites in History* (University of California Press, 1991), pp. 56–58.

Class P 11: Minor Planet/Asteroid

A minor planet is one class of small rocky bodies of the Solar System orbiting the Sun, as distinct from the classes of major planets (P 2 – P 4), dwarf planets (P 9), and satellites (P 6). Also known as asteroids, they are distinguished from comets (P 12) and other small bodies of the Solar System such as trans-Neptunian objects (P 13), as defined by the International Astronomical Union in 2006. They range in size from about 600 miles in diameter down to pebbles. The dividing line between minor planets, dwarf planets (P 9), and classical planets on the upper end of the mass scale, and minor planets and other small bodies of the Solar System down to meteoroids (P 10) on the lower end of the mass scale, has been the subject of much controversy over the last two centuries.

More than 800,000 minor planets are known to exist in our Solar System, most of them in the asteroid belt between the orbits of Mars and Jupiter, or in other asteroid groups (P 19) such as the near-Earth and Trojan asteroids. The latter are located where the gravitational pull of the Sun and a planet are balanced, a point known as a Lagrangian point. Jupiter's Lagrangian points hold by far the most Trojan asteroids. Minor planets undoubtedly exist around other stars as well as the Sun. In fact, two "interstellar asteroids" have been observed in the last few years in our Solar System. Oumuamua (Hawaiian for "messenger" or "scout") was discovered with the Pan-STARRS telescope in Hawaii in October 2017 and will pass Neptune's orbit in 2022 on its way out of the Solar System. Illustrating the difficulties of classification, it is now believed to be a comet, since it shows an acceleration that could only be caused by cometary outgassing (unless it is an alien spaceship!). But, as we discuss below, the differences between asteroids and comets can be hard to detect.[13] Asteroid 2015 BZ509, also discovered by Pan-STARRS in 2014, is another object that has likely been in our Solar System for millions or billions of years, traveling in the orbit of Jupiter but circling in the opposite direction. A large population of such interstellar intruders may exist.

Since their discovery, main belt asteroids have been hypothesized to be the remnants of an exploded planet between Mars and Jupiter. In fact, they are more likely to be primordial material

---

[13] Greg Laughlin, "Oumuamua's Dramatic Visit," S&T, 136 (October, 2018), 20–26. Avi Loeb, the chairman of the Harvard Astronomy Department, has seriously argued the alien hypothesis at https://arxiv.org/pdf/1810.11490.pdf

prevented from forming into a planet because of Jupiter's strong gravitational influence. The total mass of the asteroids probably amounts to a body less than the mass of Pluto.

The controversy over whether minor planets constituted a new class of astronomical object began immediately with their detection. On New Year's Day in 1801, while working on a star catalogue at the observatory in Palermo, Sicily, the Italian astronomer Giuseppe Piazzi observed an 8th magnitude star that moved over subsequent nights.[14] Because Piazzi was using a specialized telescope known as a transit circle, he could only observe it for less than two minutes while it was crossing the meridian. But observation over four subsequent nights clearly showed the motion against the fixed stars. As with William Herschel and the discovery of Uranus 20 years earlier, Piazzi believed at first it was a comet. It was soon lost in the glare of the Sun, but with the help of predictions by the famous German mathematician Carl Friedrich Gauss, it was recovered again on December 31.[15] Piazzi called it "Cerere Ferdinandea" in honor of Ceres and Ferdinand, respectively the patron goddess of Sicily and the royal patron of the Palermo Observatory. Lacking the normal fuzzy appearance of comets, it was soon labeled a major planet, and after considerable controversy the name was shortened to Ceres.

In March 1802, Heinrich Wilhelm Olbers discovered a similar object, now known as Pallas, and William Herschel turned his large telescope toward both objects. He determined the sizes of Ceres and Pallas to be 162 and 70 miles respectively, and in a letter to Piazzi suggested that Piazzi had discovered "a new species of primary heavenly body," a more notable achievement than merely adding another member to the class of planets already known. In the first scientific paper on this new class of object, Herschel asserted, "we ought to distinguish them by a new name, denoting a species of celestial bodies hitherto unknown to us ... From ... their asteroidical appearance ... I shall take my name, and call them Asteroids."[16]

---

[14] Giuseppe Piazzi, "Results of the Observations of the New Star," Palermo Observatory, 1801; English translation excerpts reprinted in Bartusiak, 151–152.

[15] Eric G. Forbes, "Gauss and the Discovery of Ceres," *JHA*, 2 (1971), 195–199. For more details of the discovery of asteroids see Dick (2013), pp. 45–49.

[16] William Herschel, "Observations on the two lately discovered celestial bodies," PTRSL, 92 (1802), 213–232. See also David W. Hughes and Brian G. Marsden, "Planet, Asteroid, Minor Planet: A Case Study in Astronomical

The detection of Ceres and Pallas, followed by Juno in 1804 and Vesta in 1807, was only the beginning of the discovery of the numerous minor planets that we now know. Despite Herschel having dubbed them "asteroids," during the years that their overall number was small, the major government almanacs of the world called them planets. In 1828, most astronomy texts accepted 11 planets, including the seven known major planets plus Ceres, Juno, Pallas and Vesta. No new objects of this kind were discovered for nearly 39 years after the discovery of Vesta in 1807, but by the end of 1851, there were 15. Only when their numbers grew did astronomers create a new class of Solar System objects called minor planets (a term first used in the British Nautical Almanac in 1835) or asteroids.[17] Thus, there was a period of nearly 50 years of uncertainty during which many considered these objects to be planets.

The classification of asteroids is an excellent example of how such systems evolve with new techniques and data. Little was known of the composition of asteroids until the 20th century. In 1975, planetary astronomers Clark Chapman, David Morrison, and Ben Zellner developed a taxonomic system based on the color, albedo, and spectral features of a sample of 110 asteroids. Importantly, these properties are related to their composition.[18] Three main Types are currently distinguished, stemming from that classification. The most common are: the C-Types, dark *carbonaceous* objects representing some 75% of all asteroids and occupying the outer asteroid belt; the S-types, stony *silicaceous* objects comprising 17% of known asteroids and dominating the inner asteroid belt; and the M-types, composed of pure *metallic* nickel-iron and mainly located in the middle of the asteroid belt. Asteroids are believed to be the ultimate source of the meteorites that fall on Earth, the S-Type being the source of the most common stony meteorites, while carbonaceous chondrites originate with the C-Types and iron meteorites with M-type.

Nomenclature," JAHH, 10 (March, 2007), 21–30.

[17] James Hilton, "When did the Asteroids become Minor Planets?", http://aa.usno.navy.mil/faq/docs/minorplanets.php; David W. Hughes, "Planet, Asteroid, Minor Planet: A Case Study in Astronomical Nomenclature," *JAHH*, 10 (2007), 21–30; Dick (2013), p. 49.

[18] C. R. Chapman, D. Morrison and B. Zellner, "Surface properties of asteroids – A synthesis of polarimetry, radiometry, and spectrophotometry," *Icarus*, 25 (1975), 104 ff.

In the 1980s, David J. Tholen proposed a new classification resulting in 14 categories. In 2002, Schelte Bus expanded the Tholen system, and in 2009, Bus and Francesca DeMeo further expanded the system. Updates continue. The 2009 system is based on reflectance spectrum characteristics for 371 asteroids measured in the infrared and results in 24 classes of asteroids. These classes are not always correlated with differences in composition but have nonetheless proven useful for astronomers. For example, a recent study found that the main asteroid belt's most massive classes in the Bus-DeMeo system are C, B, P, V, and S in descending order. In one indication of the importance that planetary astronomers attach to classification, in 2018 DeMeo was awarded the distinguished Harold C. Urey Prize "in recognition of the broad foundational understanding of the study of Solar System bodies using the modern system of asteroid classification that bears her name." The classification system, the award announcement noted, is a tool for understanding the geologic structure of the asteroid belt, revealing compositional complexity, and supporting dynamical models of the mixing of bodies in the early Solar System. While it may seem confusing to have several classification systems for the same objects in common usage, they have built on each other, and each level of sophistication is useful in its own way depending on the level of research being done.[19]

Asteroids have much in common with comets. Indeed, another type of asteroid are the Centaurs, so called because they have properties of both asteroids and comets. 2060 Chiron, discovered by Charles T. Kowal in 1977, was the first Centaur recognized as a new type of asteroid. The Centaurs are icy objects that orbit between Jupiter and Neptune, and may be an intermediate stage in an object's migration from the scattered disk of the Kuiper Belt (P 21) to the inner Solar System. At least some asteroids may be extinct comets, such as 3200 Phaeton, responsible for the Geminid meteoroid stream (P 20).

---

[19] D. J. Tholen, "Asteroid taxonomic classifications," *Asteroids II; Proceedings of the Conference*. University of Arizona Press (1989), pp. 1139–1150; S. J. Bus, "Phase II of the Small Main-belt Asteroid Spectroscopy Survey: A feature-based taxonomy," *Icarus*, 158 (2002): *146*. F.E DeMeo and B. Carry, "The taxonomic distribution of asteroids from multi-filter all-sky photometric surveys," *Icarus*, 226 (2013), 723–741, https://arxiv.org/pdf/1307.2424.pdf

Fig. 4.3. Gaspra, a mosaic of two images taken by the Galileo spacecraft from a range of 3,300 miles (5,300 km), some 10 minutes before closest approach on October 29, 1991. The portion illuminated in this view is about 11 miles (18 kilometers) from lower left to upper right. A striking feature of Gaspra's surface is the abundance of small craters. More than 600 are visible here. Credit: NASA/JPL

The 1990s proved a landmark decade for asteroid studies, due in part to the Space Age. In 1991, the Galileo spacecraft approached within 1,000 miles of 951 Gaspra (Fig. 4.3), revealing it to be an irregular object with dimensions 12 × 7 × 7 miles and numerous small craters, rotating every 7 hours, and composed of metal-rich silicates and perhaps blocks of pure metal. Two years later, Galileo approached 243 Ida and found a similarly elongated but larger object, and confirmed the first asteroid moon, named Dactyl and only about a mile across. Numerous asteroid moons are now known. In the mid-1990s the Hubble Space Telescope imaged Vesta, and in 1997 the near-Earth asteroid mission known as NEAR-Shoemaker flew within 750 miles of asteroid 253 Matildhe, which proved to be rounder in shape and relatively large at 41 miles in diameter. The asteroid was rich in carbon, darker than charcoal, and sported a huge crater.

NEAR-Shoemaker reached minor planet Eros in December, 1998, was inserted into orbit on Valentine's day, 2000, and after mapping Eros actually landed on the asteroid a year later, returning a treasure trove of information. A variety of geological features were observed, including craters, most likely made by impacts rather than internal activity. In late 2005, the Japanese Hayabusa (peregrine falcon) spacecraft achieved rendezvous with the tiny near-Earth asteroid 25143 Itokawa, a 600-meter-sized, potato-shaped S-type asteroid. The spacecraft briefly touched down on the surface of the asteroid and, after a harrowing journey, returned to Earth in 2010 with a surface sample consisting mainly of rocky, iron-rich olivine, and pyroxene. In 2018, Hayabusa2 dropped several probes onto the surface of the asteroid Ryugu, and was scheduled to make three touchdowns to collect samples for return to Earth.

NASA's Dawn mission orbited Vesta from August 2011 to May 2012, then orbited the dwarf planet Ceres and returned data from 2015 to 2018, when it ran out of thruster fuel (see P 9 and Fig. 4.2). Dawn's instruments showed both objects to be highly cratered, with a variety of other surface features ranging from troughs to pits. In addition, Vesta is thought to have a metallic iron-nickel core, one of the few asteroids with a differentiated interior. Fragments of Vesta have impacted Earth in the form of meteorites, aiding our study of the asteroid.

Altogether, since 1991 spacecraft have visited eleven asteroids either as flybys, orbiters, or landers, and many of them have been of the S silicaceous Type. In 2018, the Japanese Hayabusa2 and NASA's Osiris-REx studied two near-Earth asteroids dubbed Bennu and Ryugu, the first time these carbon-rich "carbonaceous" C-type of asteroids have been studied. Their ultimate goal is to bring small samples of these bodies back to Earth. In part because of spacecraft data, Phobos and Deimos, the two moons of Mars, are believed to be captured asteroids, and some of the smaller moons of the gas giant planets may be also. Meanwhile, in addition to these in situ spacecraft, the Gaia satellite has also observed about 14,000 asteroids, most of them already known but enabling astronomers to refine their orbits.[20]

---

[20] On Hayabusa and Osiris-Rex see Dan Durda, "Space Rock Rendezvous," S&T, 135 (June, 2018), 22–28.

Scientists have long known that minor planets are of more than academic interest. In the future, they may be the source of rich minerals and the site of mining operations. More immediately, the close approaches of some of them threaten the Earth. It therefore behooves us to know as much as possible about the size and structure of these roaming celestial neighbors. As the near-Earth object team at the Jet Propulsion Laboratory has stated, "whether looking for the richest source of raw materials or trying to nudge an Earth threatening object out of harm's way, it makes a big difference whether we're dealing with a 50-meter-sized fluff ball or a one-mile slab of solid iron." For this reason, systematic search programs have been established for near-Earth asteroids, such as the Lincoln Near-Earth Asteroid Research (LINEAR) program undertaken by NASA, the US Air Force, and Lincoln Laboratory. We know asteroid impacts have happened in the past, and we do not know if the next hit will be in a few years or many years. Asteroids can also be classified by their orbital characteristics, and studies of asteroid groups may help to understand the probabilities of Earth impact as well as a variety of other research problems.

While Gaia is primarily a star and galaxy program, the Gaia Solar System program is described at http://sci.esa.int/gaia/58706-gaia-turns-its-eyes-to-asteroid-hunting/, including a map of its asteroid detections. NASA's Jet Propulsion Laboratory provides a detailed Small Body database at https://ssd.jpl.nasa.gov/sbdb.cgi and https://ssd.jpl.nasa.gov/sbdb_query.cgi. Thanks to spacecraft observations, numerous images of asteroids are now available. For example, images of Eros from NEAR are found at https://nssdc.gsfc.nasa.gov/planetary/mission/near/near_eros.html. Images of Ceres and Vesta are available at the Dawn mission website https://dawn.jpl.nasa.gov. In 2018, NASA, along with the Office of Science and Technology Policy, the Federal Emergency Management Agency, and several other US governmental agencies, released "The National Near-Earth Object Preparedness Strategy and Action Plan" to coordinate NEO efforts for the next 10 years. The plan can be accessed at https://www.whitehouse.gov/wp-content/uploads/2018/06/National-Near-Earth-Object-Preparedness-Strategy-and-Action-Plan-23-pages-1MB.pdf.

Class P 12: Comet

Aside from solar eclipses and meteor storms, comets are considered to be among the most spectacular astronomical phenomena, known to cultures from ancient times. For most of history their true nature was a mystery, but today comets are known to be small objects formed in the outer Solar System with nuclei composed of ice, rock, and dust, interspersed with frozen gases and other volatiles that sublimate (vaporize) as the comet enters the inner Solar System, producing a coma and sometimes a spectacular tail. Their nuclei typically range in size from 100 meters to 50 km. Their comas may be several thousand kilometers in diameter, and their tails may span millions or tens of millions of kilometers. In 2006, the International Astronomical Union classified comets as "small Solar System bodies," along with minor planets (P 11) and trans-Neptunian objects (P 13) such as those in the Kuiper belt (P 21), where some comets may originate. Several thousand comets are known—only a fraction of the hundreds of thousands believed to exist. According to some theories, comets may evolve into minor planets after they have lost their volatiles.

Comets are usually classified not by composition but by their more observable orbital characteristics, specifically as long period (greater than 200 years) and short period, less than 200 years. A comet with a period less than 20 years will typically travel to the orbit of Jupiter and is thus called a Jupiter-family comet, while those between 20 and 200 years belong to the Halley family. A few so-called main belt comets have been discovered in nearly circular orbits in the asteroid belt. Comets that come within a few million miles of the Sun are designated "Sungrazing" and may have broken up from one large comet. They are known as the "Kreutz group" after the German astronomer Heinrich Kreutz, who first identified them as related and following a similar orbit called the Kreutz path.

Following Aristotle, for most of history comets were believed to be atmospheric phenomena. The Danish astronomer Tycho Brahe (along with Thaddaeus Hagecius and Cornelius Gemma) was among the first to identify comets as a class of astronomical objects when he obtained a distance for the comet of 1577 based on its parallax, placing it in the celestial regions some 230 Earth radii distant. From the 19th century onward, comets were believed to be dusty sandbanks that released gases as they approached the Sun. Harvard astronomer Fred Whipple revolutionized comet science in 1950 when he suggested that comet nuclei were "a conglomerate of ices" including water, ammonia, methane, carbon dioxide, or carbon monoxide combined with meteoritic materials.[21] About the same time, the Dutch astronomer Jan Oort suggested that comets originated from a vast reservoir of objects at the outer edges of the Solar System, now known as the Oort cloud (P 22).[22] That is likely true of long-period comets, but shorter period comets appear to originate from the nearer Kuiper belt (P 21).

Today, Whipple's "dirty snowball" model has been largely verified by spacecraft observations. With the return of Halley's comet in 1986, the Soviet Vega 1 and Vega 2 spacecraft took the first images of a cometary nucleus. A few days later, on March 14, the European Space Agency's Giotto approached within 370 miles of the nucleus, revealing a dark body with hills and ridges whose dimensions proved to be 16 by 18 km (Fig. 4.4). Among the gases

Fig. 4.4. The nucleus of Halley's comet as seen by the Giotto spacecraft in 1986, from a distance of about 2,000 km, the first close encounter with a comet. Outbursts of gas and dust from the comet's nucleus are caused by the heat from the Sun toward the upper left. Credit: ESA/MPS

---

[21] Fred L. Whipple, "A Comet Model I. The Acceleration of Comet Encke," ApJ, 111 (1950), 375–394, reprinted in part in Bartusiak, 445–448. Dick (2013), pp. 202–206.

[22] Jan H. Oort, "The structure of the cloud of comets surrounding the solar system and a hypothesis concerning its origin," BAIN, 11 (1950), 91–110, reprinted in part in Bartusiak, 441–445.

in the coma were water vapor, the dust was indeed meteoric, and large amounts of the organic building blocks, including carbon, hydrogen, oxygen, and nitrogen, were found.

Since Halley's 1986 apparition, a variety of spacecraft have voyaged to these denizens of the heavens. After a journey of 171 days and 268 million miles, on July 3, 2005, NASA's Deep Impact spacecraft released an 820-pound impactor on a course for Comet Tempel 1. The following day it impacted the comet's 14-km-long nucleus at 23,000 miles per hour, producing a spectacular flash of light and a crater of undetermined depth. Analysis of the ejection plume showed large amounts of organic material, confirming that during its history the Earth might have been infused with organics from similar comets. In addition, images from three cameras showed what appear to be impact craters, never before seen on a comet. Other data indicates that the nucleus is extremely porous, a fluffy structure weaker than powdered snow. Continuing its journey, in November 2009 Deep Impact flew within 435 miles of Comet 103P/Hartley 2, and revealed an irregular peanut-shaped body 1.4 km long, spraying jets from both ends and featuring a smooth, perhaps dust-covered middle. Contrary to expectations, the dust-laden jets were being driven by solar heating of subsurface frozen carbon dioxide, rather than by water. The comet was also shedding fluffballs of material a foot or so in diameter, composed of grains 1 to 10 microns across. Meanwhile, in 2006 the Stardust mission returned actual samples of Comet 81/P Wild 2, yielding insights into not only comets but also interplanetary dust (P 15).

The European Space Agency's Rosetta spacecraft was one of the most dramatic comet missions due to both its successes and failures. It orbited comet Churyumov-Gerasimenko beginning in August 2014, and on November 12 its offshoot Philae successfully landed on the comet's surface—the first comet landing (the first minor planet landing had occurred in 2001 with NEAR-Shoemaker). However, Philae's batteries lasted only two days, severely limiting the scientific return. Rosetta itself crashed into the comet in September, 2016. Subsequent analysis of dust grains emitted from the comet as it neared the Sun found that 45% of the comet by mass is organic material and 55% is non-hydrated minerals, meaning they lack water compounds, probably because frozen water prevented mixing. The chemical composition of the comet included 30% each for carbon, oxygen, and hydrogen, with small amounts of iron, magnesium, and other elements.

These observations culminated a long association between comets and culture. The first comet recorded was around 1059 BC in China; since then, comets have often been associated with famous historical events.[23] For example, the daylight comet of 44 BC appeared just after the assassination of Julius Caesar, including the seven days during which Augustus Caesar held games in honor of his predecessor and adopted father. This comet was interpreted as a sign of Julius Caesar's apotheosis, but the same comet had most likely first been seen by the Chinese.[24] In 1066, just before William the Conquerer's invasion of England, a comet appeared that is famously depicted on the Bayeux Tapestry. The *Anglo-Saxon Chronicle* describes the event: "Easter was then on the sixteenth day before the calends of May. Then was over all England such a token seen as no man ever saw before. Some men said that it was the comet-star, which others denominate the long-hair'd star. It appeared first on the eve called 'Litania major', that is, on the eighth before the calends of May; and so shone all the week." This turned out to be what would later be known as the famous Halley's comet, marking one of its 76 year appearances.

Based on their brightness and duration in a dark sky, comets are sometimes designated "Great Comets." Ronald Stoyan's *Atlas of Great Comets* specifies 30 Great Comet apparitions since 1471, including seven apparitions of Halley's comet. Sometimes they are associated with historical events. The great comet of 1811, also known as comet Flaugergues, was visible for 260 days and is sometimes called "Napoleon's comet." It appeared at the peak of Napoleon's power in Europe and was believed to presage his entry into Moscow. Leo Tolstoy wrote in *War and Peace* of the "huge, bright comet of the year 1812 [sic] – surrounded, strewn with stars on all sides, but different from them in its closeness to the earth, its white light and log, raised tail – that same comet

---

[23] A catalog of naked-eye comets reported through 1700 AD is found in Yeomans, *Comets*, pp. 361–424. On comets and culture see Sara J. Schechner, *Comets, Popular Culture, and the Birth of Modern Cosmology* (Princeton: Princeton University Press, 1997).

[24] John T. Ramsay and A. Lewis Licht, *The Comet of 44 B.C. and Caesar's Funeral Games* (Atlanta: Scholars Press, 1997).

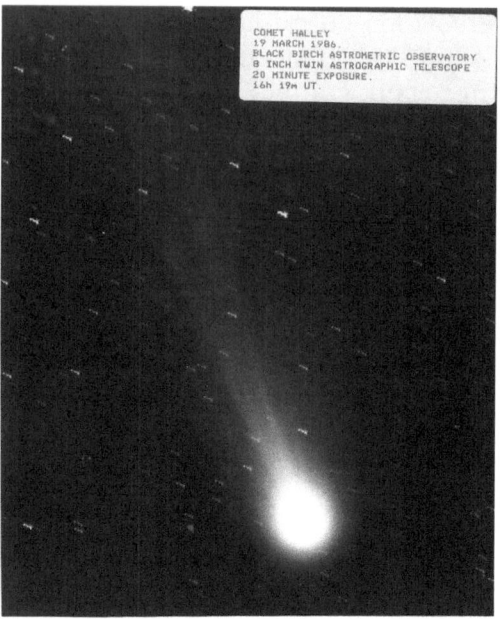

COMET HALLEY
19 MARCH 1986.
BLACK BIRCH ASTROMETRIC OBSERVATORY
8 INCH TWIN ASTROGRAPHIC TELESCOPE
20 MINUTE EXPOSURE.
16h 19m UT.

Fig. 4.5. Halley's comet on its last return in 1986, taken March 19 at Black Birch Astrometric Observatory near Blenheim, New Zealand. Credit: US Naval Observatory

which presaged, as they said, all sorts of horrors and the end of the world."[25] Gradually, comets were seen as more benign, and most people are familiar with recent appearances like Comet Halley in 1986, a disappointment in the northern hemisphere but quite spectacular from southern hemisphere locations such as New Zealand (Fig. 4.5). Among the most recent great comets are Hyakutake (1996), Hale-Bopp (1997), and McNaught (2007). Hale-Bopp induced 39 members of the Heaven's Gate cult to commit suicide in San Diego, in the hope that their souls would hitch a ride on a spaceship they believed accompanied the comet.

A unique cometary event occurred in 1994 when Shoemaker-Levy 9 broke into fragments and its "string of pearls" crashed into Jupiter. Most telescopes on Earth were trained on the event, proving in a more positive way than Hale-Bopp that comets still

---

[25] Leo Tolstoy, *War and Peace*, translated by Richard Pevear and Larissa Volokhonsky (New York: Alfred A. Knopf, 2007), p. 600.

affect culture. More than that, Shoemaker-Levy 9 may have demonstrated what has happened many times on Earth. Impacting comets may originally have brought complex organic molecules and the beginnings of life to our planet. One theory even holds they brought fully formed cells and bacteria to Earth.[26] On the other hand, they may have been responsible for some of the mass extinctions of organisms throughout history, justifying their reputation as "bringers of life, bringers of death."

Comets are often discovered by amateur astronomers and given the names of their discoverers, as in Shoemaker-Levy 9, named after geologists Carolyn and Eugene Shoemaker and ace observer David Levy. That comet broke up in a "string of pearls" prior to impacting Jupiter (Fig. 4.6). Some amateurs are prolific comet discoverers, including Robert McNaught (82 comets), the Shoemakers (32), and Levy (22), mostly discovered photographically. But no one has come close to outdoing systematic search

Fig. 4.6. Fragment R of Comet Shoemaker-Levy 9 impacts Jupiter on July 21, 1994. Composite image taken with the Hubble Space Telescope Planetary Camera. Credit: Hubble Space Telescope Comet Team and NASA

---

[26] Janaki Wickramasinghe, Chandra Wickramasinghe and William Napier, *Comets and the Origin of Life* (Hackensack, NJ: World Scientific, 2010). This volume is the latest in a series of books and articles making this claim, dating to the 1970s, and involving astronomer Fred Hoyle.

programs such as those undertaken with the SOHO spacecraft, the Lincoln Laboratory Near-Earth Asteroid Research project (LINEAR) mentioned above, and the Catalina Sky Survey. By 2015, SOHO had discovered its 3,000[th] comet, marking it as the greatest comet finder of all time.

Comets have been celebrated in art, literature, and poetry. In addition to the Bayeux Tapestry, comets appear in Giotto's Adoration of the Magi of 1303, Durer's Melancolia I engraving of 1514, and numerous paintings of the 19th and 20th centuries. Shakespeare famously wrote in the second Act of *Julius Caesar* (1599), "When beggars die there are no comets seen/the heavens themselves blaze forth the death of princes." Comets are featured in science fiction in works ranging from H. G. Wells's *In the Days of the Comet* (1906) to Gregory Benford and David Brin's *Heart of the Comet* (1986). The drama of a possible cometary impact on Earth has been visualized in films such as *Deep Impact* (1998). Comets are also associated with meteor showers, often caused when the Earth passes through meteoroid streams (P 20) composed of particles released from the previous passage of comets. These spectacular phenomena too have been the subject of woodcuts and art.

The International Astronomical Union's Minor Planet Center maintains a list of coordinates of currently observable comets at https://www.minorplanetcenter.net/iau/lists/CometLists.html as well as forthcoming close approaches and much more information on comets. Images and videos of Sun-grazing comets may be found at the SOHO site https://sohowww.nascom.nasa.gov/gallery/Movies/comets.html.

Class P 13: Trans-Neptunian Objects

While minor planets (P 11) and comets (P 12) mostly orbit the Sun closer than Neptune, many even larger objects, and with somewhat different physical characteristics, orbit beyond Neptune. Such bodies are quite logically termed trans-Neptunian objects, or TNOs for short, and are likely remnants from the accretion process in protoplanetary disks (P 1) during the early Solar System. Until 1992, Pluto and its moon Charon were the only known TNOs (and Pluto for only part of its orbit); since then, more than 2,500 have been discovered. They have become objects of a rich program of exploration of the outer Solar System, spearheaded by ground-based telescopes detecting extremely faint objects, and supplemented by space telescopes and the New Horizons spacecraft. Aside from New Horizons *in situ* imagery from the Pluto system, because of their relatively small sizes and large distances, we lack spectacular images of TNOs. They nevertheless represent a revolution in our knowledge of the Solar System.

Just as galaxies may be distinguished from galactic systems, stars from stellar systems, and planets from planetary systems, so trans-Neptunian objects can be distinguished from trans-Neptunian systems. The majority of TNOs exist in two such systems: the Kuiper belt (P 21) and the more distant Oort cloud (P 22). In addition to these two systems, TNOs include some of the dwarf planets (P 9), which are the largest known trans-Neptunian objects. The International Astronomical Union officially designates TNOs as part of the Subfamily of Small Bodies of the Solar System, along with minor planets and comets. They deserve class status not only for this reason, but also because they are believed to differ in physical composition from minor planets and comets. Although this represents some redundancy in the classification system (Pluto being simultaneously a dwarf planet, a TNO, and a member of the Kuiper belt), it reflects the IAU decision, the literature on the subject, and our incomplete knowledge of trans-Neptunian objects themselves. Such redundancy is common in classification systems.[27]

---

[27] M. Barucci et al., "The Solar System Beyond Neptune: Overview and Perspectives," in M. A. Barucci et al., eds., *The Solar System Beyond Neptune*, (Tucson: University of Arizona Press, 2008), pp. 3–10.

Despite the difficulties of their detection, a long history of speculation exists about objects beyond Neptune. Theories of planetary accretion, as well as comets and asteroids known to exist in the inner Solar System, indicated that similar bodies should be found in the outer Solar System. The technology of the time, however, made them very difficult to detect, as the painstaking discovery of Pluto in 1930 proved. Still, this did not prevent speculation. Both the American astronomer Gerard Kuiper and the Irish astronomer Kenneth Edgeworth noted in the 1940s and 1950s that there was no reason to expect that the Solar System ended with Neptune or Pluto. Quite aside from possible major planets, Edgeworth theorized that comets might originate in a reservoir beyond Pluto, and Kuiper argued that such objects would be scattered to the Oort cloud and be perturbed back into the inner Solar System as comets.[28]

Harvard astronomer Fred Whipple and others developed these ideas in the 1960s, and by 1980 a comet belt was advanced as the source of short-period comets, as was the Oort cloud for long-period comets. The short-period belt was first named the "Kuiper belt" in 1988, a term that gained rapid usage thereafter, although "Edgeworth-Kuiper Belt" is occasionally used, especially in Europe.[29] Detection of any such objects, however, remained elusive, since a given asteroid was 10,000 times fainter when moved from 3 to 30 astronomical units.

All this changed with the arrival of CCD technology in the 1990s, which enabled detection of extremely faint objects that we now recognized as Kuiper belt objects, such as 1992 QB1 (Fig. 4.7), along with a few possible inner Oort cloud members. The largest trans-Neptunian objects (Fig. 4.8) now known are those officially

---

[28] John K. Davies et al., "The Early Development of Ideas Concerning the Transneptunian Region," in M. A. Barucci et al., eds. (2008), pp. 11–23. K. E. Edgeworth, "The Evolution of our Planetary System," JBAS, 53, (1943), 181–188: 186; Edgeworth, "The origin and evolution of the solar system," MNRAS, 109 (1949), 600–609; and Gerard Kuiper, "On the origin of the solar system," in *Astrophysics: A Topical Symposium*, J. A. Hynek, ed., (New York: McGraw-Hill Book Co. 1951), pp. 357–424:402.

[29] F. Whipple, "Evidence for a comet belt beyond Neptune, PNAS, 51 (1964), 711–718; J. Fernandez, "On the existence of a comet belt beyond Neptune," MNRAS, 192 (1980), 481–491.

M. Duncan et al., "The origin of short-period comets," ApJ, 328 (1988), L69–L73. On the nomenclature controversy see http://www.cfa.harvard.edu/icq/kb.html; (Dick, 2013), pp. 57–62.

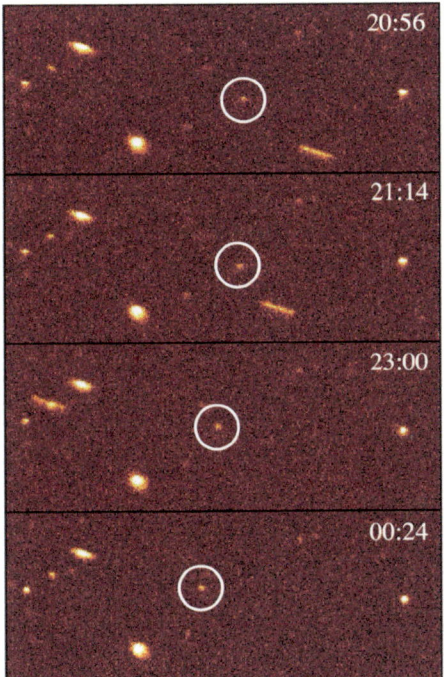

Fig. 4.7. Discovery image of 1992 QB1, the first Kuiper belt object, aside from Pluto itself, detected by David Jewitt and Jane Luu. The extremely faint 22nd magnitude object (circled) was detected with the University of Hawaii's 2.2-meter telescope, equipped with a charge-coupled device (CCD) detector. Courtesy David Jewitt

designated dwarf planets (P 9), including Pluto (Fig. 4.1), Eris, Makemake, and Haumea, and those in the Kuiper belt, including Orcus, Quaoar, Varuna, and Ixion. Most are much smaller; the smallest found to date, only 3,200 feet across, was detected by the Hubble Space Telescope and announced in December 2009. At a distance of 4.2 billion miles and 35th magnitude, it is too faint to be imaged even by Hubble, and was detected by occulting another star.[30] Like the other small bodies of the Solar System, and as originally suspected by Edgeworth, Kuiper, and others, TNOs may shed light on the nature of the original protoplanetary disk of our Solar System.

---

[30] H. E. Schlichting et al., "A single sub-kilometre Kuiper belt object from a stellar occultation in archival data," *Nature*, 462 (2009), 895; HST Release, "Hubble Finds Smallest Kuiper Belt Object Ever Seen," December 16, 2009, http://hubblesite.org/newscenter/archive/releases/2009/33/

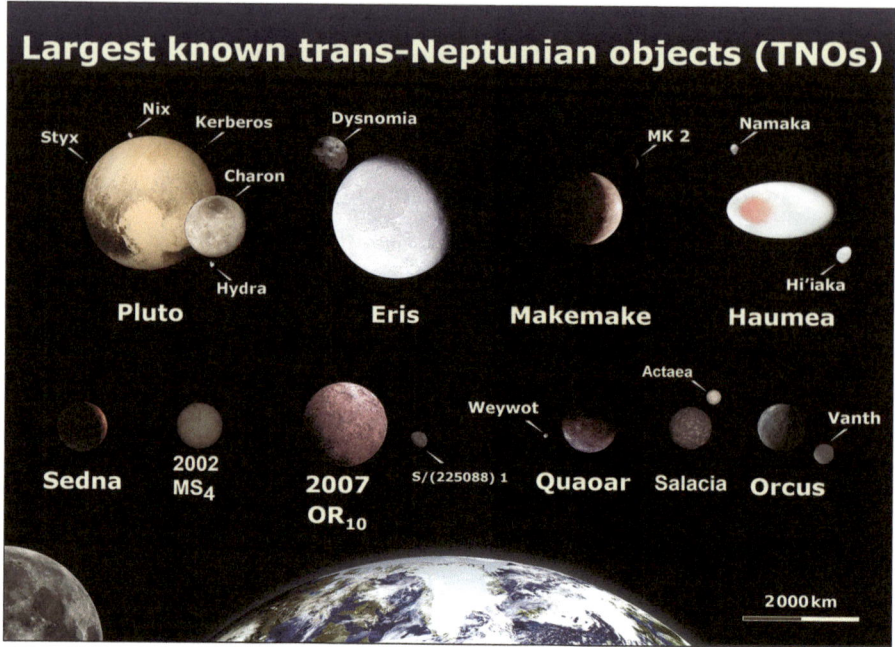

Fig. 4.8. Comparison of the eight brightest TNOs: Pluto, Eris, Makemake, Haumea, Sedna, 2007 OR10, Quaoar, and Orcus. Credit: GSFC/NASA

No TNOs have been unambiguously detected in the much more distant Oort cloud, which remains a theoretical but highly likely location of millions more. However, some astronomers, including original discoverer Michael Brown, have argued that the object known as Sedna is an inner Oort cloud object. The IAU's Minor Planet Center still classifies it as a scattered disk object, a subset of the Kuiper belt.[31] Discovered on March 15, 2004, Sedna is three times farther than Pluto at about 90 astronomical units—the most distant known object in the Solar System. It takes 10,500 years to orbit the Sun, compared to 248 years for Pluto. Its likely diameter is between 1,200 and 1,600 km. Some astronomers prefer to call it simply a "detached object" in the sense that it is

---

[31] Michael E. Brown, Chadwick Trujillo and David Rabinowitz, "Discovery of a Candidate Inner Oort Cloud Planetoid," ApJ, 617 (2004), 645–649; http://www. gps.caltech.edu/%7Embrown/papers/ps/sedna.pdf. Dick (2013), pp. 57–62.

detached from Neptune's influence. Sedna is named after the Inuit goddess of the sea. Other detached objects have also been found closer than Sedna, but none have been found farther, most likely due to the difficulties of detection. Many more are sure to be found in the future.

Sophisticated attempts have been made to develop a taxonomy of trans-Neptunians largely on the basis of color, probably a reflection of their physical characteristics.[32] All categories of trans-Neptunian objects observed so far appear to have roughly the same low densities and composition of frozen ices such as water and methane. However, the so-called scattered disk objects tend to have a more white or grey appearance, compared to reddish color for the classical Kuiper belt objects; this may be due to methane having frozen over the entire surface as opposed to Pluto, for example, where the methane may have frozen only at higher elevations, giving a lower surface brightness. This feature does not seem to be enough to distinguish scattered disk objects as a new class of objects, at least in the opinion of many astronomers. Dynamical properties would be a stronger argument for class status separate from KBOs, but even on those grounds, astronomers have declared scattered disk objects to be part of the Kuiper belt. Like so many classes of objects, it is a matter of definition—one that becomes clearer if physical composition is the defining criterion for class status, and which in this case is reinforced by dynamical considerations.

An official list of trans-Neptunian objects, updated daily, is found at http://www.cfa.harvard.edu/iau/lists/TNOs.html, and a more readable but incomplete list with more information is at https://en.wikipedia.org/wiki/List_of_trans-Neptunian_objects

---

[32] M. Fulchignoni et al., "Transneptunian Object Taxonomy," in M. A. Barucci et al., eds., *The Solar System Beyond Neptune*, (Tucson: University of Arizona Press, 2008), pp. 181–192.

5. The Interplanetary Medium Family

Class P 14: Gas

Like the interstellar medium, the interplanetary medium is composed of gas and dust (P 15), as well as energetic particles from the solar wind (P 16) and cosmic rays (S 30). The solar wind creates the heliosphere, a bubble within the more general interstellar wind, and while most of the material in the heliosphere emanates from the Sun itself, a small part consists of neutral gas that makes its way into the Solar System rather than flowing outward from the Sun. Thus, whereas both the solar wind and most interplanetary dust originate within the Solar System, interplanetary gas is a rarefied neutral gas flowing through the Solar System at a speed of about 25 km/sec. More precisely, the Solar System itself is moving through what is sometimes called the "local interstellar cloud" at that speed, so space scientists measure the wind as would a motorcyclist speeding down the highway. The particles become charged as they approach the Sun, and eventually may become Anomalous Cosmic Rays (P 17).

Interplanetary gas is thus a rarefied interstellar wind that differs from the much denser solar wind, which is composed of energetic charged particles emanating from the Sun at very high speeds. The local interstellar wind is composed mostly of helium and perhaps 10% hydrogen; near Earth, about one atom of neutral interstellar gas exists per ten cubic centimeters, compared to ten atoms of ions from the solar wind in the same ten cubic centimeters. Put another way, interplanetary helium is a billion trillion times less dense than the Earth's atmosphere at its surface. Interplanetary gas may also be in the form of plasma—ionized gas from the solar wind.

The existence of diffuse interstellar gas had been known at least since 1930, and the existence of interplanetary plasma in the form of solar wind was observed since the late 1950s. However, observations of any neutral gas in the interplanetary medium depended on ultraviolet detectors, because neutral gaseous components of the medium effectively absorb far ultraviolet light. The so-called

© Springer Nature Switzerland AG 2019
S. J. Dick, *Classifying the Cosmos*, Astronomers' Universe,
https://doi.org/10.1007/978-3-030-10380-4_5

Lyman alpha transition in hydrogen, the same transition that can be observed in intergalactic gas and Lyman alpha blobs (G 14 and G 15), is a prime example. Such observations could only begin when rockets or spacecraft could carry ultraviolet detectors above Earth's atmosphere.

In the 1960s, detectors aboard the Russian Zond 1 and Venera spacecraft, as well as the American Orbiting Geophysical Observatory (OGO) and Mariner spacecraft, found hydrogen in interplanetary space based on observations of solar Lyman alpha scattering. OGO-5 was able to make more global measurements. In reporting results from the OGO-5 in 1971, French physicists Jean-Loup Bertaux and Jacques Blamont concluded, "Our results give strong support to the theory that in its motion towards the apex, the sun crosses neutral atomic hydrogen of interstellar origin, giving rise to an apparent interstellar wind . . . it can be inferred that hydrogen coming from the interstellar medium does penetrate deep in the heliosphere; the density and the direction of flow can be deduced from the Lyman-alpha emission." Similarly, University of Colorado astronomers G.E. Thomas and R.F. Krassa, also reporting OGO-5 results in an adjacent article, concluded, "The most likely origin of the observed emission is scattering of solar Lyman alpha from interstellar hydrogen that is swept into the vicinity of the earth from the direction of the apparent apex of motion of the Solar System."[1]

Interest in the interplanetary medium greatly increased after the detection of interplanetary hydrogen in the form of Lyman alpha emission, and in 1960 and 1961 Herbert Friedman and J. C. Brandt, both space rocket pioneers, predicted that a diffuse glow of neutral helium might also be detected from the interplanetary medium.[2] In March 1970, Francesco Paresco, Stuart Bowyer, and their colleagues detected radiation emitted by this neutral helium when a Nike Tomahawk rocket flew three extreme-ultraviolet photometers to an altitude of 264 km to study the 584 Angstrom terrestrial nightglow. "We believe these measurements to be the first observation of this radiation. We advance the possibility that a major portion of the 584 Angstrom nightglow has an astrophysi-

---

[1] J. L. Bertaux and J. E. Blamont, "Evidence for a Source of an Extraterrestrial Hydrogen Lyman Alpha Emission: The Interstellar Wind," A&A, 11 (1971), 200-217: 200, 216; G. E. Thomas and R. F. Krassa, "OGO 5 Measurements of Lyman Alpha Sky Background," A&A, 11 (1971), 218-233. Sections VII and VIII of the latter article provide references for earlier sporadic detections.

[2] Herbert Friedman, in *Physics of the Upper Atmosphere*, ed. J. A. Ratcliffe (New York and London: Academic Press, 1960), pp. 202-206; J. C. Brandt, "Interplanetary Gas VI. On Diffuse Extreme Ultraviolet Helium Radiation in the Night and Day Sky," ApJ, 134 (1961), 975-980.

cal rather than a geophysical origin," the authors wrote in their 1973 paper. A second paper the following year explicitly stated that the observed radiation "is emitted by neutral helium in the interplanetary medium."[3]

Such rocket observations were complicated by the terrestrial atmosphere and limited observation time, and detailed observation were left for a variety of satellites in the 1970s, including the Naval Research Laboratory's STP 72-1, OGO-8, Solrad 11-B, Mariner, and the Russian Prognoz 5 and 6 spacecraft. Dedicated ultraviolet spacecraft, including Copernicus and the International Ultraviolet Explorer, provided even more detail, and the Hubble Space Telescope has also recorded the spectrum of the interplanetary hydrogen Lyman alpha glow.[4] The same ultraviolet photometry and spectroscopy used for the study of planetary atmospheres was the key technology for early studies of the interplanetary medium.

General properties of the interplanetary medium, including its gaseous component, have also been studied more recently with many other satellites, including ACE (Advanced Composition Observer), POLAR, SOHO, Wind, Ulysses, and IBEX.[5] ACE observations of the "helium-rich breeze" from the stars flowing into the Solar System from the direction of the constellation Ophiuchus are described at http://science.nasa.gov/science-news/science-at-nasa/2004/17dec_heliumstream/, including diagrams to help explain the flow. "We see strange gusts, ebbs and flows" likely caused by interactions with the solar wind, said George Gloecker, one of the lead co-investigator for ACE. A diagram of the local interstellar cloud can be found at http://apod.nasa.gov/apod/ap020210.html. This Astronomy Picture of the Day site contains much succinct information relevant to different classes of objects in space.

---

[3] The discovery paper for neutral interplanetary helium is Francesco Paresce, Stuart Bowyer et al, "Evidence for an Interstellar or Interplanetary Source of Diffuse He I 584 Angstrom Radiation," ApJ, 183 (1973), L87-L90, and "Observations of He I 584 Angstrom Nighttime Radiation; Evidence for an Interstellar Source of Neutral Helium," ApJ, 187 (1974), 633-639.

[4] F. Dalaudier, J. L. Bertaux et al, "Characteristics of interstellar helium observed with Prognox6 58.4 nm photometers," A&A, 134 (1984), 171-184. On Hubble observations see R. Lallement et al., "The GHRS [Goddard High-Resolution Spectrograph] and the Heliosphere," *Science with the Hubble Space Telescope II*, P Benvenuti et al., eds. http://www.stsci.edu/stsci/meetings/shst2/lallementr.html

[5] E. Mobius et al., "Diagnosing the Neutral Interstellar Gas Flow at 1 AU with IBEX-Lo," *Space Science Reviews* 146(2009),149-172; V. G. Kurt et al, "Ultraviolet Studies of the Interplanetary and Local Interstellar Medium," *Cosmic Research*, 47 (Feb 2009), 1-13.

Class P 15: Dust

Interplanetary dust (IDP), sometimes called cosmic dust, consists of particles ranging in diameter from 10 to 200 micrometers— about the width of a single human hair. In general, meteoroids (P 10) are larger particles, though at the micrometeoroid level astronomers consider the boundary between them and IDPs indistinct and a matter of semantics. While cosmic dust was long believed to originate with collisions among asteroids, comets, and other small bodies of the Solar System, recent computer models suggest that 90% of interplanetary dust particles originate with Jupiter-family comets, which (unlike asteroids) range far from the plane of planetary orbits, thus accounting for the observed breadth of distribution of the dust particles above and below the ecliptic plane. The dust particles especially span the inner Solar System from Mercury to Jupiter. More than 100 tons of cosmic dust grains fall onto Earth each day.[6]

In contrast to interplanetary gas, which is the local interstellar medium the Solar System happens to be passing through, most IDPs originate in the Solar System itself. In aggregate, IDPs are sometimes referred to as the interplanetary dust cloud, or "zodiacal cloud." Their collective mass might be equal to a small asteroid perhaps 15 km in diameter. Along with meteorites and icy trans-Neptunian objects (P 13), IDPs are important for studying the earliest stages of Solar System formation (P 1), including the formation of minerals and organic matter. IDP astronomers have likened themselves to archaeologists studying human origins through four- million-year old fossils, except they are studying Solar System origins through 4.6 billion-year-old fossils in the form of IDPs.

Interplanetary dust particles scatter sunlight and produce the naked-eye phenomenon known as zodiacal light, a faint white glow seen only in the darkest skies (Fig. 5.1). Although the zodiacal light was undoubtedly first seen in the dark skies of antiquity, the French/Italian astronomer Giovanni Domenico Cassini first investigated this phenomenon in a paper dated 1683. The following year, the Swiss mathematician Nicolas Fatio de Dullier explained

---

[6]David Nesvorny, Peter Jenniskens, Harold F. Levison et al., "Cometary Origin of the Zodiacal cloud and Carbonaceous Meteorites: Implications for Hot Debris Disks," ApJ, 713 (2010), 816.

Fig. 5.1. Zodiacal light caused by interplanetary dust is seen as the verti-cal white glow in this image from the European Southern Observatory's La Silla Observatory in Chile. The image was taken just after sunset. Credit: ESO/Y. Beletsky

the phenomenon was due to the reflection of sunlight by particles circling the Sun.[7] Zodiacal light has been studied ever since.

The greatest progress has been made during and after the second half of the 20th century with quantitative measurements and spaceborne observations. Optical properties such as albedo and polarization have been measured for different parts of the zodiacal cloud, as well as physical properties including the porosity, size distribution, and bulk density of various popula-tions of dust. In addition to observations in the visual part of the spectrum, thermal emission in the infrared region has also been observed from IDPs. Nothing, however, is known about the

---

[7]Giovanni Domenico Cassini, "Nouveau phenomenone rare et singulier d'une Lumiere Celeste, qui a paru au commencement du Printemps de cette annee 1683," *Journal des Scavans*, June 10, 1683, and Cassini "Decouverte de la lumiere celeste qui paroiste dans le zodiaque," French Academie des Sciences, 1693; Hugo Fechtig et al., "Historical Perspectives," in E. Grun et al., eds., *Interplanetary Dust* (Berlin: Springer-Verlag, 2001), pp. 1-56.

temporal evolution of the cloud.[8] Zodiacal light experiments on the Pioneer 10 and 11 and Helios A and B spacecraft determined the overall distribution of the interplanetary dust cloud. Observations of comet Halley flyby spacecraft, as well as infrared measurements by IRAS and COBE, have also contributed substantially to the field. Conferences were held as early as the 1960s on the subject and have been continued under the auspices of COSPAR, the International Astronomical Union, and other international bodies.

In addition to astronomical observations from the ground and in space, IDPs can be collected for study in the laboratory. The ability not only to observe but also to collect dust particles constituted a landmark in the study of interplanetary dust. The first attempt to collect particles was made using V-2 rockets in 1950. In the mid-1970s, University of Washington astronomer Don Brownlee and his colleagues pioneered modern collection techniques when they began collecting IDPs in the stratosphere (20 km altitude) using balloons and U-2 aircraft. The particles, 2–30 micrometers in size, could be returned to laboratories for study with electron microscopes and all the advanced techniques available to science, the first task being to determine that they were indeed extraterrestrial.

Analysis of 300 interplanetary dust particles recovered in this way showed them to be primitive Solar System material similar to so-called C1 meteorites, but with their own unique features.[9] Antarctic micrometeorites (AMMs), which constitute the upper end of the interplanetary dust mass range and the dominant mass of the material accreted by the Earth, have also been collected and subject to detailed analysis. For example, the noble gases He, Ne, Ar, Kr, and Xe have been measured in 27 individual Antarctic micrometeorite in the size range 60 to 250 micrometers that were collected at the Dome Fuji Station. Their composition implies that carbonaceous chondrite-like objects may be the source of these particles (see P 11).[10]

---

[8] A. Ch. Levasseur-Regourd, "Zodiacal light, certitudes and questions," *Earth, Planets, Space*, 50 (1998), p. 610. *Interplanetary Dust and Zodiacal Light*, H. Elsässer and H. Fechtig, eds. (Springer-Verlag, Heidelberg, 1976); *Physics, Chemistry, and Dynamics of Interplanetary Dust*, B. A. S. Gustafson and M. S. Hanner, eds. (Astronomical Society of the Pacific: San Francisco, 1996).

[9] D. E. Brownlee et al., "An Atlas of Extraterrestrial Particles Collected with NASA U-2 Aircraft: 1974-1976, NASA TM X-73152, 47 pp; D. E. Brownlee et al., "Interplanetary dust; A new source of extraterrestrial material for laboratory studies," *Proc. Lunar Sci. Conf.* 8 (1977), 149-160; D. Brownlee "Cosmic Dust: Collection and Research," AREPS, 13 (1985), 147-183.

[10] Takahito Osawa and Keisuke Nagao, "Noble gas compositions of Antarctic micrometeorites collected at the Dome Fuji Station in 1996 and 1997," *Meteoritics and Planetary Science*, 37 (2002), 911-936.

The most ambitious dust-collecting project either on the ground or in space has been NASA's Stardust mission to the Jupiter-family comet 81P/Wild 2. Like other comets, Wild 2 is believed to have formed in the Kuiper belt and only recently entered the Solar System; as such, it is an emissary from a distant, inaccessible region of the system. Launched in 1999, Stardust encountered the comet on January 2, 2004 at a distance of only 234 km, collecting particles from the coma in the 1- to 300-micron-diameter range. The sample capsule returned to Earth January 15, 2006.

Analysis showed the comet dust was similar to materials found in primitive meteorites. "What we found was remarkable," Stardust principal investigator Donald Brownlee noted. "Instead of rocky materials that formed around previous generations of stars, we found that most of the comet's rocky matter formed inside our Solar System at extremely high temperature. In contrast to its ice, our comet's rocky material had formed under white-hot conditions... We now know that comets are really a mix of materials made by conditions of both fire and ice." The larger particles are composed of several minerals, including chondrules and calcium aluminum inclusions. The total mass of the approximately 10,000 cometary particles collected by Wild 2 was equivalent to several hundred thousand of the generally smaller nanogram IDPs studied in the laboratory over the previous 35 years.[11] The first decade of Stardust research resulted in 142 research papers. Study of these particles, which involved more than 200 investigators around the world, and of other interplanetary dust particles previously collected by other means, continues to provide important information on the origin of the Solar System as well as circumstellar disks (S 15).

Beginning in 1981, cosmic dust was studied at the Cosmic Dust Laboratory of NASA's Johnson Space Center in Houston. The Laboratory, which holds several thousand cosmic dust particles and distributes them to researchers, closed in 2009 and transitioned to a new laboratory to process the materials returned by the Stardust mission.[12] Based on the results of combined research

---

[11] Donald Brownlee et al., "Comet 81P/Wild 2 Under a Microscope," *Science*, 314 (2006), 1711-1716, a special issue on the Stardust mission; Donald Brownlee, "Stardust: A Mission with Many Scientific Surprises," October 29 2009, at https://stardust.jpl.nasa.gov/news/news116.html. A comprehensive review after a decade of Stardust analysis is A. J. Westphal et al., "The Future of Stardust Science," *Meteoritics and Planetary Science*, at https://arxiv.org/pdf/1704.01980.pdf

[12] NASA, "Cosmic Dust," http://curator.jsc.nasa.gov/dust/ and NASA, "Astromaterial Acquisition and Curation Office," https://curator.jsc.nasa.gov/index.html

of many scientists around the world, IDPs are now often classified by size: those collected in the stratosphere are generally less than about 50 microns, Antarctic micrometeorites that have fallen to Earth are generally 20 to hundreds of microns in diameter, and those from Stardust's comet Wild 2 ranged up to 300 micrometers. IDPs may also be classified by composition.

The most general studies of interplanetary dust and their relation to dust in the interstellar medium have been carried out with spacecraft. In particular, the Ulysses and Galileo spacecraft showed that a small component of interplanetary dust originates outside the Solar System. In addition to the regular dust grains, the spacecraft also detected relatively large dust grains between 0.2 and 6 micrometers in diameter. Because they flow with the same velocity and direction as the local interstellar wind that is the source of interplanetary neutral hydrogen gas, they are inferred to originate from the same source.

Along with spacecraft studies and samples, progress continues to be made with nearby samples. Recent analysis indicates that certain types of interplanetary dust containing silicate grains known as GEMS, collected from the Earth's upper atmosphere and believed to originate from comets, survive from interstellar environments preceding the formation of the comet and are the original building blocks of planetary systems.[13]

Cosmic dust is curated at the Johnson Space Center in Houston as part of its Astromaterials Acquisition and Curation Office. Information on its collections from a variety of sources, including the Stardust mission, is available at the website https://curator.jsc.nasa.gov.

NASA's Curation and Analysis Planning Team for Extraterrestrial Materials (CAPTEM) coordinates matters concerning the collection and curation of extraterrestrial samples, including planning future sample return missions as well as evaluating proposals requesting allocation of all extraterrestrial samples contained in NASA collections. A paper on interplanetary dust for the planetary science decadal survey provides a survey of research issues on interplanetary dust: https://www.lpi.usra.edu/decadal/sbag/community_wp/SB_Community_WP_Final_Dust.pdf.

---

[13] Hope Ishii et al., "Multiple generations of grain aggregation in different environments preceded solar system body formation," PNAS (2018), online at http://www.pnas.org/content/early/2018/06/04/1720167115, news article at http://www.sci-news.com/space/interplanetary-particles-presolar-dust-06094.html

Class P 16: Solar Wind

Surprisingly, the Sun sends matter as well as light into space. This "solar wind" consists of a stream of charged particles, primarily electrons and protons, ejected from the Sun at speeds of 200 to 600 miles per second, about a million miles per hour. This corresponds to energies of 10 to 100 electron volts, whereas air molecules at room temperature have energies of about 1/40 of an electron volt. The acceleration mechanism is believed to be the magnetic field associated with sunspots. These solar energetic particles are also sometimes termed "solar cosmic rays," but the term "cosmic rays" also refers to much higher energy particles originating beyond the Solar System (P17, S30 and G18). By contrast, the sources of solar wind include solar flares and corona mass ejections from the Sun.[14] The solar wind also creates the Solar System's heliosphere.

The solar wind is important for a variety of reasons, both scientific and practical. It is responsible for the phenomenon of interplanetary scintillation—the fluctuation of radio intensity from compact celestial radio sources such as pulsars and quasars, analogous to "twinkling" in the optical realm. It is also a significant component of space weather. In 1964 the British science fiction writer Arthur C. Clarke published the famous short story "Sunjammer," later retitled "The Wind from the Sun," about solar sailing making use of the solar wind; the latter is also the title of a collection of his short stories first published in 1972.

The realization of solar wind as a reality came only slowly. British astronomer Richard Carrington first suggested the concept in 1859, after observing a solar flare. By the 1930s scientists realized the Sun's corona must be about a million degrees Celsius (1.8 million degrees Fahrenheit). In 1955, Sydney Chapman calculated the properties of such a hot gas and concluded that it must extend beyond Earth's orbit. Indirect observational evidence of particle emission from the Sun came in 1951 from the German scientist Ludwig Biermann, who postulated that this steady stream of particles caused a comet's tail to always point away from the Sun. James Van Allen added another piece of the puzzle with his discovery of Earth's radiation belts (P 8) using his detector on the first American satellite, Explorer 1. Van Allen

---

[14] R. A. Howard, "A Historical Perspective on Coronal Mass Ejections," *Solar Eruptions and Energetic Particles*, N. Gopalswamy, R. Mewaldt, and J. Torsti, eds (2006).

attributed these particles to the solar plasma, but proof of the source remained elusive.[15]

The landmark event in recognizing the solar wind as a class of energetic particles came in 1958 when astronomer Eugene Parker wrote a seminal paper on the subject of interplanetary gas in the *Astrophysical Journal* and coined the term "solar wind."[16] Parker's theory of the solar wind was not widely accepted at first; in fact, his paper was initially rejected and only published after the intervention of the journal's editor, Subrahmanyan Chandrasekhar. But satellites confirmed the solar wind in the 1960s, at first tentatively with the brief measurements of Luna 3 in 1959 and Explorer 10 in 1961, and then more persuasively with the electrostatic analyzer aboard Mariner 2 on its way to Venus. It was Mariner 2 that established the continuous flow of the solar wind, its composition, velocity, temperature, and density; the Pioneer spacecraft and many others have since refined our knowledge.

The solar wind was even studied from the ground using the newly discovered technique of interplanetary scintillation. Anthony Hewish, who received the 1974 Nobel Prize for the discovery of pulsars using the same technique, recalled of his work in the early 1960s:

> Since interplanetary scintillation, as we called this new effect, could be used in any direction in space, I used it to study the solar wind, which had by then been discovered by space probes launched into orbits far beyond the magnetosphere. It was interesting to track the interplanetary diffraction patterns as they raced across England at speeds in excess of 300 km/second, and to sample the behaviour of the solar wind far outside the plane of the ecliptic where spacecraft have yet to venture.[17]

---

[15] On Carrington and the 19th century history of our understanding of the Sun see Stuart Clark, *The Sun Kings* (Princeton: Princeton University Press, 2007). For subsequent solar history see Karl Hufbauer, *Exploring the Sun: Solar Science Since Galileo* (Baltimore and London: Johns Hopkins University Press, 1991).

[16] Eugene N. Parker, "Dynamics of the Interplanetary Gas and Magnetic Field," ApJ, 128 (1958), 664-676; Parker, "Cosmic-ray modulation by solar wind," *Physical Review*, 110 (1958), 1445-1449; Karl Hufbauer, *Exploring the Sun*, pp. 213 ff.

[17] Anthony Hewish, "Pulsars and High Density Physics," Nobel Lecture, December 12, 1974, pp. 174-175, online at https://www.nobelprize.org/prizes/physics/1974/hewish/lecture/; James Van Allen, "Interplanetary Particles and Fields," SciAm (September, 1975), reprinted in *The Solar System* (W. H. Freeman: San Francisco, 1975), 125-134.

The solar wind has been the subject of much research since that time, in part because it is a component of space weather, and its effects include geomagnetic storms that disturb power grids on Earth, as well as aurora and other solar-terrestrial phenomena. The Solar Wind Composition Experiment was one of the first experiments employed during Apollo 11, the first manned lunar landing in 1969. It was repeated on Apollo 12, 14, 15, and 16. It consisted simply of an aluminum foil sheet, 1.4 by .3 meters, deployed on a pole facing the Sun. This was exposed to the Sun for periods ranging from 77 minutes on Apollo 11 to 45 hours on Apollo 16. The embedded particles were then returned to Earth for analysis. Among the gases detected were helium-3, helium-4, neon-20, 21, and 22, and argon 36. But the Solar Wind Spectrometer on Apollo 12 and 15 confirmed that more than 95% of the solar wind is in the form of equal numbers of electrons and protons.[18]

Dedicated spacecraft such as the Solar and Heliospheric Observatory (SOHO) studied the solar wind conditions in much more detail (Fig. 5.2). The Advanced Composition Explorer,

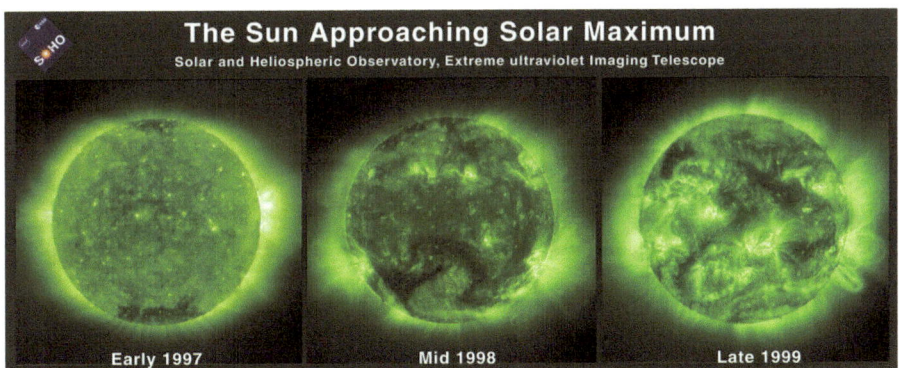

Fig. 5.2. A comparison of three images almost three years apart illustrates how the level of solar activity increases significantly as the Sun attains its expected sunspot maximum. The progression towards more active regions and the number/size of magnetic loops is unmistakable. Many more sunspots, solar flares, and coronal mass ejections occur during the solar maximum, and so the solar wind increases accordingly. Credit: SOHO/NASA/ESA

---

[18] Lunar and Planetary Institute, "Apollo 11 Mission," http://www.lpi.usra.edu/lunar/missions/apollo/apollo_11/experiments/swc/

launched in 1997 and placed at the gravitationally stable L1 point between the Earth and the Sun, still monitors real-time solar wind conditions at the spacecraft. The Genesis spacecraft actually collected solar wind particles for return to Earth. And in 2018 a new solar probe, appropriately named the Parker Solar Probe, was launched to study the solar wind and the dynamics of the solar corona. It will orbit the Sun 24 times during its seven-year mission, passing within nine solar radii of the Sun on December 19, 2024. Needless to say, it sports one of the strongest heat shields ever constructed in order to counter the 2,500° F temperature it will face.[19]

Ludwig Biermann's suggestion—that the solar wind is also partly responsible for the observed fact that a comet's tail always points away from the Sun—has proven true. It has been estimated that the solar wind removes about two billion pounds of matter from the Sun every second; even so, during its 5 billion year history, only .01% of the Sun's mass has been lost through the solar wind. Some stars have much stronger stellar winds (S 29), including the massive Wolf-Rayet stars.

The Space Weather Prediction Center at NOAA gives current space weather predictions, including the state of the solar wind speed and magnetic fields, at https://www.swpc.noaa.gov and https://www.swpc.noaa.gov/phenomena/solar-wind, with numerous further links. Much more information about the current state of the Sun is available at the Solar Dynamics Observatory and SOHO websites https://sdo.gsfc.nasa.gov and https://sohowww.nascom.nasa.gov/home.html. For information and updates on the Parker Solar Probe, see http://parkersolarprobe.jhuapl.edu and https://www.nasa.gov/content/goddard/parker-solar-probe.

---

[19] NASA Marshall Space Flight Center, "Solar Physics," http://solarscience.msfc.nasa.gov/SolarWind.shtml

Class P 17: Anomalous Cosmic Rays

Anomalous cosmic rays (ACRs) are energetic particles of interstellar origin that travel into the Solar System near the speed of light. They have energies of about one thousand electron volts (one KeV, ten times the solar wind energies) to 350 million electron volts (350 MeV). They are not to be confused with lower energy particles, called solar energetic particles, from solar flares and coronal mass ejections that originate from our Sun and constitute the solar wind (P 16). Solar energetic particles are also considered cosmic rays, in the general sense that they are charged particles that bombard the Earth from beyond the Earth's atmosphere. But ACRs are very different from that more local source of energetic particles. ACRs are also distinct from higher energy galactic cosmic rays (S 30) or extragalactic cosmic rays (G 18), although those classes of cosmic rays can and do also enter the Solar System from more distant regions. Because anomalous cosmic rays and extragalactic cosmic rays were discovered much later than galactic cosmic rays, in common usage "cosmic rays" usually refers to galactic cosmic rays as discussed in Class S 30.

Anomalous Cosmic Rays are termed "anomalous" because of their unusual composition by comparison with the other two classes of cosmic rays. They are composed chiefly of helium and the heavier elements nitrogen, oxygen, argon, and neon. By contrast, most incoming higher energy galactic and extragalactic cosmic rays are simple protons. The heavier elements comprising ACRs are believed to represent the local interstellar medium at the boundary of the heliopause, the edge of the Solar System where the strength of the stellar wind (S 29) begins to overtake the solar wind. These elements are thought to have originated as neutral atoms in the interstellar medium that have been ionized by solar ultraviolet radiation after entering the Solar System, then accelerated at the termination shock of the heliosphere. The mechanisms of acceleration are still uncertain, but the solar wind helps in the acceleration process. After acceleration, they diffuse toward the inner heliosphere.

Anomalous cosmic rays were first discovered during 1972–1973 at the 10 MeV level by detectors aboard the Explorer spacecraft IMP-5 and IMP-7, and by the Pioneer 10 spacecraft. Many more spacecraft have detected them since then, including the SAMPEX Polar Orbiter and Ulysses. These spacecraft have detected ions of hydrogen, helium, carbon, nitrogen, oxygen, neon, silicon, sulfur, argon, and iron, singly, doubly, or triply ionized and at energies ranging from several MeV to 100 MeV (10^6 to 10^8 eV).[20] By contrast, the world's highest energy particle accelerator, the Large Hadron Collider, can now produce energies of 7 TeV, almost 10^{13} eV. The abundances of these energetic particles have been found to differ in different parts of the heliosphere.

By the time anomalous cosmic rays were discovered in the early 1970s, cosmic ray research had already been underway for 60 years. The Austrian physicist Victor F. Hess first discovered cosmic rays in 1912 during the ninth of a series of daring balloon flights.[21] These dangerous hydrogen-gas balloon flights carried Hess and three ionization chambers to great altitudes, culminating at 17,500 feet (5,350 meters). The ionization chambers were greatly improved over those invented by the Jesuit priest, Theodore Wulf, who during an Easter visit to Paris in 1910 had made measurements at the top of the Eiffel Tower in an attempt to determine how ionizing radiation changed as a function of distance from the surface of the Earth. Hess found that ionizing radiation fell off at an altitude of about 6,000 feet (2,000 meters), which made sense for radiation emanating from Earth due to the recently discovered property of radioactivity. But surprisingly he found that radiation levels increased above that. This he ascribed to "radiation of very great penetrating power [that] impinges on our atmosphere from above and still evokes in the lowest layers a part of the ionization observed in closed vessels."[22] For his work, Hess received the Nobel Prize in Physics in 1936.

By 1926, CalTech physicist Robert Millikan agreed that the Hess's phenomenon originated beyond Earth, and in one of the great misnomers of science, he coined the term "cosmic rays." A great debate took place in 1932 between Millikan and the physicist Arthur Holly Compton at the Christmas meetings of the

---

[20] Roger Clay and Bruce Dawson, *Cosmic Bullets: High Energy Particles in Astrophysics* (Reading, Mass.: Addison Wesley, 1997); Michael Friedlander, *A Thin Cosmic Rain: Particles from Outer Space* (Cambridge, Mass.: Harvard University Press, 2000), pp. 70-71.

[21] Friedlander, pp. 8-9.

[22] V. F. Hess, "Über Beobachtungen der durchdringenden Strahlung bei sieben Freiballonfahrt". *Physikalische Zeitschrift* **13** (1912) 1084–1091; translated as "Concerning Observations of Penetrating Radiation on Seven Free Balloon Flights," in Lang and Gingerich, 13-20.

American Association for the Advancement of Science in Atlanta, with Millikan arguing the observed phenomenon consisted of rays, photons similar to light created when hydrogen or other elements fused in interstellar space. Compton, on the other hand, argued they were charged particles. In the previous months he had mounted a worldwide campaign to observe cosmic rays and himself traveled 50,000 miles in search of a latitude effect, which he and others found and interpreted as an uneven distribution of charged particles due to the influence of the Earth's magnetic field. Millikan, on the other hand, saw no such effect. The argument between the two Nobelists was closely followed by the press, notably the *New York Times*.[23]

We now know that Compton was right: cosmic "rays" are not light rays but atomic or subatomic particles. We also know that because primary cosmic rays with energies below 1 MeV do not reach the Earth, and because very high-energy cosmic rays are rare, Hess, Millikan, and Compton were likely observing showers of secondary particles produced when galactic cosmic rays hit molecules in the atmosphere. The different categories of cosmic rays based on energy, composition, and source were only gradually separated out over time, ACRs being the last discovered. In the process, cosmic ray physics gave rise to high-energy physics, which in the 1950s began to accelerate its own charged particles in increasingly large accelerators.

NASA's website on anomalous cosmic rays in the context of the ACE spacecraft is found at http://helios.gsfc.nasa.gov/acr.html#acrnews. Though no longer updated it still contains useful information. The ACE mission website is accessible at http://www.srl.caltech.edu/ACE/.

---

[23] Robert A. Millikan and G. H. Cameron, "The Origin of the Cosmic Rays," *Physical Review*, 32 (Oct., 1928), 533-557; A. H. Compton, "A Geographic Study of Cosmic Rays," *Physical Review*, 43 (1933), 387-403. For context see Daniel Kevles, *The Physicists* (Vintage Books New York, 1971), 179.

6. The Planetary Systems Family

Class P 18: Planetary Systems/Exoplanets

Because of their mechanism of formation, planets tend to exist in systems, each bound to its central star. Although "free floaters" are known to roam interstellar space, in general planets as we know them live a social rather than a solitary life, mutually interacting not only gravitationally but also when material is exchanged between them, as in the case of the famous Mars rock ALH84001. In common usage, a planetary system consists not only of the planets themselves but also everything associated with them—what we term in this book the entire Kingdom of the Planets. At the Class level in this entry, we consider the planets as a system of objects orbiting a star. This is in contrast to "clustering," as stars and galaxies tend to do under the action of gravitation. Until recently, there were no known analog planet "clusters" in the absence of a parent star, but in 2018 the Hubble Space Telescope did report one such binary planet with no star.[1] For most of our history, our Solar System was the only planetary system known. Only in the last quarter-century have thousands of exoplanets been discovered around other stars, hundreds of them known to exist in multiple planet systems.

The existence of other planetary systems had long been conjectured by analogy in works such as Bernard le Bovier de Fontenelle's *Entretriens sur la pluralite des mondes [Conversations on the Plurality of Worlds*, 1686] and Christiaan Huygens' *Cosmotheoros* (1698), two of the most popular works of the 17th century. But planetary systems beyond the Sun, today termed "extrasolar planets," were much too faint to be detected directly. In the early 20th century, unaware of the scale of the universe, many astronomers believed the spiral nebulae were Solar Systems in formation.

---

[1] The binary planet cluster is reported at http://hubblesite.org/news_release/news/2018-03/42-brown-dwarfs

© Springer Nature Switzerland AG 2019
S. J. Dick, *Classifying the Cosmos*, Astronomers' Universe,
https://doi.org/10.1007/978-3-030-10380-4_6

"Tidal" theories of planetary formation in the early 20th century postulated that material was tidally pulled from stars during close encounters with other stars and indicated that planets might be rare, since such encounters were believed to be rare. But a turning point came in the 1940 and 1950s with a return to the nebular hypothesis of planetary coalescence from protostars (S 1), believed to be a normal part of stellar evolution. There were also claims that actual planets had been detected in the 1940s. By 1958, Harvard astronomer Harlow Shapley proclaimed the possibility of billions of planetary systems in the Galaxy, and in 1963 astronomer Peter van de Kamp claimed an actual detection of a planet around Barnard's star, based on intricate observations of the wobbling position of the star due the gravitational effect of a planet. Such claims were difficult to dispute, even as more claims were being made.[2]

It was only in the late 20th century that empirical evidence became available directly bearing on the frequency of planetary systems. In 1983 circumstellar debris disks (S 15) were found, now believed to be the remnants of post-planetary formation. In 1992, protoplantary "proplyds" (P 1) were imaged and pulsar planets (P 5) were detected. The claim of actual planets detected around solar-type stars was more ambiguous at first and intertwined with the search for brown dwarfs (S 22)—those substellar objects that range in mass from 13 to 80 times the mass of Jupiter, equal to .08 of the mass of the Sun. Radial velocity techniques, which relied on an object gravitationally tugging its parent star and resulting in measurable periodic variations in the radial velocity of the star, were being refined at this time. Using this technique, in 1989 Harvard astronomer David Latham and his colleagues reported a companion to the star HD 114762, which they calculated could have a mass as small as .001 of the Sun. Not only was this value smaller than the .08 traditional dividing line between brown dwarfs and stable hydrogen-burning stars, but it was also smaller than the proposed dividing line between a brown dwarf and a planetary companion. Because the title of the paper claimed a brown dwarf rather than a planet, Latham and his colleagues seldom get credit for discovering the first planet around a Sun-like star, though some astrono-

---

[2] For this history see Steven J. Dick, *The Biological Universe: The Twentieth Century Extraterrestrial Life Debate* (Cambridge: Cambridge University Press, 1996); Dick, *Life on Other Worlds* (Cambridge: Cambridge University Press, 1998).

mers still claim they did in fact do so. Only a determination of the inclination of the orbit, possible with refined measurements in the future, will settle the matter.[3]

It was by applying this same radial velocity technique that in 1995 the Swiss team of Michel Mayor and Didier Queloz first unambiguously detected an extrasolar planet around a Sun-like star, 51 Pegasi. Their observation was confirmed, and soon supplemented, by the American team consisting of Geoff Marcy and Paul Butler. This discovery was followed by many more using the ground-based radial velocity method that they and the Swiss team pioneered and that other teams soon perfected.[4]

In the early 21st century, it was spacecraft observations that opened the floodgates of exoplanet discovery. The two early primary planet-hunting spacecraft were CoRoT (Convection, Rotation, and planetary Transits), a product of the French and European Space agencies active between 2006 and 2012, and NASA's Kepler spacecraft, which operated from 2009 to 2018. Both employed a photometric methodology, the dimming of starlight by a transiting planet. Unlike ground-based detections, these spacecraft can detect planets between 1,500 and 6,000 light years from Earth, making ground-based follow-up very difficult— a source of some frustration for astronomers using telescopes on Earth. In May 2007, the CoRoT team announced its first detection of an extrasolar planet, a hot Jupiter dubbed CoRoT-1b; by early 2009, it had found the first Earth-like extrasolar planet, CoRoT-7b, a possible super-Earth. Altogether, CoRoT discovered 24 planets. These numbers were soon dwarfed by NASA's Kepler spacecraft, the brainchild of NASA astronomer Bill Borucki. Between 2009 and 2018 it detected more than 2,600 planets, as well as several thousand candidate planets, most of which are expected to be confirmed. Kepler observed only one quarter of one percent of the sky

---

[3] David W. Latham et al., "The unseen companion of HD114762 - A probable brown dwarf," *Nature*, 339 (1989), 38-40; Dick (2013), pp. 111–112. It has been proposed that the dividing line between giant planets and brown dwarfs, traditionally .08 solar masses or 13 Jupiter masses, could actually be set at 4 to 10 Jupiter masses, based on the observations of 146 planetary systems. Kevin C. Schlaufman, "Evidence of an Upper Bound on the Masses of Planets and Its Implications for Giant Planet Formation," ApJ, 853 (2018), 14 pp.

[4] Michel Mayor and Didier Queloz, "A Jupiter-mass companion to a solar-type star," *Nature*, 378 (1995), 355–359; Geoff Marcy, Paul Butler et al., "The Planet around 51 Pegasi," ApJ, 481 (2007), 926–935.

out to a distance of 3,000 light years, a very small sample that is nevertheless an indication of the abundance of planetary systems.[5]

In the midst of this cornucopia of exoplanets, ground-based discoveries continued apace, led by the High Accuracy Radial-velocity Planet Searcher (HARPS) instrument, which began observations in Chile in 2003, and the international Wide-Angle Search for Planets (WASP). Together they have discovered over 300 exoplanets. Other programs include the European Southern Observatory's EXPRESSO instrument mounted on the Very Large Telescope at the Paranal Observatory in Chile. Many other groups around the world are using spectrometers to hunt for exoplanets.

In addition to radial velocity and photometric methods, gravitational microlensing has also been used to discover exoplanets. Microlensing occurs when the gravitational field of a star or galaxy acts like a lens, magnifying the light of a more distant object. Exoplanets using this technique were reported in our own Galaxy beginning in 2002 as part of the Optical Gravitational Lensing Experiment (OGLE). In 2006, astronomers announced a planet using this technique of only five Earth masses, 22,000 light years away. Since then, at least six exoplanets have been reported as part of the OGLE project, and several dozen have been detected altogether in our Galaxy. Astonishingly, in 2018 two astronomers studying the quasar RXJ1131-1231 via an elliptical lensing galaxy 3.8 billion light years away reported 2,000 free-floating "rogue" exoplanets in the elliptical galaxy, with masses ranging from the Moon to Jupiter. Such quasar-galaxy lensing systems hold promise as the only technique for finding planets beyond our own Galaxy.[6]

These ground-based efforts are complementary to new planet-hunting spacecraft such as NASA's Transiting Exoplanet Survey Satellite (TESS) launched in April 2018. The latter is an all-sky survey over two years, expected to reveal many more rocky plan-

---

[5] William J. Borucki et al., "Kepler Planet-Detection Mission: Introduction and First Results," *Science*, 327 (2010), 997 ff.

[6] B. Scott Gaudi, "Microlensing Surveys for Exoplanets," ARAA, 50 (2012), 411–453; J. P. Beaulieu et al., "Discovery of a cool planet of 5.5 Earth masses through gravitational microlensing," *Nature*, 439 (2006), 437–440; Xinyu Dai and Eduardo Guerras, "Probing Extragalactic Planets Using Quasar Microlensing," ApJ Letters, 853, preprint at https://arxiv.org/abs/1802.00049. A list of exoplanets discovered through microlensing with OGLE is at https://en.wikipedia.org/wiki/Optical_Gravitational_Lensing_Experiment

ets along with other (and perhaps new) classes. These planets, circling stars only 30 to 300 light years from Earth, are much closer than previous planet discoveries and thus are ripe for follow-up from ground-based observatories. The European Space Agency's Gaia spacecraft, launched in 2013, is measuring the orbits and inclinations of thousands of extrasolar planets using astrometric techniques. ESA also plans to launch its CHEOPS (Characterizing Exoplanets Satellite) spacecraft in 2019. The latter will look at exoplanets already discovered and reveal in more detail the nature of those planets, including density and composition. In 2018, ESA announced another exoplanet mission known as ARIEL, the Atmospheric Remote-sensing Infrared Exoplanet Large-scale mission. It will launch in 2028 and study exoplanet atmospheres, focusing on planet formation and evolution.

In short, today the search and discovery of exoplanets is a robust field in astronomy. As of 2018 there are more than 3,700 known planets based on both ground-based and spacecraft detections. Among these, more than 600 multiplanet systems are known to exist as of 2018, most of them wildly different from our own system in terms of planetary size and orbits. Many of the newly discovered planets are "hot Jupiters," gas giants (P 3) orbiting very close to their host star. Others are in highly eccentric orbits. A small percentage of exoplanets are rocky "super-Earths," perhaps the precursor to finding Earth-sized and Earth-like terrestrial planets (see P 2). The Kepler spacecraft alone discovered some 30 rocky planets. Only a few planets have been directly imaged, notably Fomalhaut b within its debris disk (see Fig. 9.1 in entry S 15) and HR 8799 b, c, d, and e (the letter nomenclature for exoplanets represents the discovery order, not distance from the parent star). More planets will be imaged and the spectroscopic study of their atmospheres, especially for any "biosignatures," is already becoming a robust new subfield.[7]

The nearest known exoplanet is Proxima Centauri b, circling our nearest star, the red dwarf Proxima Centauri 4.2 light years distant. Announced in 2016, the planet circles only 7.5 million kilometers, or .05 astronomical units, from its parent star, compared to .39 AU for Mercury. It is only 1.3 Earth masses. Because the parent star is a red dwarf, the planet is in the habitable zone,

---

[7] Kevin Stevenson et al., "Possible thermochemical disequilibrium in the atmosphere of the exoplanet GJ 436b," *Nature*, 464 (2010), 1161–1164; Jasmina Blecic et al., "The Atmosphere of WASP-14b Revealed by Three Spitzer Eclipses," in press.

but flares from this kind of star make life on the planet unlikely since they would have stripped the planet of any atmosphere. A dusty debris disk (S 15) has also recently been discovered outside the planet's orbit.[8]

The detection of exoplanets has driven theory in several directions. The "core accretion theory," whereby a planet starts out with a solid core and can accrete gas suddenly when the core grows beyond a critical mass, is the most widely accepted theory of planetary formation. The existence of hot Jupiters has spawned the theory that these gas giants formed further from their parent star and migrated inward, since several factors including high temperatures inhibit planet growth so close to the star. No one yet knows how some exoplanets acquired their highly eccentric orbits, nor how planets such as HAT P-7b, WASP-17b, and dozens of others thus far detected acquired their retrograde orbits, except by some general planet-planet or star-planet scattering.[9] The planetary orbits themselves are being studied with ever more subtle and elaborate techniques, including the so-called "transit time variations" (TTVs), a new photometric dynamical tool that allows the study of planet-planet interactions.

Already in 1997, a connection was found between a star's chemical composition—namely its high fraction of elements heavier than helium (its "metallicity," see S 36)—and the likelihood that it will have planets, a finding supported by subsequent exoplanet surveys and by a 2009 survey revealing that 160 metal-poor main sequence stars had no detectable planets.[10] Most planets detected so far circle either main sequence stars (S4 and S5, mainly F, G, and K dwarfs) or subgiants (S 6), but they are also found

---

[8] On TESS see Sara Seager, "TESS: The Transiting Exoplanet Hunter," S&T, 135 (March 2018), 22–27. On Proxima Centauri's planet see Anglada-Escude et al., "A terrestrial planet candidate in a temperate orbit around Proxima Centauri," *Nature*, 536 (2017), 437–440.

[9] D. N. Lin, P. Bodenheimer, and D. C. Richardson, "Orbital migration of the planetary companion of 51 Pegasi to its present location." *Nature*, 380 (1996), 606-607; A. Pal et al., "HAT-P-7b: An Extremely Hot Massive Planet Transiting a Bright Star in the Kepler Field," APj, 680 (2008), 1450-1458; D. R. Anderson et al., "Wasp-17b: An Ultra-Low Density Planet in a Probable Retrograde Orbit," ApJ, 709 (2010), 159.

[10] G. Gonzalez, "The stellar metallicity-giant planet connection," MNRAS, 285 (1997), 403–412; Debra Fischer and Jeff Valenti, "The Planet-Metallicity Correlation," ApJ, 622 (2005), 1102–1117; A. Sozzetti et al., "A Keck HIRES Doppler Search for Planets Orbiting Metal-Poor Dwarfs. II. On the Frequency of Giant Planets in the Metal-Poor Regime," ApJ, 697 (2009), 544–556.

around M dwarfs. Although the latter is surprising, it is important because M dwarfs are the most common class of star in the Galaxy.

Theory has also been stretched because planets are sometimes found in the most unusual places, pulsars being a prime example but not the only one. In 2010, astronomers reported a planet orbiting the post-red giant star HIP 13044, a 10th magnitude yellow giant 2,000 light years away in Fornax. The star is on the horizontal branch of the HR diagram, having already undergone helium burning and again begun contraction and reheating. The planet has a minimum mass of 1.25 Jupiters, and its host star has the lowest metallicity of any star yet detected with planets, only 1% that of the Sun. Because the parent star is a member of the Helmi stream, a shredded dwarf galaxy that forms a ring of stars (G 10) around the Milky Way Galaxy, both the planet and its star are likely of extragalactic origin, originating in the galaxy that fell into the Milky Way 6 to 9 billion years.[11] This discovery is yet another indication that planets are more common that once believed, even surviving the red giant stage in this case. Because "red horizontal branch" planets may have a different composition than planets around stars on the main sequence, like pulsar planets they may prove to be a new class of planets. The same is true of the few planets found thus far around subdwarfs, the hot cores of evolved massive stars.

The characterization of these thousands of planets, individually and statistically, is an ongoing and hot topic in planetary astronomy. But it is notable that many of them are known to exist in systems that are diverse when compared to our Solar System. In 2017, astronomers announced the discovery of seven Earth-sized planets circling the nearby red dwarf star known as TRAPPIST-1 (Fig. 6.1). The star is named after the Transiting Planets and Planetesimals Small Telescope (TRAPPIST) in Chile, which discovered two of the planets, followed by the Spitzer Space Telescope's confirmation and discovery of five additional ones. The star, only 40 light years away, has only 8% of the Sun's mass and 12% of its diameter. All of the planets orbit in a compact system with periods ranging from 1.5 to 18.8 days, thus much shorter than even Mercury's "year." But because of the size of the star, all of the planets were within the habitable zone of the star. One study concluded that some of these planets could be 5% water, containing more than 250 times the water found on Earth.

[11]Johny Setiawan et al., "A Giant Planet Around a Metal-Poor Star of Extragalactic Origin," *Science*, 330 (2010), 1642–1644.

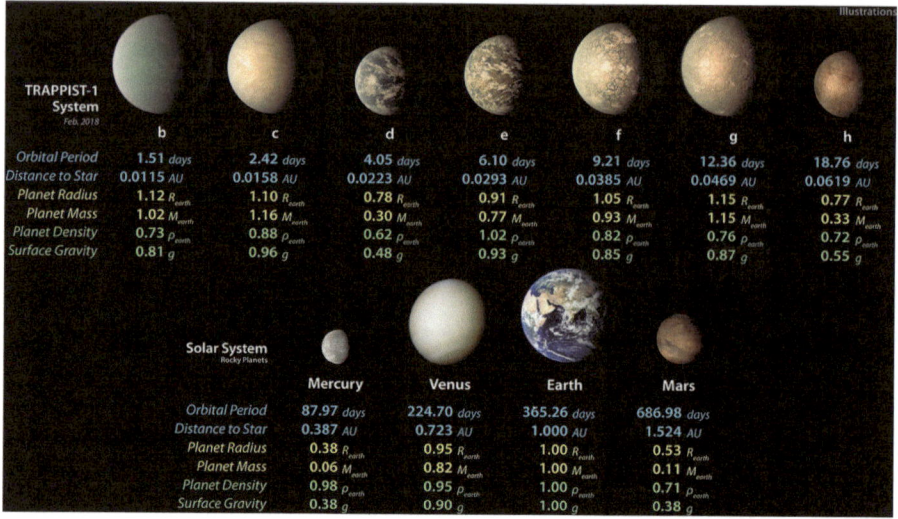

Fig. 6.1. Artist's conception of the TRAPPIST system of planets. All are believed to be rocky terrestrial planets. Credit: NASA/JPL-Caltech

Two other stars are known to have seven-planet systems: HD 10180 and Kepler 90. In 2017, another planet (dubbed Kepler 90-i) was discovered around the latter, making it the only other system to have eight planets like our own Solar System. Kepler 90, some 2,500 light years from Earth, consists of six rocky inner planets and two outer gas giants, but all orbit at a distance that would be inside Earth's orbit. Kepler 90-i, and previously Kepler 80-g, are notable for being the first exoplanets discovered using a neural network, an artificial intelligence machine learning algorithm that identifies pattern in the data.

The discovery of so many exoplanets raises questions of their classification, just as in earlier periods stars and galaxies had to be classified. Most of the extrasolar planets found so far fall in the gas giant class (P 3) with Types such as hot Jupiters, and some fall in the terrestrial planet class (P 2) as in the TRAPPIST-1 system. But from early on it was recognized that one of the Kepler planets (Kepler 7b) has the density of styrofoam, and others such as WASP-17b and WASP 107b are being discovered with very low densities. WASP 17b has 1.6 Saturn masses but 1.5 to 2 Jupiter radii, yielding a density of 6–14% that of Jupiter.[12] Only further observations will

---

[12] D. Latham et al., "Kepler-7b: A Transiting Planet with Unusually Low Density," ApJ Letters, 713 (2010), L140-L144; D. R. Anderson et al., "Wasp-17b: An Ultra-Low Density Planet in a Probable Retrograde Orbit," ApJ, 709

determine if these are different enough from the others to constitute a new class, and even then it will be a subjective call.

As mentioned in entry P 2, already attempts are being made to classify extrasolar planets based on their physical characteristics. In fiction, Star Trek fans will recall that planets were classified as A through Y, with an M designating a terrestrial-class planet like Earth. The Planetary Habitability Laboratory at the University of Puerto Rico at Arecibo offers a classification by mass and another one by thermal habitability. Another study finds that the extrasolar planets discovered thus far across five orders of magnitude in mass can be accommodated by the three classes already known in our own Solar System: terrestrial planets, gas giants, and ice giants. Yet another study by Italian astronomer S. Marchi distinguishes five classes based on planetary mass, semimajor axis, eccentricity, stellar mass, and stellar metallicity.[13] Whether the latter are truly new classes of objects or Types of a particular Class remains to be seen and is dependent on the definition of "Class."

Yet another possible class is carbon planets, which are believed to form if the material in a protoplanetary disk has more carbon than oxygen, and silicates are replaced with carbon and diamond. WASP-12b, reported in 2010 at a distance of 1,200 light years from Earth, may be an example. In any case, these pioneering efforts are likely to expand as more and more planets are discovered around other stars, and as instruments such as the James Webb Space Telescope begin to analyze exoplanet composition. Meanwhile, rather ad hoc practical classification systems are being devised to meet the needs of specific missions.[14]

(2010), 159. One recent attempt to classify exoplanets based on density is Soham Kulkarni and Shantanu Desai, "Classifying Exoplanets with Gaussian Mixture Model," preprint at https://arxiv.org/abs/1708.00605

[13] On Star Trek classification of planets see http://www.sttff.net/planetary-class.html. Planetary Habitability Laboratory, "A Mass Classification for both Solar and Extrasolar Planets," http://phl.upr.edu/library/notes/amass-classificationforbothsolarandextrasolarplanets. J. J. Fortney et al., "Planetary Radii across Five Orders of Magnitude in Mass and Stellar Insolation: Application to Transits," ApJ, 659 (2007), 1661-1672; S. Marchi, "Extrasolar Planet Taxonomy: A New Statistical Approach," ApJ, 666 (2007), 475-485. An attempt to classify planets into thermal habitability classes using data from the Planetary Habitability Laboratory and machine learning is Suryoday Basak et al., "Habitability Classification of Exoplanets: A Machine Learning Insight," https://arxiv.org/abs/1805.08810

[14] Ravi Kopparapu et al, "Exoplanet Classification and Yield Estimates for Direct Imaging Missions," https://arxiv.org/pdf/1802.09602.pdf

These discoveries have a direct bearing on one of the great motivations of astronomy—the search for life in the universe. A broad enough sample is now available to pin down one of the parameters of the Drake Equation, which attempts to estimate the number of communicative technological civilizations in the Galaxy. Astronomers have estimated that at least 17% of Sun-like stars have gas giants orbiting at distances closer than Uranus in our own system, and that 30% of G and K stars have rocky or icy planets up to the mass of Neptune.[15] Since in common usage planets are believed to be necessary for life, one of the primary conditions is now in place, with terrestrial-class exoplanets being discovered at in increasing pace. Many more conditions need to be fulfilled for actual life to exist. Determining the existence of those conditions elsewhere in the universe is the primary mission of the science of astrobiology.

Given the daily discoveries of exoplanets, it can be difficult to keep up with both individual discoveries and the statistical analysis of their nature. The NASA Exoplanet Archive, a service of the NASA Exoplanet Science Institute, is very helpful in tracking the latest discoveries and can be accessed at https://exoplanetarchive.ipac.caltech.edu. NASA's exoplanet website is at https://exoplanets.nasa.gov. Individual programs also maintain their own websites. Notably, the Kepler spacecraft website contains the latest news on its discoveries at https://www.nasa.gov/mission_pages/kepler/main/index.html, and TESS at https://tess.gsfc.nasa.gov. WASP results are at https://wasp-planets.net. Recent news on the TRAPPIST system is at https://www.jpl.nasa.gov/news/news.php?feature=7052. A list of multiplantary systems is at https://en.wikipedia.org/wiki/List_of_multiplanetary_systems. A list of possible rogue planets may be found at https://en.wikipedia.org/wiki/Rogue_planet and http://www.space.com/scienceastronomy/oldest_planet_030710-1.html. The Planetary Habitability Laboratory at the University of Puerto Rico at Arecibo maintains a list of habitable planets at http://phl.upr.edu/projects/habitable-exoplanets-catalog. As of mid-2018, some 55 exoplanets were on the list.

---

[15] Andrew Cumming et al., "The Keck Planet Search: Detectability and the Minimum Mass and Orbital Period Distribution of Extrasolar Planets," PASP, 120 (2008), 531–554; M. Mayor et al., "The HARPS search for southern extra-solar planets XVIII. An Earth-mass planet in the GJ 581 planetary system," A&A, 507 (2009): 487–494.

Class P 19: Asteroid Groups

Asteroid groups are populations of minor planets (P 11) with similar orbits, probably related to their origins. While the terms "minor planet" and "asteroid" are technically interchangeable for individual objects, for historical reasons the latter term in practice is usually used for groups, including the main asteroid belt located between Mars and Jupiter. Other important groups are the Trojans, most of which share the orbit of Jupiter at the Sun-Jupiter Lagrangian points; the Centaurs, orbiting between Jupiter and Neptune; and the Amors, Apollos, and Atens. The latter three groups are known as near-Earth asteroids (NEAs) because of their proximity to Earth. The term near-Earth object (NEOs) is broader, including not only near-Earth asteroids but also near-Earth comets and larger meteoroids. Asteroid Groups are not to be confused with asteroid Families, which are subsets of asteroid groups consisting of a few hundred asteroids with very similar orbits. About 30 asteroid families have been identified, many of them in the main asteroid belt, where one-third of its members can be identified as families when their orbital inclination is plotted versus orbital eccentricity.

Giuseppe Piazzi discovered the first object in the main asteroid belt, Ceres, on the first day of 1801. But the realization that a "belt" of objects existed between Mars and Jupiter came only gradually as new asteroids were discovered and their distances determined. As early as 1850, an English translation of Alexander von Humboldt's *Kosmos* mentioned in passing the "belt of asteroids," and the idea gradually came into common usage in the 1850s and 1860s, by which time 100 asteroids were known.[16] Main belt asteroids include the largest known minor planets: 1 Ceres, 2 Pallas, 4 Vesta, and 10 Hygiea comprise more than half the belt's mass (the numbers indicate order of discovery). Ceres is so large (about 950 km in diameter) that it was designated a dwarf planet (P 9) by the International Astronomical Union in 2006. The JPL Small-Body Database lists more than 670,000 known objects in this belt (Fig. 6.2).

For many years, the main asteroid belt was believed to have originated from an exploded planet. This idea stemmed from the 18th century Titius-Bode law, whereby planetary distances from the Sun were seen to follow a numerical sequence, except for a gap between Mars and Jupiter. After the discovery of minor planets in

---

[16] Alexander von Humboldt, *Cosmos: A Sketch of a Physical Description of the Universe*, vol. 1 (New York: Harper & Brothers (1850), p. 44.

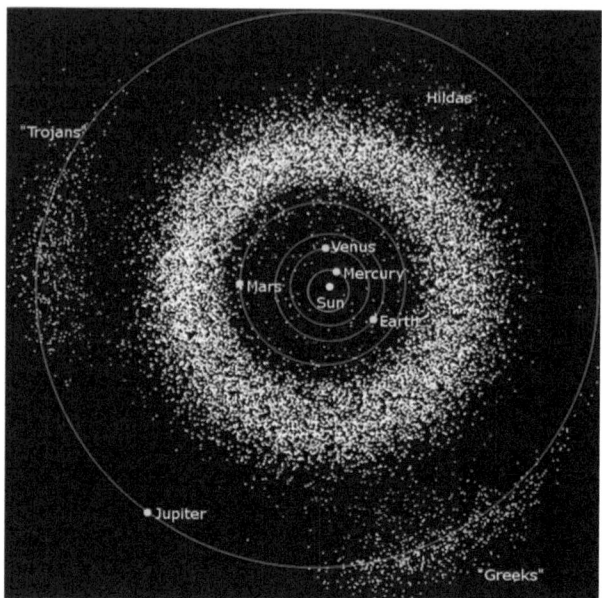

Fig. 6.2. Plotted diagram showing the main asteroid belt (white), Trojans and Greeks (green) at the L4 and L5 points, and near-Earth asteroids (red). The Hildas (orange) are a separate dynamical group in a 3:2 orbital resonance with Jupiter. Based on data from JPL Development Ephemeris 405. Credit: Mdf at English Wikipedia (Transferred from en.wikipedia to Commons.) Public domain via Wikimedia Commons

this region, the concept of an exploded planet became a popular idea. This theory has been discredited based on the small mass of the combined asteroids (about 4% of the Moon's mass), as well as what is now known about their chemical composition. Instead, it is believed that Jupiter's gravitational field prevented objects in the asteroid belt from ever coalescing in the first place.

The realization of more asteroid groups came more than a century after the discovery of the first main belt object. The first objects in a new asteroid group, known as the Trojans, centered around Jupiter and its so-called Lagrangian points, where the gravitational pull of the Sun and Jupiter are balanced (green in Fig. 6.2). In 1906, the German astronomer Max Wolf discovered 588 Achilles at the L4 Lagrangian point of the Jupiter-Sun system. This was followed in the next few years by the discovery of 624 Hektor, also at the Jovian L4, and 617 Patroclus at L5, hinting that this was a real group. Eleven Jovian Trojans were known by 1938, and as of 2017 more than 4,200 are known at L4 and 2,300 at L5 (Fig. 6.2). More are constantly being discovered.

Hektor is the largest known Jovian Trojan, an elongated object measuring 125 by 230 miles (270 by 300 km), and is one of only two Trojans with a known companion Skamandrios at 7.5 miles (12 km). The other is Patroclus, whose companion is almost the same diameter as Patroclus itself. Aside from Hektor and Patroclus, the names of Trojan War heroes have consistently been used at L5, and Greek names at L4. Officially, the Greek and Trojan "camps" at L4 and L5 are collectively referred to as the Trojan asteroids. Each Lagrangian region is located 60° ahead and 60° behind Jupiter itself, and each stretches about 26°. In 2021, NASA is set to launch a Discovery-class spacecraft, dubbed Lucy, to study the composition and geology of the Trojan asteroids. It should arrive by 2027. From Earth, the Trojans appear to be a curious mix of spectral type, color, and size, and Lucy aims to determine their origin. All planets except Mercury and Saturn are known to have their own Trojan asteroids located at their respective Lagrangian points with respect to the Sun, including 18 at Neptune.

The near-Earth asteroids (red in Fig. 6.2) were the next asteroid group to be discovered, in several subgroups. The first of the subgroups was heralded by 433 Eros, discovered on August 31, 1898 by Gustav Witt in Berlin and Auguste Charlois at Nice. Because of its proximity to Earth, its parallax could be determined, and it played an important role in determining an accurate value for the astronomical unit (the Earth-Sun distance).[17] Eros was visited by the NEAR-Shoemaker probe in the year 2000, and landing took place in 2001 (see P 11). Although Eros is now classed in the Amor subgroup, it was only in 1932 that German astronomer Karl Wilhelm Reinmuth discovered the prototypes of what are now recognized as two of the subgroups of the near-Earth asteroids, 1221 Amor and 1862 Apollo. In 1976, Eleanor F. Helin discovered 2062 Aten, the beginning of the third subgroup of NEAs now known as the Atens. The Atens have semi-major axes less than one astronomical unit and so lie largely within Earth's orbit. The Apollos are at distances slightly larger than the Earth's orbit, ranging from 1 to 1.017 AU. The Amors, which are classed as Earth-approaching rather than Earth-crossing, have orbits between Earth and Mars, up to about 1.3 AU. More than

---

[17] Luisa Pigatto and Valeria Zanini, "The 1900-1 opposition of 433 Eros, the solar parallax, and the contribution of Padova Observatory," JAHH, 5 (2002); R. d'E Atkinson, "The Eros Parallax, 1930-31" JHA, 13 (1982), 77–82.

15,000 near-Earth asteroids are known, up to 1,000 of them over one kilometer in diameter.

Near-Earth asteroids are the subject of much concern because of their proximity to Earth. For example, the largest Apollo asteroid, 1866 Sisyphus, has a diameter of about 10 km. Should such an object hit Earth, the result would be catastrophic. In 2004, the Aten asteroid known as 99942 Apophis gained brief notoriety when it was predicted it might impact the Earth in 2029. Subsequent calculations pushed the encounter to 2036 and significantly reduced the likelihood of an encounter even then. On October 7, 2008, a small near-Earth asteroid known as 2008 TC3, estimated to be no more than a few meters in diameter, entered Earth's atmosphere and caused a fireball above northern Sudan. The asteroid was discovered the day before and for the first time allowed JPL astronomers to accurately predict the impact location of an asteroid. NASA's Jet Propulsion Laboratory maintains a near-Earth object program to detect, track, and characterize potentially hazardous asteroids and other near-Earth objects. Thanks to this program, which includes the Catalina Sky Survey and Pan-STARRS in Hawaii, roughly 30 new discoveries are being added each week.[18]

The last of the major groups to be discovered was the Centaurs, so called because they behave as half asteroid, half comet. Their orbits lay between Jupiter and Neptune. Chiron was the first centaur discovered and recognized as part of a distinct population in 1977.[19] The largest Centaur is 10199 Chariklo, at 260 km in diameter. Although only 180 Centaurs have been catalogued, it has been estimated that as many as 44,000 may exist. They may originate in the scattered disk of the Kuiper belt.

For more on near-Earth asteroids see Class P 11 (Minor Planets/Asteroids). The JPL Small Body database can be accessed at https://ssd.jpl.nasa.gov/sbdb.cgi, and more advanced searches at https://ssd.jpl.nasa.gov/sbdb_query.cgi.

---

[18] On JPL's Near Earth Object program see http://neo.jpl.nasa.gov/

[19] C. T. Kowal, W. Liller, and B. G. Marsden, B.G., "The Discovery and Orbit of (2060) Chiron," *Dynamics of the solar system*, R. L. Duncombe, ed. (Tokyo: Reidel Publishing Co., 1979), pp. 245–250.

Class P 20: Meteoroid Streams

A meteoroid stream is a collection of meteoroids (P 10) believed to be shed from a short-period comet (P 12) during its perihelion passage, or (more rarely) from an Earth-crossing asteroid. The stream is thus in orbit around the Sun and results in a meteor shower if the Earth intersects its orbit. Though the term "meteor stream" is sometimes used even by astronomers, "meteoroid stream" is the correct usage adopted by the International Meteor Organization and the International Astronomical Union, since meteoroid streams reside in space and have not entered the Earth's atmosphere.

Because particles are ejected from the parent comet in different directions and different velocities, a meteoroid stream has a width that usually increases with age. Among the meteoroid streams with indisputable parent comets are the Perseid stream associated with comet 109P/Swift-Tuttle, the Taurids associated with comet 2P/Encke, the Leonids associated with comet 55P/Temple-Tuttle, and the Eta Aquarids and Orionids associated with comet 1P/Halley. Fred Whipple, the pioneer of comet studies, was the first to notice that the Geminids are in the same orbit as an extinct comet, the Apollo asteroid known as 3200 Phaethon, discovered by the IRAS spacecraft in 1983. This is believed to be one of several meteoroid streams caused by near-Earth asteroids (P 11).

The "November meteors" (later known as the Leonids) as well as the August Perseids proved important in demonstrating that meteoroid streams existed and that their source was cometary debris (Fig. 6.3). After studying numerous cycles of the Leonids, in 1863 Yale astronomer Hubert A. Newton predicted the November meteors would storm in 1866. When the storm indeed occurred, astronomers in four European countries were inspired to solve the riddle of the orbit of the meteor stream and its parent body. The largely independent and almost simultaneous work in 1866–1867 of John Couch Adams in England, Giovanni Schiaparelli in Italy, U.J.J. Le Verrier in France, and Theodor von Oppolzer and C.F.W. Peters in Germany established that orbit.

As Adams recounted in 1867, Schiaparelli, the Director of the Milan Observatory, showed that the orbits of meteoroid streams around the Sun are very elongated, as are those of the comets, and that "… both these classes of bodies originally come into our system from very distant regions of space." More specifically,

Fig. 6.3. Leonid meteor storm in the early morning hours of November 13, 1833. This depiction first appeared in April, 1888, painted by the Swiss artist Karl Jauslin and engraved by Adolf Vollmy, based on eyewitness descriptions. Courtesy Seventh Day Adventist Church

in a letter dated December 31, 1866, Schiaparelli remarked on the very close agreement in the orbital elements of the August meteors (now known as the Perseids) and Comet II 1862. To the Italian astronomer thus goes the credit of showing that the comet now known as 109P/Swift-Tuttle is the source of the Perseids. One month later, Peters noticed that the orbital elements for the November Leonids agreed closely with comet I 1866, now known as comet 55P/Tempel-Tuttle, as determined by Oppolzer. Thus in the space of only one month, the connection that we accept today between comets and meteors was established beyond doubt.[20]

---

[20] J. C. Adams, "On the orbit of the November meteors," MNRAS, 27 (1867), 247–252; G. V. Schiaparelli, "Intorno al corso ed all'origine probabile delle telle meteoriche," Bulletin meteorologico dell'osservatorio del Collegio

Because the annual meteor showers are due to Earth-crossing meteoroid streams, historical observations of these showers are important for determining how the stream may have changed over the years, and even for improving the orbit of the parent comets. The Perseid showers have been observed for about 2,000 years, while the first known recorded Leonid event was in 902 AD. The discovery of meteor events in historical literature remains a promising field, as witnessed by the fact that in 1863 Hubert Newton had uncovered only six historical Leonid events, while we now can be reasonably certain of at least 58 such events, about half of them storms.[21]

Following in the footsteps of Newton, historical observations continue to be used for modern scientific purposes. For example, NASA scientist Donald Yeomans has used the full range of Leonid meteor shower data from the period AD 902–1969 to map the distribution of dust surrounding the parent comet Tempel-Tuttle, to predict the strength of the Leonid event in 1998–1999, and to re-determine the orbit of Tempel-Tuttle. Similarly, J.W. Mason found that "Of the 58 Leonid displays researched, 27 of the 35 outstanding showers, and all 23 meteor storms, have occurred between 750 days before and 1750 days after the parent comet's passage through the descending node ...." Of those, "A total of 19 meteor storms took place between 250 days before and 750 days after the comet's nodal passage." The study also showed that most of the dust was concentrated outside the comet's orbit, where 41 of the 58 documented Leonid events, including 18 of the 23 storms, occurred.[22]

The Taurids are an important annual meteoroid stream associated with comet Encke. The associated debris is the largest stream in the inner Solar System. Increased knowledge of the Taurids has given rise to "coherent catastrophism" concept of British astronomers Mark Bailey, Victor Clube, and Bill Napier. Their books

Romano, 5 (1867), 8–12 and vol. 6, 2; C. F. W. Peters, Letter dated January 29, *Astronomische Nachrichten*, 68 (1867), 287–288; Steven J. Dick, "Observation and interpretation of the Leonid meteors over the last millennium," JAHH, 1, (1998), 1–20: 5–6; Dick (2013), 208–211; 306–307.

[21] Dick, 1998, 11–15.

[22] D. K. Yeomans, "Comet Tempel-Tuttle and the Leonid meteors," *Icarus*, 47 (1981), 492–499; J. W. Mason, "The Leonid Meteors and comet 66P/Tempel-Tuttle, JBAA, 105 (1995), 219–235.

The Cosmic Serpent (1982) and *The Cosmic Winter* (1990), along with a number of articles, argue that objects from meteoroid streams have impacted Earth during timescales relevant to human history, and with strong patterns rather than randomly. In particular, they view the Taurid Complex—including comet Encke, its debris, and several asteroids captured by the inner Solar System about 20,000 years ago—as having produced remnants that have impacted Earth in historical times still recalled in mythology.[23]

While there is no doubt that numerous impacts have occurred over Earth's geologic history—some of them perhaps causing mass extinctions—most astronomers consider the claim of impacts influencing recent historical events to be unsupported by the scientific evidence. On the other hand, some specialists have made a case that spectacular meteor showers have influenced historical events. For example, a meteor shower observed on the evening of April 10, 531 AD in the Mayan regional political center of Caracol may have prompted the crowning four days later of king K'an I, known as Lord Jaguar. The authors of these claims convincingly demonstrate that this shower, perhaps better characterized as a meteor storm, occurred when the Earth encountered particles from three previous passages of comet Halley in 295, 374, and 451 AD—what we now term the Eta Aquariid stream. Because these passages were so recent, the debris stream would still have been concentrated. This may have been the first meteor shower recorded in the Western Hemisphere, though the Chinese likely recorded Eta Aquariids as far back as 74 BC. Furthermore, comet Halley itself may have been responsible for the royal ascension of the Teotihuacan ruler Spearthrower Owl, who ascended to the throne about one month after the comet's very close approach on April 1, 374 AD.[24]

---

[23] S. V. M. Clube and W. M. Napier, *The Cosmic Serpent* (Universe Publishers, 1982); Clube and Napier, *Cosmic Winter* (Oxford: Basil Blackwood, 1990); Clube, "The nature of punctuational crises and the Spenglerian model of civilization," *Vistas in Astronomy*, 39 (1996), 673–698; M. E. Bailey, "Recent results in cometary astronomy: Implications for the ancient sky," *Vistas in Astronomy*, 39 (1996), 647–671; M. E. Bailey, S. V. M. Clube and W. M. Napier, *The Origin of Comets* (Oxford: Pergamon, 1989).

[24] James Romero, "Halley's Comet and Maya Kings," *S&T*, April 2018, pp. 36–40.

While the Perseids, Leonids, Eta Aquariids, and Taurids are among the most impressive meteoroid streams, many others exist, and some of their parent bodies are known. 112 meteor showers are known to exist and hundreds of others have been proposed, but parent bodies are known for only 32 of them. Research to identify others is important because meteoroid streams and their associated meteor showers provide one of the few methods for determining past cometary activity. [25]

Lists of meteor showers and meteoroid streams can be found at http://en.wikipedia.org/wiki/List_of_meteor_showers and http://www.astro.sk/~ne/IAUMDC/STREAMLIST/, as well as at the IAU's Meteor Data Center at https://www.ta3.sk/IAUC22DB/MDC2007/. Peter Jenniskens and Ian Webster at the SETI Institute have created spectacular meteoroid stream visualizations at https://www.meteorshowers.org.

---

[25] A study of some 50 meteoroid streams is at http://adsabs.harvard.edu/abs/1994A&A...287..990J; A recent review is Peter Jeniskens, "Meteor Showers in Review," *Planetary and Space Science*, Volume 143 (2017), p. 116–124

Class P 21: Kuiper Belt

Kuiper belt objects (KBOs) include those trans-Neptunian objects (P 13) that reside in stable orbits around the Sun in the space extending from 30 to approximately 55 astronomical units (AU) from the Sun, as well as those less dynamically stable objects known as scattered-disk objects that may extend as far as 2,000 AU. They comprise, as astronomer Alan Stern has suggested, the "third zone" of the Solar System, beyond the terrestrial and giant planets.[26]

As mentioned in the entry on trans-Neptunian objects, they are sometimes called Edgeworth-Kuiper belt objects, because the Irish astronomer Kenneth Edgeworth briefly suggested the existence of such a belt of objects in 1943 and 1949, before the American astronomer Gerard Kuiper independently made a similar claim in 1951.[27] "Kuiper belt" remains the most commonly used term in the United States, while "Edgeworth-Kuiper belt" is sometimes used in Europe. That "Kuiper belt" is more commonly used is possibly a case of wrongful nomenclature, since Edgeworth proposed such a belt still existed while Kuiper proposed it once existed. However, things are never so simple. Harvard astronomer Brian Marsden has argued that others, such as Harvard astronomer Fred Whipple, more nearly described the nature of the Kuiper belt, and that it should be named after him. David Jewitt, one of the pioneers in the field, argued that Uruguayan astronomer Julio Fernandez deserved credit for predicting the Kuiper belt "based on clear statements and physical reasoning" set forth in a 1980s paper. "Just hand-waving on the possibility that there might be something 'beyond Pluto' is not sufficient," he argued.[28]

---

[26] S. A. Stern, "The Third Zone: Exploring the Kuiper Belt," S&T, 106 (2003), 31–36.

[27] K. E. Edgeworth, "The Evolution of our Planetary System," JBAS, 53, (1943), 181-188: 186; Edgeworth, "The origin and evolution of the solar system," MNRAS, 109 (1949), 600–609; and Gerard Kuiper, "On the origin of the solar system," in *Astrophysics: A Topical Symposium*, J. A. Hynek, ed., (New York: McGraw-Hill Book Co. 1951), pp. 357–424:402. Edgeworth's manuscript, preserved in the National Library of Ireland in Dublin, actually dated to 1938. On this controversy see also entry P 13.

[28] David Jewitt, "Why Kuiper Belt?," http://www2.ess.ucla.edu/~jewitt/kb/gerard.html

The nomenclature problem is not new in science; we refer to Darwinian evolution rather than Wallace evolution, even though A.R. Wallace had the idea of natural selection first (though not the term) and it was jointly presented in Wallace's absence to the Linnaean Society of London in 1858. In any case, the Kuiper belt controversy has led some astronomers to prefer the more generic term trans-Neptunian objects (TNOs), but this obscures the fact that the Kuiper belt is one of two systems harboring trans-Neptunian objects, clearly distinguishable from the more distant Oort cloud (P 22). Some astronomers categorize the so-called "scattered disk" as a separate class of objects due to their high eccentricity orbits; others have argued there really is no defined scattered disk. It may sound confusing, but this is how science works.

A dynamical analysis of the known population of KBOs adopted 2,000 astronomical units as "the formal (somewhat arbitrary) beginning of the inner Oort cloud (and thus the end of the Kuiper belt in terms of its scattered disk objects)."[29] The authors emphasized that this makes other claimed classes of TNOs, such as scattered disk objects, smaller populations of the Kuiper belt. They likely have no major compositional differences. While the physical distinction between KBOs and Oort cloud objects is unknown because of the unknown nature of Oort cloud objects, the former are composed largely of ices of methane, ammonia, and water, and the Oort cloud objects likely are also so. This distinguishes both classes from the class of minor planets consisting largely of rock and metal. It has been estimated that 100,000 Kuiper belt objects exist, with a combined mass about 10% of Earth (Fig. 6.4).

The first unambiguous Kuiper belt object, dubbed 1992 QB1, was announced on August 30, 1992 after a five-year search by astronomer David Jewitt and his student Jane Luu (Fig. 4.7). It was detected as a very faint 22nd magnitude object with slow retrograde motion, using the University of Hawaii's 2.2-meter telescope on Mauna Kea. Six months later they found a similar object, named 1993 FW. With the advance of telescope detector technology, such objects have been discovered in increasing numbers ever since.[30]

---

[29] Brett Gladman, Brian G. Marsden and Christa VanLaerhoven, "Nomenclature in the Outer Solar System," in M. A. Barucci et al., eds., *The Solar System Beyond Neptune*, (Tucson: University of Arizona Press, 2008), pp.43–58.

[30] David Jewitt and Jane Luu, "Discovery of the candidate Kuiper belt object

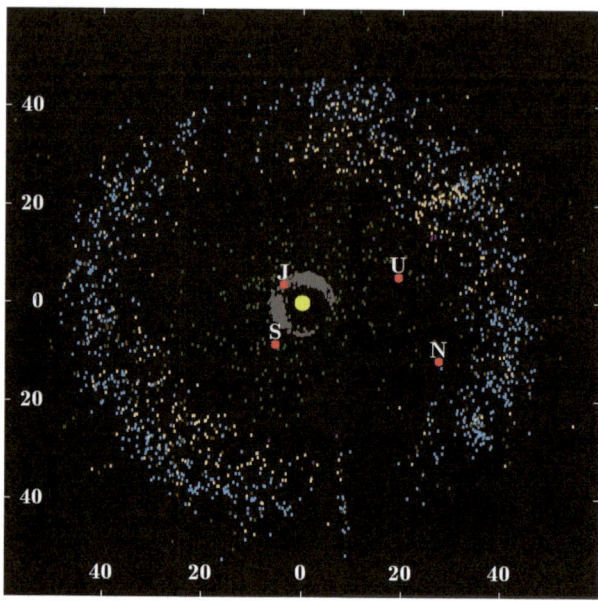

Fig. 6.4. Plot of Kuiper belt objects (blue) within 60 astronomical units of the Sun. For perspective, the gas and ice giant planets are shown in red and the Jupiter Trojans and Greeks in grey. Orange indicates scattered-disc objects, and purple the Neptune Trojans. Distances are to scale, but not sizes. Credit: Minor Planet Center, www.cfeps.net

By 1995, the Hubble Space Telescope saw evidence of the KBO population. Since the International Astronomical Union declared Pluto a dwarf planet in 2006, and further declared that a dwarf planet is subplanetary in nature, it is now considered a Kuiper belt object. Aside from Pluto, the largest KBOs not (yet) declared dwarf planets are about 850 miles in diameter—half that of Pluto (Fig. 6.5). The population of trans-Neptunian objects in the Kuiper belt with diameters larger than 100 km is likely to exceed 100,000, with tens of millions more down to 1 km size.

1992 QB1," *Nature*, 362 (1993), 730–732.

Fig. 6.5. Largest known Kuiper belt objects. Xena is now known as Eris, 2003 EL61 as Haumea, and 2005 FY9 as Makemake. All three are dwarf planets. Credit: NASA

On October 9, 1996 during a wide area search for TNOs, Jewitt, Luu, and their colleagues also discovered the first object in the so-called scattered disk, dubbed 1996 TL66. [31] As with 1992 QB1, they found the object using a CCD detector on the 2.2-meter telescope of the University of Hawaii located on Mauna Kea. Data from the Spitzer Space Telescope indicates that its diameter is about 575 km and that it may be a candidate for dwarf planet status. The same survey found three more scattered KBOs in 1999. 1996 TL66 has a perihelion distance of 35 AU and a diameter of about 600 km. Scattered disk objects are distinguished from more stable KBO orbits in that they have orbital eccentricities as high as .8 (where zero is perfectly circular), inclinations as high as 40° (where zero is the plane of the ecliptic), and perihelia extending as close as 30 AUs and aphelia beyond 100 AUs. Thus, while the closest approach of scattered-disk objects to the Sun may overlap those objects with more stable orbits in the Kuiper belt disk, at the most distant points of their orbits, they are two to three times the distance of Pluto. Because of their high eccentricities, their

---

[31] Jane Luu, Brian G. Marsden, David Jewitt et al., "A new dynamical class of object in the outer Solar System," *Nature*, 387 (1997), pp. 573–575.

semi-major axes may exceed 50 AU. Such pioneers in the field as Jewitt himself have called them "scattered Kuiper belt objects;" whether they should be considered a separate class of objects is open to definition and interpretation, but based on the criterion of physical composition, they are KBOs.

By the year 2000, astronomers had distinguished three types of KBOs based on their orbital characteristics:[32]

1) *classical KBOs*, composing about two-thirds of the objects known at that time, have small eccentricities and semi-major axes between 41 and 46 astronomical units. Because the first classical KBO detected was QB1, this type is sometimes called "cubewanos" after its prototype. Varuna, discovered by the Spacewatch project at the University of Arizona in 2000 and with a diameter around 500 km, falls in this category. It is also a dwarf planet candidate.

2) *resonant KBOs*, trapped in 2:3 resonance with Neptune (2 orbits for every 3 of Neptune), have semi-major axes at about 39.4 AUs. Because Pluto is the prototype for this type of KBO, they are often referred to as "plutinos." The first plutino, other than Pluto and its moon Charon, was found in 1993 and named 1993 RO. Other important plutinos are Orcus and Ixion, with diameters 946 km and 650 km respectively, and several others in that size range. About one-fourth of known KBOs are plutinos.

3) *scattered KBOs*, with large eccentricities and perihelia near 35 AU, are thought to originate from Neptune scattering. Since the discovery of 1996 TL66, over 100 scattered disk objects have been found, including the dwarf planet Eris. Neptune's large satellite Triton with its retrograde orbit is likely captured from the Kuiper belt, perhaps from the scattered disk.

Research suggests that the centaur asteroids (half comet, half asteroid in terms of physical characteristics) are scattered-disk objects that have been perturbed inward by Neptune. Chiron, first discovered by Charles T. Kowal in 1977, is a primary example. Based on physical characteristics, Saturn's distant moon Phoebe may be a captured centaur, and thus a captured scattered disk object. Imaged by the Cassini spacecraft in 2004, Phoebe has a diameter of 220 km (140 miles), shows a heavily cratered surface,

---

[32] D. C. Jewitt and J. X. Luu, "Physical Nature of the Kuiper Belt," in *Protostars and Planets IV*, V. Manning, A. Boss et al, eds., (Tucson: University of Arizona Press, 2000), p. 1201–1229.

and a dark coloring. It is believed that scattered-disk objects may be perturbed closer to the Sun and become periodic comets, and they may be perturbed outward toward the Oort cloud.

It has been estimated that the number of known KBOs will increase from 1,000 to 20,000 in the next decade, due largely to two large surveys, the Panoramic Survey Telescope and Rapid Response System (Pan-STARRS) located in Hawaii, and the Large Synoptic Survey Telescope (LSST) in Chile, scheduled for science first light in 2021.[33] More than 100,000 KBOs greater than 62 miles (100 km) in diameter are thought to exist. The arrival of NASA's New Horizons spacecraft at Pluto in 2015 allowed the first in-depth study of a Kuiper belt object that had also been designated a dwarf planet (see P 9). New Horizons continued past Pluto to rendezvous with Kuiper belt object 2014 MU69 (Ultima Thule) in January 2019, revealing it to consist of two joined bodies. That object is at a distance of 43.4 astronomical units from Earth. New Horizons plans to visit many more such objects. In the Planetary Science Vision Workshop for 2015, Alan Stern and colleagues called for "a series of flyby missions to explore the diversity of phenomenology and origins of the objects found in these vast, primordial reservoirs," meaning both the Kuiper belt and the Oort cloud, and including the Pluto system.

Hundreds of KBO-type disks are now known in other star systems. Based on far-infrared surveys, they are estimated to exist around 20% of nearby main sequence stars, where they are also known as debris disks (S 15), one of several classes of circumstellar matter. As Jewitt has pointed out, "The main importance of the Kuiper belt is that it provides a very local example of a highly evolved circumstellar disk. Processes that are probably general to the circumstellar disks of solar-mass stars can be discerned already in the size and orbital element distributions of the Kuiper belt." After a large object was observed being shredded in the atmosphere of a white dwarf star (S 12), Kuiper belts have even been inferred around those degenerate stars. The observation of exo-Kuiper belts is therefore a robust area of research.[34]

---

[33] Chadwick A. Trujillo, "Future Surveys of the Kuiper Belt," in in M. A. Barucci et al., eds., *The Solar System Beyond Neptune*, (Tucson: University of Arizona Press, 2008), pp. 573–585.

[34] David C. Jewitt, "The Kuiper Belt as an Evolved Circumstellar Disk," in *The Origin of Stars and Planets: The VLT View* (Berlin: Springer-Verlag, 2002), 405–415. The shredded Kuiper Belt object around a white dwarf is HST

Class P 22: Oort Cloud

The Oort cloud, occasionally called the Öpik-Oort cloud, is a spherical reservoir of trans-Neptunian objects (P 13) surrounding the Solar System. If the Kuiper belt (P 21) is the Solar System's "third zone" beyond the terrestrial and gas giants, the Oort cloud is its fourth zone, and the one containing most of its objects. They are believed to be the primordial remnants of the formation of the Solar System. The Oort cloud extends from about 50,000 astronomical units outward 100,000 astronomical units from the Sun, nearly half the way to the nearest known star, Proxima Centauri. By contrast, the scattered disk component of the Kuiper belt, the other reservoir of trans-Neptunian objects, extends to about 2,000 astronomical units and is not spherical but disk-shaped. Both classes likely have similar composition, in which case there would be no cause to grant them separate class status based on that criterion. But since they are so dynamically distinct from the Kuiper belt, and because their true composition remains unknown, they are given separate class status here. Some astronomers believe there may be an inner Oort cloud, known as the Hills cloud, beginning at about 2,000 astronomical units.

Whereas many objects have been observed in the Kuiper belt, the Oort cloud is largely inferred, with the possible exception of Sedna and a few other objects. The Estonian astronomer Ernst Öpik first suggested such a reservoir of objects in 1932, and the Dutch astronomer Jan Oort revived the idea in 1950 when he theorized that long-period comets (greater than 200 years) must be falling inward to the Solar System from a distant unseen cloud surrounding the Sun. He believed perturbations by Jupiter originally placed the comets into highly elliptical orbits at a distance where passing stars could in turn perturb some of them back into the system as long-period comets.[35]

Release, February 9, 2017, "Hubble Witnesses Massive Comet-Like Object Pollute Atmosphere of a White Dwarf,"http://hubblesite.org/news_release/news/2017-09/90-white-dwarfs

[35] E. J. Öpik, "Note on stellar perturbations of nearly parabolic orbits," *Proc. Am. Acad. Arts Science.* 67 (1932), 169–183; Jan H. Oort, "The structure of the cloud of comets surrounding the Solar System and a hypothesis concerning its origin," BAIN, 11 (1950), 91–110, reprinted in part in Bartusiak, 441–445.

Astronomers estimate the Oort cloud may contain 100 billion objects with radii greater than 1 km, and the total number of objects may exceed one trillion. Oort's theory that passing stars may perturb these objects inward, where they are seen as comets, remains valid based on what we know about dynamical considerations along with new observations from the Gaia satellite and the University of Hawaii's Pan-STARRS program. Other Oort cloud members, however, may be ejected out of the Solar System into interstellar space. The interstellar interloper known as Oumuamua (see P 11) may have been ejected from the Oort cloud of another stellar system. Perturbations from the giant planets, galactic tides, and the interstellar medium may also influence the Oort cloud, which may contain primordial materials before the Solar System was formed.

The Oort cloud was inferred long before it was observed. At present, there are only four candidates for inner Oort cloud membership. The first such claim came in 2004 for Sedna, likely about 1,000 miles in diameter. Three times farther than Pluto, it remains one of the most distant known objects in the Solar System, requiring 10,500 years to orbit the Sun. Since then, the objects known as 2000 CR_{105}, 2006 SQ_{372}, and 2008KV_{42} have been claimed as possible inner Oort cloud objects. 2006 SQ_{372} was discovered by the Sloan Digital Sky Survey. Even more distant than Sedna, it takes between 22,000 and 32,000 years to orbit the Sun and is likely much smaller than Sedna.[36]

The origin of the Oort cloud has been the subject of considerable debate. Many of its objects undoubtedly originated in the Sun's protoplanetary disk (P 1), when leftover materials from the outer planets were scattered outward, perhaps first entering the scattered disk of the Kuiper belt before they were perturbed into the Oort cloud or ejected altogether. However, based on numerical simulations, in 2010 astronomer Harold Levison and colleagues suggested that 90% or more of these objects may have been captured from other nearby stellar systems, either when the Sun was

---

[36] Michael E. Brown, Chadwick Trujillo and David Rabinowitz, "Discovery of a Candidate Inner Oort Cloud Planetoid," ApJ, 617 (2004), 645-649; Nathan Kaib et al., "2006 SQ_{372}: A Likely Long-Period Comet from the Inner Oort Cloud," ApJ, 695 (2009), 268.

still in its birth cloud during its first few million years, or (less likely) through later close encounters with other stars. Simulations published in 2017 lend support to the delayed formation of the Oort cloud, after the Sun had left its birth cluster. Because observations are so difficult, many numerical models of the Oort cloud have been undertaken and, along with long-period comets, remain an important source of speculation about these objects, pending improved observational techniques.[37]

By analogy with our Solar System, other star systems may have their own Oort clouds. A search of the infrared data from the IRAS satellite found no clear signature of such clouds for stars in the solar neighborhood.[38] But these observations are extremely difficult and will undoubtedly continue, especially with the discovery of more planetary systems (P 18).

With the Oort cloud, we take leave of the Kingdom of the Planets and enter the realm of the stars in Part II.

---

[37] Harold F. Levison, "Capture of the Sun's Oort Cloud from Stars in its Birth Cluster," *Science,* 239 (2010), 187–190; T. Nordlander et al., "The destruction of an Oort Cloud in a rich stellar cluster," A&A, 603 (2017), A112, 18 pp., https://www.aanda.org/articles/aa/pdf/2017/07/aa30342-16.pdf
[38] Gregory J. Black, "Revisiting the Search for Extrasolar Oort Clouds," AAS Abstract, January, 2010, http://adsabs.harvard.edu/abs/2010AAS...21560106B

Part II
The Kingdom of the Stars

Stars of all colors crowd the Milky Way Galaxy in this image from the Hubble Space Telescope, released in 2018. Colors range from red giants to yellow Sun-like stars and recently formed blue stars. The classification of these stars has been an ongoing process for the last 150 years. Credit: NASA, ESA, and T. Brown (STScI)

Like the wandering planets, the "fixed" stars have been known throughout history simply by glancing at the sky. But apparent differences in stellar brightness and color gave only the faintest hint of their true nature. Not until Giordano Bruno—who paid for his speculations with his life—was it seriously suggested that the stars might be suns like our own. And not until the 19th century did spectroscopy reveal the great diversity and the true natures of the stars.

With the Kingdom of the Stars we enter a more distant realm, where gaseous rather than solid objects reign, and with additional physics from the hydrostatics laws dominating the planets. Gravity and hydrostatics are still our constant companions, but here they work their will in different ways. Although the Kingdom of the Planets contains gaseous objects in the form of the gas giant planets, we are now in a realm where much more massive objects are formed that can ignite nuclear fusion, and where convection, conduction, and radiation are at play.

Like planets, stars must be born, and their birth, life, and death depend heavily on their initial masses. Some fling away their lives in only a few million years; others with lower masses while away their time in trillions of years. We can now pick out extremely young stars like the T Tauri's and Herbig's, before they settle down to adulthood on the so-called "main sequence" of the Hertzsprung-Russell diagram in the form of hydrogen-burning dwarfs and subdwarfs. And we can now distinguish their successors—the subgiants, giants, supergiants, and hypergiants—burning helium and higher elements. We can follow their evolutionary endpoints in the form of spectacular exploding supernovae, or as they enter old age as white dwarfs, neutron stars, or the amazing black holes, depending on their masses.

The German-British astronomer William Herschel likened the heavens to a luxuriant garden, where we see in succession "the germination, foliage, fecundity, fading, withering, and corruption" of plants, except that in the heavens, we see all stages at once. Whether plants or planets, sunflowers or stars, we see the lifecycle of nature's creations spread before us, awaiting understanding.

All this activity gives rise to a variety not only of stable and steady stars but also of variable stars—those whose brightness fluctuates whether for reasons intrinsic or extrinsic, as in an eclipsing binary. Although some stars are variable at the class level, such as the young and forming T Tauri stars, more often in the system laid out here, variables are Types of classes. Among these are the so-called RR Lyrae variables within the giant and bright giant classes, the pulsating Cepheid variables found mainly among the supergiants, and the luminous blue variables among the hypergiants. Novae, supernovae, flare stars, M dwarfs, and many more give rise to temporary or long-term fluctuations. All these variable phenomena are eagerly monitored by professional astronomers and amateur organizations such as the American Association of Variable Star Observers (AAVSO). They belie the constancy of the heavens proclaimed from the ancients through Shakespeare.

Like planets, stars also have a circumstellar family of objects. These include not only protoplanets and planets themselves but also debris disks left over after planetary formation; shells surrounding certain dying stars like the Mira variables and carbon stars; the beautiful but transitory planetary nebulae representing the ejected outer layers of red giants on their way to becoming white dwarfs; and nova remnants and core collapse supernovae.

More surprisingly, some stars sport jets of energetic matter and radiation being ejected from stars accreting matter, sometimes accompanied by so-called Herbig-Haro Objects formed when gas from the jet collides with gas in the interstellar medium. That medium itself consists of 99% gas, coalescing into classes of objects including cool atomic hydrogen clouds (H I objects), hot ionized hydrogen clouds (H II regions), and cooler molecular clouds of hydrogen molecules, all extremely important in the life cycle of stars. The other 1% of the interstellar medium is dust, evidenced in dark but striking nebulae like the famous Horsehead Nebula in Orion, and in beautiful and colorful reflection nebulae like the one surrounding the Pleiades cluster. And, as in the planetary Kingdom, whizzing through the whole is a component of energetic particles in the form of stellar winds and galactic cosmic rays.

Stars also like to form systems, ranging from binaries to multiples, small associations to larger open clusters, teeming globular clusters to entire distinct populations. And when they form systems even as simple as binaries, unexpected and spectacular things happen. Binary stellar black holes can spiral into each other, emitting gravitational waves, as first detected in 2015. Binary neutron stars can do the same thing, as evidenced in another detection of gravitational waves in 2017. A neutron star or black hole paired with a normal star can produce X-rays as the neutron star or black hole pulls mass into an accretion disk—what astronomers call an X-ray binary, exemplified by Cygnus X-1 (a black hole and a supergiant). A white dwarf in a binary system may explode as a supernova after it accretes more than 1.4 solar masses from a normal star in the system, the so-called Chandrasekhar limit. Even the less energetic but still impressive phenomenon of novae occurs when hydrogen from a normal star is accreted onto a white dwarf in a binary system. Binary system manifestations can also be as simple as one star passing in front of another, resulting in an eclipsing binary as in the prototypical Algol A. Additionally, binary systems can be responsible for the particularly spectacular shapes of some planetary nebulae.

The largely gaseous Kingdom of the Stars provides ample opportunity for the creation of phenomena not seen in the Kingdom of the Planets. And these stars in turn provide the basic units for the next Kingdom—the even more spectacular Kingdom of the Galaxies, which takes us into the realm of cosmology.

7. The Protostellar Family

Class S 1: Protostar

Like planets, stars are born, but they do so in considerably more spectacular fashion and at rates highly dependent on their environment. A protostar, sometimes called an embryonic star, is a molecular cloud (S 25) undergoing gravitational contraction and accretion of gas and dust prior to initiating fusion reactions. The end of accretion and the onset of a stellar wind (S 29) is often considered the dividing line between a collapsing protostar and the next stage in stellar evolution, the so-called pre-main sequence stars (S 2 and S 3) such as T Tauri or higher mass Herbig Ae/Be objects, discussed in the next two entries. But the transition is not sharp: some pre-main sequence stars are still accreting matter. Other classes of objects associated with newborn stars are jets (S 20) and Herbig-Haro objects (S 21). Protostars are also called "young stellar objects" (YSOs), but that term sometimes also encompasses pre-main sequence stars. Surprisingly, our Galaxy births only about one star per year, but "starburst" galaxies such as M82 may form stars at a thousand times that rate.

The American astronomer Lyman Spitzer was among the first to use the term "protostar" in 1949 in an article on the formation of stars, and it was used occasionally in the 1950s by astronomers such as Gerard Kuiper. The term came into widespread usage in the 1960s, especially after a review article on protostars was published in 1966 by the Japanese astronomer Chushiro Hayashi. In that article, he determined the track of a contracting protostar onto the main sequence (S 4), known today as the "Hayashi track."[1]

The lifetime of a protostar prior to becoming a pre-main sequence star depends on the mass of the protostar; the more massive the protostellar cloud, the more rapidly it collapses to a star.

---

[1] Lyman Spitzer, Jr., "The Formation of Stars," ASP Leaflets, 5 (1949), 336-343: 342; G. P. Kuiper, "On the Origin of Binary Stars," PASP, 67 (1955), 387-396; Chushiro Hayashi, "Evolution of Protostars," ARAA, vol. 4 (1966), p.171-192.

© Springer Nature Switzerland AG 2019
S. J. Dick, *Classifying the Cosmos*, Astronomers' Universe,
https://doi.org/10.1007/978-3-030-10380-4_7

Stars similar in mass to our Sun would spend several tens of millions of years in the protostar stage. The story is personal: about 4.6 billion years ago, the molecular cloud of hydrogen, helium, and a sprinkling of heavier elements that became our Sun collapsed under its own gravity, spinning faster as it shrank. While the spinning flat disk produced the planets, the central mass continued to shrink, heating up and deriving its energy from gravitational contraction until it was hot enough to initiate nuclear fusion.

The basic idea of star formation can be traced back to the nebular hypothesis of Pierre Simon de Laplace in the late 18th century, in which a cloud of gas and dust condenses into planets and a central star. The general idea of stars condensing out of gas clouds was thus in the air in the early 20th century, as when Henry Norris Russell suggested that the masses of the dark nebulae (S 27) systematically photographed by E.E. Barnard were "probably sufficient to form hundreds of stars." In the 1940s a few astronomers such as Lyman Spitzer and Fred Whipple drew attention to the possible formation of stars out of condensations in the interstellar medium, and in 1947 Bart Bok drew attention to small circular dark nebulae as possible sites of such formation.[2] Little was known of the nature of the development of potential stars until the advance of observational astronomy, and even then protostar formation was difficult to observe because it was buried within the molecular cloud from which it is forming. For this reason, protostars are best observed with infrared and radio telescopes. Only in the late 1970s and 1980s did infrared, radio, millimeter, and submillimeter observations begin to penetrate these clouds, using ground-based instruments such as the NASA Infrared Telescope Facility (IRTF) on Mauna Kea, Hawaii, and radio telescope arrays such as the Very Large Array in New Mexico. Space-based observatories, including the Infrared Astronomical Satellite (IRAS) and the Hubble and Spitzer space telescopes, also then took the lead in uncovering protostars.

---

[2] Henry Norris Russell, "Dark Nebulae," PNAS, 8, (1922), 115-118: 117; Lyman Spitzer, "The Dynamics of the Interstellar Medium. II. Radiation Pressure," ApJ, 94 (1941), 232; Fred Whipple, "Concentrations of the Interstellar Medium," ApJ, 104 (1946), 1; Bart J. Bok and Edith F. Reilly, "Small Dark Nebulae," ApJ, 105 (1947), 255-257.

The general picture that has emerged is of a large, cold molecular cloud in which a dense sub-cloud, known as the core, collapses under its own weight, triggered either by internal changes or external forces such as pressure from a supernova explosion. This core fragments into many stellar embryos, each of which accretes gas and dust. As the protostar shrinks in size, its density increases until nuclear fusion begins and the surrounding gas and dust dissipate.

If gravity were the only factor, star formation would be much faster. But gravity has its own fight in the form of highly chaotic turbulence within the gas cloud, evident in the broadening of molecular lines in the spectrum of the clouds as observed by instruments such as the Atacama Large Millimeter Array (ALMA). The same turbulence is believed to cause compressions of matter that then serve as seeds for accretion and gravitational collapse. Magnetic fields in the cloud also play a role. The initial mass of these stars plays a defining role for the rest of their lives, determining their life, evolution, and death. Most stars begin with masses less than half of the Sun, others more or less. Stars with more than 10 solar masses were thought to be rare, but recent observations of the star-forming region 30 Doradus (Fig. 11.8) in the Large Magellanic Cloud suggests that massive stars are much more common than previously thought. Stars up to 200 to 300 solar masses may not only exist, but could also be in such numbers as to hold most of the stellar mass of the universe. The distribution of stellar masses after formation is known as the initial mass function (IMF).[3]

By the 1980s, enough observations had been made that astronomers could divide young stellar objects into three classes (or Types as we would term them in the Three Kingdom system) depending on the amount of circumstellar material: Class I are still embedded in their molecular clouds, Class II are T Tauri Stars,

---

[3] Christoph Federrath, "The Turbulent Formation of Stars," *Physics Today*, 71 (2018), pp. 38-42; F. R. Schneider et al., "An excess of massive stars in the local 30 Doradus starburst," *Science*, 359 (2018), 69-71. For a recent review of the history and importance of the star formation rate and initial mass function see Patrick Francois and Danielle Briot, "Rate of formation of stars suitable for the development of intelligent life, R*, 1961 to the present," in Douglas A. Vakoch and Matthew F. Dowd, eds., *The Drake Equation* (Cambridge, CUP, 2015), pp. 38-52.

and Class III have lost most of their circumstellar material. These stages can be observed based on their spectral energy distribution from near- to mid-infrared wavelengths and roughly reflect their evolutionary sequence. Class I constitutes what is here termed a protostar prior to the pre-main sequence star. As observational capabilities increased, an even earlier Class 0 was added in 1993 based on submillimeter observations of the rich star-forming core dark cloud Rho Ophiuchi A. Arguing that "An unambiguous example of a protostar remains to be found ... in spite of repeated searches for over two decades," radio astronomer Philippe Andre and his colleagues described their radio observations and wrote that "We ... suggest that these objects define an entirely new class of YSOs, which we call 'Class 0' to indicate their extreme youth."[4] Class-0 and Class I objects are believed to be about 10,000 and 100,000 years old respectively. In the Class-0 phase, a dense molecular cloud and heavy accretion of gas onto the newly forming star shroud the region.

By 1995, the Hubble Space Telescope showed newborn stars emerging from EGGS (evaporating gaseous globules) in the famous image of a tiny portion of the Eagle Nebula (Fig. 7.1). In this image, Hubble caught protostars in the act of being uncovered by the process known as photoionization, as hot stars blow away the materials the protostar is feeding on. The Eagle Nebula itself is an enormous H II region (S 24) about 7,000 light years distant, which includes a large open cluster of stars (S 34).[5]

In the year 2000, astronomers using the Chandra Observatory detected activity in this early phase for the first time, in the form of highly attenuated X-rays coming from Class-0 protostars in the Orion Molecular Cloud (OMC-3), at 1,400 light years from the closest star-forming region to Earth. The discovery of X-rays in such a nascent setting was very surprising. "Far beyond our imagination, a star immediately after birth at the center of a cold molecular core at temperatures of only a few tens of Kelvin [-400° F] frequently generates very hot plasma with 10 to 100 million Kelvin,"

---

[4] Philippe Andre et al, "Submillimeter continuum observations of Rho Ophiuchi A - The candidate protostar VLA 1623 and prestellar clumps," ApJ, 406 (1993), 122-141; On Classes I, II and III, see C. J. Lada, in *Star Forming Regions*, M. Peimbert and J. Jugaku, eds. (Dordrecht, 1987), p. 1.

[5] HST Press Release, November 2, 1995, "Embryonic Stars Emerge from Interstellar "Eggs," http://hubblesite.org/newscenter/archive/releases/1995/44/text/

Fig. 7.1. EGGS emerge from the nest in the Eagle Nebula. The three columns are molecular hydrogen gas and dust protruding from the molecular cloud. The EGGS, approximately 100 astronomical units across, are the denser regions at the tips of the columns, some of which will form embryonic stars. This iconic image was taken on April 1, 1995 with the Hubble Space Telescope. Credit: NASA, ESA, STScI, J. Hester and P. Scowen (Arizona State University)

astronomer Katsuji Koyama of Kyoto University stated. Koyama speculated that these X-ray flares might be generated from stellar spin and convection prior to the onset of hydrogen fusion.

The transition to a Class-1 protostar is characterized by changes in the infrared spectrum of the protostar and the gas as the dust envelope diminishes. The Chandra spacecraft also detected X-rays emanating from 17 of these Class-1 protostars in the famous molecular cloud rho Ophiuchi 460 light years from Earth, first studied by E.E. Barnard a century earlier. "Virtually all the Class 1 protostars in the rho molecular cloud may emit X rays with extremely violent and frequent flare activity," said Kensuke Imanishi, lead investigator of the rho Ophiuchi observation. "The X-ray fluxes in the flares we saw were up to 10,000 to 100,000 brighter than those in our Sun's flares."[6] Other Class 0 objects have been found; one of the best-studied is IRAS16293-2422, where formaldehyde and other complex molecules have been detected.

Over the last decade, both Spitzer Space Telescope and Herschel Space Observatory teams have carried out surveys of young stellar objects, including the Orion and Taurus molecular clouds as well as 18 other molecular clouds within 1,500 light years of Earth. They found massive infrared dark clouds, 100 to 100,000 times the mass of the Sun, which may be the link between molecular clouds and protostars. These infrared clouds are three times more massive than the stars formed from them; the rest is returned to the interstellar medium. Both Spitzer and the Chandra X-ray Observatory have also found evidence of multiple triggering mechanisms for star formation, including radiation from a bright massive star in a region known as Cepheus B (Fig. 7.2). Researchers using the Hubble Space Telescope have recently surveyed 50 local galaxies to analyze their star-forming regions in the ultraviolet region of the spectrum, part of an international program known as the Legacy ExtraGalactic UV Survey (LEGUS). By "local," astronomers mean space stretching as far as 58 million light years away to the Virgo cluster of galaxies![7]

Theoretical work on protostars has proceeded hand-in-hand with observations, building on the work of such pioneers in the field as Hayashi, Spitzer, and Frank Shu. Spitzer concluded his detailed review of the subject in 1968 with the words "There are so many uncertainties in this picture that at present we do not really have a theory of star formation. In a general way, however, we can

---

[6]Chandra Release, November 8, 2000, "X-Ray Snapshots Capture the First Cries of Baby Stars," http://chandra.harvard.edu/press/00_releases/press_110800.html; Yohko Tsuboi et al, "Discovery of X-rays from Class 0 Protostar Candidates in OMC-3," ApJ, 554 (2001), 734-741; NASA Release, March 1, 2005, "Baby Star is Way Ahead of its Time," http://www.nasa.gov/vision/universe/starsgalaxies/xmm_magnetic_starbirth.html

[7]Michael M. Dunham et al., "The Evolution of Protostars: Insights from Ten Years of Infrared Surveys with Spitzer and Herschel," Protostars and Planets VI, Henrik Beuther, Ralf S. Klessen, Cornelis P. Dullemond, and Thomas Henning, eds., (Tucson: University of Arizona Press, 2014), pp.195-218, preprint at https://arxiv.org/abs/1401.1809; Spitzer Release, August 12, 2009, "Trigger-Happy Cloud," http://spitzer.caltech.edu/images/2719-ssc2009-17a-Trigger-Happy-Cloud

Trigger-Happy Cloud

Spitzer Space Telescope • IRAC / Chandra X-ray Observatory • ACIS

NASA / CXC / JPL-Caltech / PSU / CfA   ssc2009-17a

Fig. 7.2. Composite image combining data from NASA's Chandra X-ray Observatory and Spitzer Space Telescope, showing the star-forming cloud Cepheus B, located in our Milky Way galaxy about 2,400 light years from Earth. Credit: NASA/CXC/JPL-Caltech/PSU/CfA

at least visualize some of the important physical processes that act in the formation of stars from interstellar clouds."[8] Fifty years later, not only have we "visualized" these processes with real telescopic images, but the theory has also advanced significantly.

Although many details have been filled in during the last half-century, important questions remain unresolved. These questions range from the origin of the molecular clouds and the cause of their collapse to the details of formation of multiple stars (S 32) and very massive stars. Despite the uncertainties in our current understanding, star formation seems to be a robust, if slow, process.

---

[8] Lyman Spitzer, "Dynamics of Interstellar Matter and the Formation of Stars," in *Nebulae and Interstellar Matter*, ed. B. M. Middlehurst and L. H. Aller (Chicago: University of Chicago Press, 1968), pp. 1-63; Frank Shu, "Self-similar collapse of isothermal spheres and star formation," ApJ, 214 (1977), 488-497.

For example, in 2010 the Herschel Infrared Space Observatory found a star ten times the mass of the Sun still drawing on an additional 2,000 solar masses of material surrounding it, set to turn into one of the most massive stars in the galaxy.[9] Surprisingly, stars have also been reported forming even very near the center of our Galaxy (within 10 light years), where one would have thought the supermassive black hole (S 14) might have disrupted the process. Even more surprising, in recent years 200 young massive stars were reported only a fraction of a light year from this central black hole, known as Sgr A* (pronounced "Sagittarius A star").[10]

Star formation is studied well beyond the bounds of our parochial Milky Way. Indeed, one of the most vibrant areas of research in astrophysics is starburst galaxies, where all or part of a galaxy is undergoing star formation a thousand times higher than in normal galaxies such as our own. Such starbursts may be triggered as a result of interactions with other galaxies (G 20), and usually last only a few hundred million years. Starburst galaxies include the prototypical M82 (the Cigar Galaxy, Fig. 14.9); NGC 1569 (Fig. 14.8); the Antenna Galaxies (Fig. 18.3); the Sculptor Galaxy; and Centaurus A, a rare case of an elliptical starburst galaxy (Fig. 14.13). Even more exotic examples include luminous infrared galaxies (LIRGs) such as the lenticular NGC 5010 in Virgo, and ultra-luminous infrared galaxies (ULIRGs) such as the irregular NGC 6240 in Ophiuchus. The latter is the result of a collision between two galaxies. Starburst galaxies are often accompanied by supergalactic winds (G 17), the galactic equivalent of solar and stellar winds (P 16 and S 29).[11]

A video zooming in on the EGGS of the Eagle Nebula can be found at http://hubblesite.org/video/100/news_release/1995-44. For a European Southern Observatory image of the entire nebula, see https://en.wikipedia.org/wiki/Eagle_Nebula#/media/File:Eagle_Nebula_from_ESO.jpg. Hubble's LEGUS program results are described at http://hubblesite.org/news_release/news/2018-27.

---

[9] ESA Herschel Press Release, May 6, 2010, "Herschel Reveals the Hidden Side of Starbirth," http://www.esa.int/SPECIALS/Herschel/SEM7N7KPO8G_0.html

[10] Andrea Thompson, "Surprise Star Formation Found Near Black Hole, January 5, 2009, http://www.space.com/scienceastronomy/090105-aas-stars-galactic-center.html reporting on Elizabeth Humphreys et al, "Active Star Formation Within Two Parsecs of Sgr A*?," BAAS, 41 (2009), 226; Camille Carlisle, "Near the Pit," S&T, 136 (2018), 22-29.

[11] Richard Jakiel, "Star Factories," S&T (March, 2018), 28-33.

8. The Star Family

Class S 2: T Tauri

T Tauri stars represent a very early stage in the history of a low-mass star, when the protostar (S 1) emerges from the surrounding molecular cloud (S 25), but before sustained core hydrogen burning on the main sequence of the Hertzsprung-Russell diagram (S 4). The energy is derived from gravitational collapse (as well as some deuterium burning beginning at about one million degrees Kelvin), since the star is not yet hot enough for the proton-proton reaction converting hydrogen to helium, for which a minimum temperature of four million K is required. T Tauri stars are therefore termed "pre-main sequence" objects, and are always young (one to ten million years) and less than two solar masses. Higher mass pre-main sequence stars are known as Herbig Ae/Be stars (S 3).

As would be expected, because of their dynamic nature, T Tauri stars are variable and produce very powerful stellar winds (S 29). They are traditionally divided into two Types: classical T Tauri stars with extensive disks resulting in strong emission lines, and weak ("naked") T Tauri stars that have almost completely lost their nebulous disk, perhaps because planetesimals—the building blocks of planets—have formed in the protoplanetary disk (P 1). T Tauri stars are therefore important for the study of planetary system formation. Spectral types range from F to M and show lines of lithium, also a sign of stellar youth since lithium does not survive long inside a star any more than a snowflake would inside a furnace. T Tauri stars are also associated with optical jets (S 20) and Herbig-Haro objects (S 21).

The nearest T Tauri stars are found in the Taurus Molecular Cloud and the Rho Ophiuchus Cloud, both about 460 light years away. The dark cloud known as the Taurus-Auriga complex (Fig. 8.1) contains a large number of T Tauri stars, as does the more distant Orion Nebula at 1,500 light years. T Tauri stars are copious emitters of X-rays, probably due in part to accretion activity. Such X-ray emission was first reported in 1981 using the Einstein (HEAO-2) spacecraft and has since been observed with numerous

© Springer Nature Switzerland AG 2019
S. J. Dick, *Classifying the Cosmos*, Astronomers' Universe,
https://doi.org/10.1007/978-3-030-10380-4_8

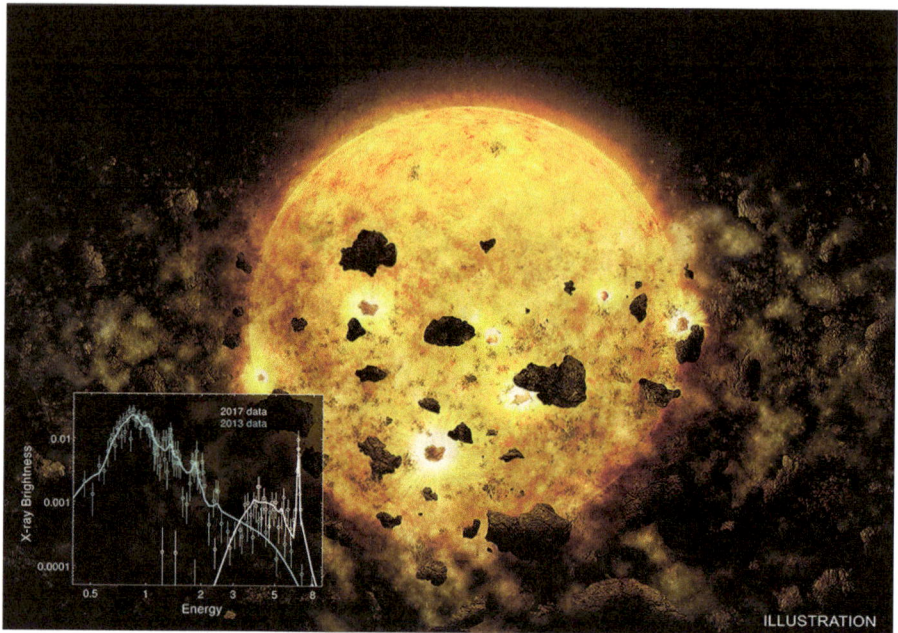

Fig. 8.1. The young T-Tauri star known as RW Auriga A in the Taurus-Auriga Dark Clouds may be destroying one or more planets, based on the X-ray spectrum (inset) from Chandra, indicating the presence of iron from the planets. Illustration credit: NASA/CXC/M.Weiss; X-ray spectrum: NASA/CXC/MIT/H.M.Günther

other spacecraft. Ground-based follow-ups to observe unidentified X-ray sources in the ROSAT spacecraft's All-Sky Survey revealed hundreds of new T Tauri stars. The Chandra X-ray Observatory's Orion Ultradeep Project studied nearly 600 T Tauri X-ray sources in the Orion Nebula Cluster.[1]

The astronomer John Russell Hind first discovered the eponymous T Tauri star as a 10th magnitude object in 1852. In 1890, S.W. Burnham found a small nebula surrounding T Tauri, which by that time had dimmed to 14th magnitude. In 1945, Mt. Wilson astronomer Alfred Joy first systematically studied the star and

---

[1] Thomas Preibisch et al., "The Origin of T Tauri X-Ray Emission: New Insights from the Chandra Orion Ultradeep Project," ApJ Supplement Series, 160 (2005), 401-422, including a good summary of X-ray observations of T Tauris.

chose T Tauri to be the prototype for a class of 11 other stars then observed to have similar behavior. "Eleven irregular variable stars have been observed whose physical characteristics seem much alike and yet are sufficiently different from other known classes of variables to warrant the recognition of a new class whose prototype is T Tauri," he wrote. His four criteria for inclusion in the class were irregular light variations of about 3 magnitudes, spectral types F5 through G5 with emission lines resembling the solar chromosphere, low luminosity, and association with dark or bright luminosity. Though the magnitude variations and spectral range were gradually extended with observations of more class members, his criteria remain remarkably accurate.[2]

Only in the 1950s did astronomers begin to consider T Tauri stars as precursors of solar-type stars. In 1960, in investigating the possibility of a class of higher-mass counterparts to T Tauri stars that would eventually bear the name Herbig Ae/Be stars (S 3), George Herbig wrote:

> Over the past decade the accumulation of observational material, together with a growing understanding of some of the processes of stellar evolution, have led many astronomers to the belief that the T Tauri stars are young stars still in the stage of gravitational contraction toward the main sequence. If this belief is correct, then the luminosities of the T Tauri stars indicate that they are objects of small to intermediate mass that will in time become main-sequence stars of type F and later.

Herbig himself played a considerable role in demonstrating this was indeed true.[3]

In 1981, T Tauri itself was found to be a binary, and most T Tauri stars may also be binary systems; two surveys in the 1990s found 74 binaries or multiples with a total of 85 companions in 174 T Tauri systems.[4] Beginning in the 1990s, T Tauri itself has

---

[2] Alfred Joy, "T Tauri Variable Stars," ApJ, 102 (1945), 168-195

[3] G. H. Herbig, "The Spectra of Be- and Ae-Type Stars Associated with Nebulosity," ApJ Supplement, 4 (1960), 337-368: 337; G. H. Herbig, "The properties and problems of T Tauri stars and related objects," Adv. Astr Astrophys, 1 (1962), 47.

[4] R. Kohler and C. Leinert, "Multiplicity of T Tauri stars in Taurus after ROSAT," A & A, 331 (1998), 977-988.

been observed in considerable detail at several wavelengths, and the results have prompted some astronomers to warn that "observations at infrared wavelengths over the last twenty years have demonstrated the danger in adopting a single astronomical source as a paradigm or representative member of a class of objects ... it is now evident that T Tau is very unusual and perhaps unique." They pointed out, for example, that T Tauri is one of the few stars in its class with detectable molecular hydrogen. It also has an unusual extremely red companion and "exhibits an unusually complex system of multiple jets and outflows unlike that seen in any other TTS."[5] Although the point is well taken and applies to other class prototypes as well, the T Tauri class is entrenched among astronomers, who can increasingly point to differences between class members and the prototype as technology improves.

Millimeter and near-infrared surveys have shown that many Sun-like pre-main sequence stars have disks ranging from one thousandth to one tenth of a solar mass. An early survey in 1990 showed disks around 42% of 86 T Tauri stars observed in the Taurus dark cloud. Further, the authors found that "The mass, size and angular momentum of these disks are similar to those associated with the solar nebula during the first stage of planet formation," and concluded that this evidence supported the idea that planetary systems are common in the Galaxy. Simple molecules including carbon monoxide (CO), CN, and HCN have been detected in disks around both T Tauri and Herbig Ae/Be stars.[6] In 2003, the twin 10-meter Keck telescopes on Mauna Kea were linked together to make their first interferometric observations by observing the relatively faint T Tauri object DG Tau, revealing a disk orbiting some 18 million miles from the star. Many more interferometric observations of resolved disks have since been made. Surprisingly, planetary mass objects have even been detected around several T Tauri stars, including HD 106906 in the

---

[5] T. M. Herbst, S. V. W. Beckwith et al., "A Near-Infrared Spectral Imaging Study of T Tau," AJ (1996), 2403-2414

[6] Steven V. W. Beckwith, Annelia I. Sargent et al, "A survey for circumstellar disks around young stellar objects," AJ, 99 (1990), 924-944; W.-F. Thi et al, "Organic molecules in protoplanetary disks around T Tauri and Herbig Ae stars," A &A, 425 (2004), 955-972.

constellation Crux about 300 light years from Earth, and Gliese 674b only 15 light years away in the constellation Ara.[7]

T Tauri stars may go through a stage of particular instability when they are observed as the so-called F U Orionis stars, giving rise to an increase in brightness of three to five magnitudes, perhaps due to the accretion of material from the surrounding disk. F U Orionis stars, also known as "Fuors," are classified as eruptive variables and are quite rare; only 11 stars have been observed going through this stage. While sometimes considered a separate class of pre-main sequence star, most astronomers consider F U Orionis stars a stage of T Tauri stars.

More on the RW Auriga A discovery is available at http://chandra.harvard.edu/press/18_releases/press_071818.html. ROSAT observations of the spatial distribution of T Tauri stars in the constellation Chamaeleon are mapped at http://heasarc.gsfc.nasa.gov/docs/rosat/gallery/stars_ttauri.html. An artist's conception of an F U Orionis star with its surrounding disk can be found at http://hubblesite.org/image/3746/category/87-variable-stars.

---

[7] M. Colavita, R. Akeson et al., "Observations of DG Tauri with the Keck Interferometer," ApJ, 592 (2003), L83-L86. Bailey, Vanessa; Meshkat, Tiffany et al. "HD 106906 b: A planetary-mass companion outside a massive debris disk," ApJ Letters (Jan, 2014), 780 (1): L4

Class S 3: Herbig Ae/Be

Herbig Ae/Be stars, sometimes referred to as HAeBe stars, are a class of the pre-main-sequence Subfamily that are higher mass than T Tauri stars (S 2). Their ages range from half a million to 20 million years before the onset of hydrogen nuclear fusion. Like T Tauri stars, they derive most of their energy from gravitational collapse rather than core hydrogen fusion. But whereas members of the T Tauri class become F, G, K, and M stars less than 2 solar masses, Herbig Ae/Be stars are the progenitors of the hotter A and B spectral types (thus their designation) and have masses that range from two to ten solar masses. Like many T Tauris, they show strong emission lines in their spectra (thus the suffix "e"). The emission lines in both cases are due to the shell or disk of material that surrounds the star, but the spectra of T Tauri stars and Ae/Be stars are quite different in other ways.

Some Herbig Ae and Be stars are not associated with gaseous nebulosity, but are characterized by an infrared excess due to surrounding dust. Still other initial candidates turn out to be "ordinary" main sequence Ae or Be stars with emission lines. The identification of true Herbig Ae/Be pre-main sequence stars is therefore difficult. As one researcher noted, "even after more than 3 decades of investigations, it is still difficult to define a unique set of observable quantities that can distinguish intermediate mass PMS [pre-main sequence] objects from more evolved stars."[8] In addition to circumstellar disks (P 1 and S 15), and envelopes, Herbig Ae/Be stars may exhibit stellar winds (S 29), bipolar outflow jets (S 20), Herbig-Haro objects (S 21), and accretion phenomena.

Despite the difficulties, Herbig Ae/Be stars are readily classified in the MK system discussed in the Introduction. In a revealing statement about classification, astronomers have also emphasized that "The distinction between T Tau stars and HAeBe stars is not merely a matter of semantics...the evolution of intermediate-mass PMS stars is qualitatively different from that of lower-and higher-mass stars owing to the differences in stellar and circum-

---

[8] P. S. The et al., "A New Catalogue of members and candidate members of the Herbig Ae/Be (HAEBE) stellar group," A&A Supplement series, 104 (1994), 315-339: 316; U. Bastian et al., "The definition of T Tauri and Herbig Ae/Be stars," A&A, 126 (1983), 438-439.

stellar physics, as well as in time scales." In particular, pre-main sequence stars in excess of two solar masses follow fully radiative tracks once contraction has ended, while those more massive than 10 solar masses spend their entire pre-main sequence lives as obscured objects.[9] About a dozen Herbig Ae/Be stars had been confirmed by 1972, several dozen were known by the 1980s, and over 100 are now known.

The American astronomer George Herbig first identified Ae/Be stars as a potential class of pre-main-sequence stars in a seminal paper in 1960.[10] In this paper, Herbig explicitly acknowledged that in the previous decade T Tauri stars were shown to be likely pre-main-sequence stars of small to intermediate mass. Inspired by that work, he inquired whether pre-main-sequence stars of larger mass existed. Herbig did this by examining the spectroscopic characteristics of 26 Be and Ae stars "that both lie in obscured regions and illuminate nearby nebulosity." Although the evidence was inconclusive as to their true nature, subsequent observations demonstrated about half of these were high-mass pre-main-sequence stars. Herbig is therefore credited with the discovery of this new class of stars. He also realized that such stars would be rarer than T Tauris, since higher mass stars are rarer in general than lower mass stars, and their higher mass means their contraction times are shorter. Stephen Strom and colleagues confirmed the pre-main sequence nature of these objects in 1972.[11]

HAeBe stars have now been observed in wavelengths ranging from X-ray to radio. In particular, the brighter members of the Ae/Be class have been identified as strong infrared sources. As early as 1971, Gillett and Stein suggested that this is due to circumstellar dust clouds, and HAeBe stars have since been studied for

---

[9] L. B. F. M. Waters and C. Waelkens, "Herbig Ae/Be Stars," ARAA, 36 (1998), 233-266: 234; on their place in the MK system, Richard O. Gray and Christopher J. Corbally, *Stellar Spectral Classification* (Princeton and Oxford: Princeton University Press, 2009), pp. 201 ff.

[10] G. H. Herbig, "The Spectra of Be- and Ae-Type Stars Associated with Nebulosity," ApJ Supplement, 4 (1960), 337-368. S. E. Strom et al, "The Nature of the Herbig Ae- and Be-Type Stars Associated with Nebulosity," ApJ, 173 (1972), 353 found that about half of Herbig's sample of 26 were actually pre-main sequence stars.

[11] S. E. Strom et al., "The Nature of the Herbig Ae- and Be-Type Stars Associated with Nebulosity," ApJ, 173 (1972), 353-366.

their possible links to both protoplanetary disks and debris disks such as Beta Pictoris and Vega-type stars. Since the 1980s, infrared spectroscopy from the ground and from spacecraft, including the Infrared Space Observatory and the Spitzer Telescope, have demonstrated that protoplantary dust is composed of amorphous silicates including carbon and oxygen, and that about 20% show evidence of polycyclic aromatic hydrocarbons. Detailed characteristics of HAeBe stars have recently been revealed by observations at the ALMA telescope, the Very Large Telescope Interferometer in Chile, and other telescopes with adaptive optics.

Since the Infrared Astronomical Satellite (IRAS) all-sky infrared survey in the mid-1980s, evidence has also accumulated that Herbig Ae/Be stars are found outside nebular regions, perhaps because of ejection or perhaps due to small offshoots of nebular gas. They have been called "isolated" Herbig Ae/Be stars and are the subject of continuing research. This discovery has caused a revision in the "working definition" of HAeBe stars as "(a) spectral type A or B with emission lines, (b) infrared (IR) excess due to hot or cool circumstellar dust or both, and (c) luminosity class III to V."[12] A well-known example of this class is R Coronae Borealis, a variable supergiant that is the prototype for an entire Type of eruptive variables.

The fact that the class of T Tauri pre-main-sequence stars is not called by analogy Joy Ge/Ke stars (after Alfred Joy, who first studied them closely and identified them with those spectral types) is another example of the vagaries of astronomical nomenclature.

A catalogue of Herbig Ae/Be stars is available at http://www.eso.org/~mvandena/haebecat.html. The Herschel spacecraft has recently imaged the G-305 high-mass star-forming complex at http://sci.esa.int/herschel/60349-hidden-secrets-of-a-massive-star-formation-region/

---

[12] Waters and Waelkens, 235.

Class S 4: Dwarf

Main sequence stars, also called "dwarfs," are those stars in their core hydrogen-burning stage, so called because they form a clear band or "sequence" of stars when stellar absolute magnitudes or luminosities are plotted against stellar surface temperatures. The astronomers Ejnar Hertzsprung and Henry Norris Russell independently developed the ideas behind such a plot in the decade from 1905 to 1915, both using spectra and brightness data obtained from the Harvard College Observatory.

The resulting diagram, clearly showing a band of stars running from upper left to lower right, is called the Hertzsprung-Russell (HR) diagram (Fig. 8.2).[13] In its fully elaborated forms, it is the key framework graphically depicting stellar evolution onto the main sequence and beyond. The time when a star joins main sequence is known as the "zero age main sequence" for that star. The position of a star in the diagram is then mainly determined by its initial mass, chemical composition, and age, but other factors also play a role. The details of the HR diagram are continuously being refined, most recently with data from the European Space Agency's Gaia spacecraft, which in 2018 released the positions, motions, and other data for 1.7 billion stars.[14]

Main sequence stars are called "dwarfs" by comparison with the much brighter giants (S7 and S8) and supergiants (S 9) that also stand out as distinct groups on the upper right of the HR diagram, not to mention the amazingly huge hypergiants (S 10). About 90% of stars in our Galaxy are on the main sequence, which is luminosity class V in the so-called MKK (Morgan-Keenan-Kellman) system developed by W.W. Morgan, Philip Keenan, and Edith Kellman at the Yerkes Observatory in 1943. After revisions in 1953, the system became known as the MK system, since Kellman had moved from the Yerkes Observatory staff to become a teacher during

---

[13] Bengt Strömgren first named it the "Hertzsprung-Russell Diagram" in his article "On the Interpretation of the Hertzsprung-Russell-Diagram," *Zeitschrift für Astrophysik*, 7 (**1933**): 222–248. On the history of the diagram see David DeVorkin, "Stellar Evolution and the Origin of the Hertzsprung-Russell Diagram," *Astrophysics and Twentieth-Century Astronomy to 1950*, Part A, Owen Gingerich, ed. (Cambridge, 1984), 90-108.

[14] Gaia collaboration, "Gaia Data Release 2: Observational Hertzsprung-Russell diagrams," A&A, vol. 616, http://adsabs.harvard.edu/abs/2018arXiv 180409378G

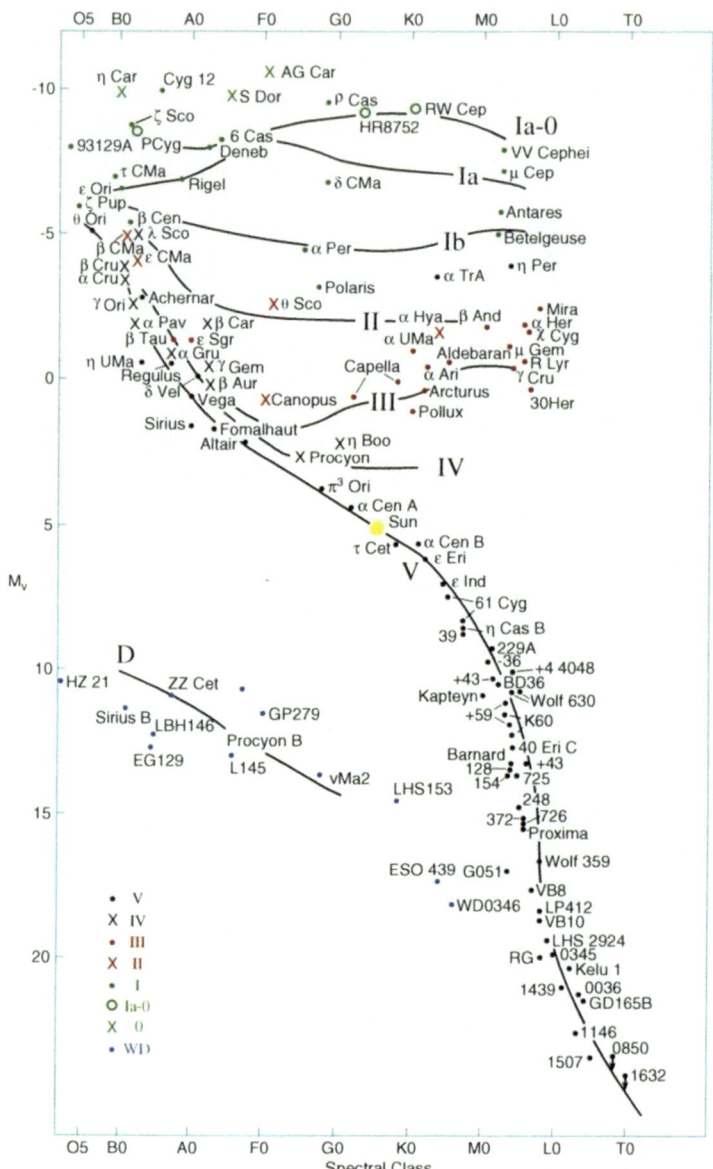

Fig. 8.2. The Hertzsprung-Russell Diagram, plotting absolute visual magnitude and Harvard spectral types for the brightest and closest stars. The different classes of stars are clearly shown, ranging from main sequence dwarfs (luminosity class V) through subgiants (IV), giants (II and III), and supergiants (I). The Diagram is the key to much of stellar astronomy. From James B. Kaler, *Encyclopedia of the Stars* (Cambridge University Press, 2006), Fig. 6.12, p. 110

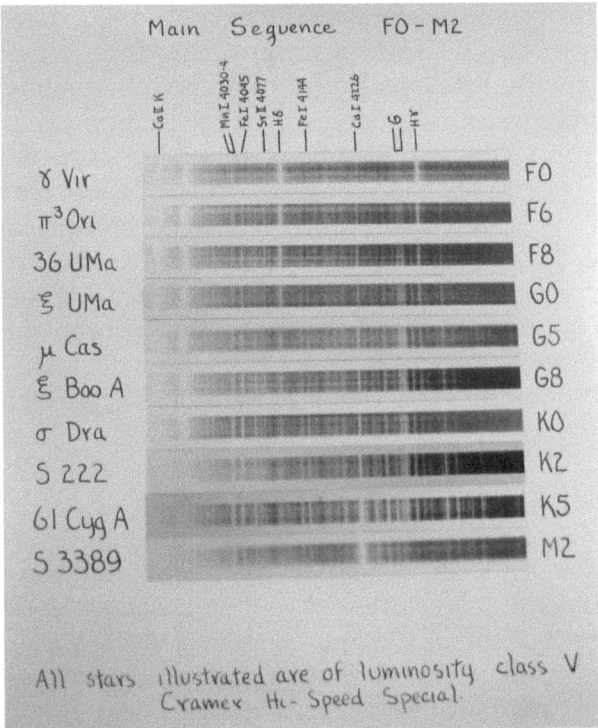

Fig. 8.3. The MKK system defined standards for luminosity classes, which represent physical classes such as giants and dwarfs. Here the standards are shown for luminosity class V (dwarfs), across the temperature sequence from F0 to M2. This is one of 55 such prints from W.W. Morgan, P. C. Keenan, and E. Kellman, *Atlas of Stellar Spectra with an Outline of Spectral Classification* (Chicago: University of Chicago Press, 1943)

World War II. Each luminosity class in this system across a range of temperatures is defined by very specific characteristics in the spectra of standard stars, as seen in Fig. 8.3. Combined with spectral types arranged in a temperature sequence from hot O stars to cool M stars according to the system devised at the Harvard College Observatory in the 1880s (the horizontal axis in Fig. 8.2), a star can be described both in terms of its temperature and its place on the HR diagram on or off the main sequence. The Sun in this system is a G2 V star.[15]

---

[15] The story of the Harvard classification has been told many times, but none so well for a popular audience as in Dava Sobel's *The Glass Universe: How the Ladies of the Harvard Observatory Took the Measure of the Stars* (New York: Viking, 2016).

In terms of their place on the main sequence, dwarf stars may be divided into two chief Types: upper and lower main sequence stars, distinguished physically by their fusion-generating mechanism. An upper main sequence star is located on the upper left of the HR diagram (Fig. 8.2) and consists of hot stars of more than about 1.5 solar masses that generate most of their energy by converting core hydrogen to helium by the carbon-nitrogen-oxygen (CNO) cycle. This cycle begins at an interior temperature of about 13 million degrees Kelvin, whereas lower main sequence stars are less than 1.5 solar masses and operate on the "proton-proton" chain reaction, which begins at around 4 million degrees. The physicists Hans Bethe and Carl von Weizsäcker independently worked out the CNO fusion reaction in 1938, and in the same year Bethe and Charles Critchfield (a graduate student of George Gamov) worked out the details of the proton-proton chain, which fuses four protons into a helium nucleus.[16] Bethe, von Weizsäcker, and Critchfield can thus can be said to have separated the upper main sequence from the lower main sequence based on their respective core nuclear reactions.[17] Compared to lower main sequence stars, upper main sequence stars are large and relatively rare, constituting only a few percent of stars on the main sequence. Their lives are brief but spectacular, and include giants and supergiants, some of the brightest stars in the sky.

Upper main sequence stars as usually defined have spectral types O, B, and A, along with some F types. In addition to being hot and massive they are also fast rotators. Lower main sequence stars are slow rotators, having lost much of their angular momentum, and they include some F spectral types, all G, K, and M spectral types, and, since the 1980s, the L spectral type. The Sun (G2 V), Alpha Centauri (also G2 V), and M dwarfs are lower main sequence stars, the latter constituting the great majority of stars in the universe because they have such long lifetimes. For example, Proxima Centauri, a red M dwarf (M 5.5 V) and the closest star to

---

[16]H. A. Bethe and C. L. Critchfield, "The formation of Deuterons by Proton Combination, "*Physical Review*, 54 (1938), 248-254; and Bethe's synthetic paper "Energy Production in Stars," *Physical Review*, 55 (March 1, 1939), pp. 434-456, reprinted in part in Bartusiak, 352-357.

[17]C. F. von Weizsäcker, "Element Transformation Inside Stars. II," Physikalische Zeitschrift, 39 (1938), 633-646, and Hans Bethe, "Energy Production in Stars," *Physical Review*, 55 (March 1, 1939), 434-456, reprinted in part in Bartusiak, 352-357.

the Sun, will live a trillion years. After their long lifetimes, lower main sequence stars will evolve into red giants (S 7 and S 8), pass through a planetary nebulae stage (S 17), and settle down as white dwarfs (S 12).

Beginning with the hottest main sequence stars at the upper left of the HR diagram, O stars are, as astronomer James Kaler puts it, "the hottest, the bluest, brightest, most massive, and rarest."[18] The surface temperatures of these hottest of upper main sequence stars range from 30,000 to 60,000 K. Only about one star in 3 million of the main sequence stars in the solar neighborhood are O stars. To put it another way, only 0.00004% of stars on the main sequence are O stars, 0.1% are B stars, 1% are A stars, and 4% are F stars. The reason for this is that massive stars burn themselves out quickly and have short lifetimes on the order of 100 million years. While the hottest O stars are giants and supergiants, main sequence O stars are also spectacular. An example is the star known as Theta-1 Orionis C (O6 V in the MK system), the brightest member of the Trapezium cluster in the Orion Nebula. It has a surface temperature of 45,000 K, is 40 times the mass of the Sun, and is 250,000 times more luminous than the Sun. It is mostly the ultraviolet light from this star that illuminates and ionizes the Orion Nebula. Another example in Orion is the Sigma Orionis system, consisting of five upper main sequence stars in orbit around each other: one O star, one A star, and three B stars.

By comparison with the O stars, B stars are cooler, with temperatures between 10,000–30,000 degrees, and constitute what Kaler calls "beacons of the skies." They have masses two to 16 times the mass of the Sun. Seven first magnitude stars in our skies are B types, and several of them are on the main sequence. Regulus, the brightest star in the constellation Leo, is a B7 main sequence star that is part of a quadruple system. Only a few hundred million years old, it has 3.5 times the Sun's mass and a rotation period of only 15.9 hours (compared to about 25 days for our Sun at the equator). Algol A, the Demon star, is a B8 main sequence star in Perseus. One of the first variable stars discovered, it is the prototype of an eclipsing binary. Acrux (B1) is the brightest star in the famous Southern Cross. One of the stars of the famous double star Albireo is a B0 main sequence star.

---

[18] James B. Kaler, *Stars and Their Spectra: An Introduction to the Spectral Sequence* (Cambridge: Cambridge University Press, 1997), 203, 221.

O stars tend to be found with B stars in loose associations (S 33) known as OB associations. Because they are young and hot, O stars also tend to be surrounded by diffuse nebula, H II regions (S 24) such as Messier 42 in Orion's sword, the Trifid Nebula in Sagittarius, the Rosette Nebula in Monoceros, and the North America and Eta Carina Nebula.

Continuing down the temperature scale on the main sequence, A stars are relatively cooler at 7,600–10,000 degrees. Sirius is the most illustrious example of an A star, which lasts no more than a billion years. Vega (A0 V), featured in Carl Sagan's *Contact* novel and movie, is another example, as well as Altair (A7 V), the brightest star in the constellation Aquila the Eagle. Some F type stars are upper main sequence and others are lower main sequence, and in some of these stars both the CNO cycle and the proton-proton chain reactions play an important role. Because of this, some astronomers distinguish an "intermediate main sequence" class of stars. Defined in this way (core energy production), it would be a relatively small class in terms of numbers, consisting of stars from about 1.5 to 2.5 solar masses. Other astronomers, however, define a larger intermediate main sequence based on stellar evolution rather than core physics. In this case, the intermediate main sequence includes those stars between about 0.85 solar masses, where stars begin to evolve off the main sequence, and 10 solar masses, which would produce neutron stars or black holes rather than white dwarfs.[19]

The study of lower main sequence stars has been greatly aided by the study of our own Sun, classified as spectral type G2 V, the closest star and consequently the best studied (Fig. 8.4). Its statistics are truly staggering, and many of its characteristics are undoubtedly shared by the multitudes of lower main sequence stars that have not yet been studied in such detail. The Sun is about 865,000 miles in diameter with a mass a million times that of the Earth. It is composed of about 70% hydrogen and 28% helium by mass (92% hydrogen and 7.8% helium by number of atoms), with traces of other elements including carbon, nitrogen, and oxygen. About 93% of its energy is generated by the proton-proton chain

---

[19]James B. Kaler, *The Cambridge Encyclopedia of Stars*, (Cambridge: Cambridge University Press, 2006), p. 250.

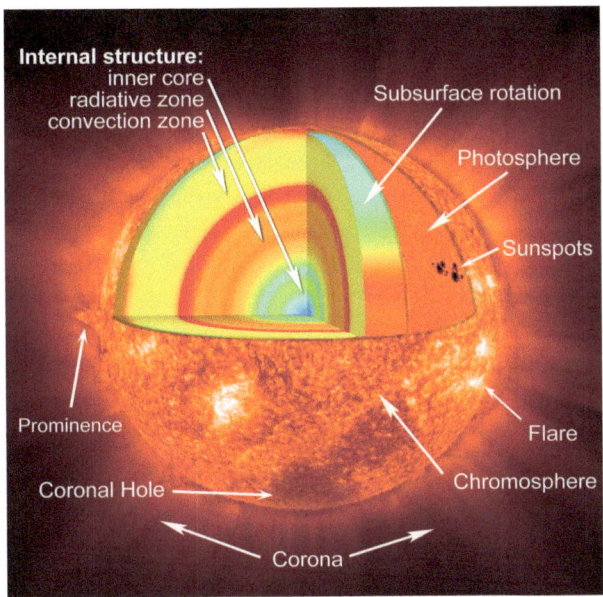

Internal structure:
inner core
radiative zone
convection zone

Subsurface rotation

Photosphere

Sunspots

Prominence

Flare

Coronal Hole

Chromosphere

Corona

Fig. 8.4. Structure of the Sun, a G2 V main sequence star. Credit: NASA/ SOHO

burning core hydrogen at a temperature of 15 million degrees K, with the remainder coming from the CNO cycle. In addition to its core, which constitutes the inner 25% of its radius, the Sun has radiative zones (to 70% of its radius) and convective zones (to near the visible surface) that transfer energy to the surface (Fig. 8.4). It loses about a million tons of mass a second in the form of solar radiation, a process that has been ongoing for 5 billion years and that will continue for another 5 billion.

The Sun radiates not only in the visible part of the spectrum but also in the infrared, ultraviolet, radio, and X-ray regions. Because it is gaseous it undergoes differential rotation; the rotation rate of the visible surface is about 25 days at the equator, 35 days at the poles. The Sun's photosphere exhibits sunspots, cooler regions that appear on an 11-year cycle, as well as flares emitting profuse X-rays and spectacular prominences. More than 1,000 miles above the photosphere lies the chromosphere at a temperature of 10,000 K, and 15,000 miles above that is the corona with a temperature of a million degrees, from which "coronal mass ejections" emanate.

In addition to the million tons per second the Sun loses in mass converted to energy, it loses a billion tons per second in the form of solar wind (P 16), mostly protons.[20]

Knowledge of the Sun, and thus of lower main sequence stars, has accumulated over 400 years of telescopic observations, beginning with Thomas Harriott's and Galileo's observations of sunspots and their use to determine Sun's rotation rate. The study of the Sun was also greatly enhanced by the phenomenon of solar eclipses, which allowed both the corona and the chromosphere to be discovered. In the early 19th century, William Wollaston and Joseph Fraunhofer observed dark lines in the solar spectrum, and beginning in the 1860s Robert Bunsen and Gustav Kirchoff interpreted these lines as the fingerprints of elements in the Sun. Spectroscopy developed in ever-more sophisticated form, and increasingly determined the characteristics and constitution of the Sun and its corona. At the turn of the century George Ellery Hale built the first spectroheliograph, allowing the Sun to be viewed in the light of specific spectral lines. In the Space Age, our knowledge of the Sun has been greatly accelerated by observations from spacecraft such as Ulysses, SOHO, TRACE, and STEREO.

Billions of solar-type G stars, including the nearby alpha Centauri, populate our galaxy alone. Billions more undoubtedly exist in other galaxies. Other types of lower main sequence stars include slightly cooler K stars such as Epsilon Eridani and Tau Ceti, and, at the bottom of the main sequence, the M stars with temperatures less than 3,600 K. Barnard's star and Gliese 581 are among the red M dwarfs, which are the most common stars in the galaxy. They are normally quite sedate, but some M stars, probably younger ones, are flare stars like the "UV Ceti" prototype (M 5.5 V). A double red dwarf star system, UV Ceti, was discovered in 1948 by Willem J. Luyten and is also designated Luyten 726-8. Proxima Centauri itself is a UV Ceti flare star that can triple its brightness in a few minutes, a phenomenon that does not bode well for its planet Proxima Centauri b.

---

[20] Among the many books on the Sun see Jack B. Zirker, *Journey from the Center of the Sun* (Princeton: Princeton University Press, 2003), and Karl Hufbauer, *Exploring the Sun: Solar Science since Galileo* (Baltimore and London: Johns Hopkins University Press, 1991).

Despite the advances, a great deal remains unexplained about the Sun and, by extension, the class of lower main sequence stars to which it belongs. Among the mysteries are the mechanisms for heating the corona to 500 times hotter than its surface, for accelerating the solar/stellar wind up to 600 miles per second, and for producing flares and coronal mass ejections.

Distinguishing the number of Types of stars on the main sequence depends on whether one wants to define its domains by the MK system (one luminosity class: V), core energy production (two classes: upper and lower main sequence), evolutionary endpoints (four classes: supernova, white dwarf, neutron star, and black hole), or Harvard spectral types (nine or more). The moral of the story is that there are many ways to classify objects, and they sometimes overlap and coexist—and persist, as long as they are useful. Stars are also sometimes exiled from their designated class—although cool subdwarfs (S 5) are also officially luminosity class V of the MK system because they are still in their hydrogen-burning phase, they are very different in other ways. As we will see in the next entry, they are in some ways so different that some of the compilers of the MK system wished to designate a separate class VI. But this idea did not catch on.

Gaia's Hertzsprung-Russell diagram, released in 2018 and plotting some four million stars within 5,000 light years of the Sun, can be viewed at http://sci.esa.int/gaia/60198-gaia-hertzsprung-russell-diagram/. A variety of solar images and movies, representing G stars, are available at NASA's SOHO site https://sohowww.nascom.nasa.gov/gallery/. On the petulant behavior of M red dwarf stars see http://hubblesite.org/news_release/news/2011-02/2-stars.

Class S 5: Subdwarf

Subdwarfs, today more accurately known as "cool subdwarfs," are an odd type of star that run about one magnitude below the main sequence in the Hertzsprung-Russell diagram (Fig. 8.2). They are usually classed as main sequence stars since they are hydrogen burning, although near the end of this phase of energy generation. The first hint of this class came in 1913, when astronomer Walter S. Adams reported that two high velocity stars, known as Lalande 5761 and 28607, had peculiar A-type spectra with weak hydrogen lines. "It is a singular coincidence that two stars, both of peculiar but similar spectra, should be characterized by such velocities," he wrote, estimating those velocities as receding at 144 and 170 kilometers per second, very close to the current measured value.[21] By 1934, Adams and Alfred Joy had found three more similar stars. Because their luminosities were fainter than the main sequence, but brighter than the white dwarfs, Adams and his team called them "intermediate white dwarfs." Because of their high velocities, subdwarfs are most often detected in proper motion surveys.

In 1939 Gerard Kuiper coined the term "subdwarf" as a better descriptor for this class of objects. Kuiper used the first month of operation of the 82-inch telescope at the MacDonald Observatory in Texas to determine accurate spectral types for large proper motion stars. In delineating a new class of objects, he commented that what Adams and Joy had called intermediate white dwarfs "extend almost along the whole main sequence. Since these stars merge into the main sequence and are much more similar to main-sequence stars than to white dwarfs (probably also in the interior), the name 'subdwarfs' is suggested for this class of stars, in analogy with 'subgiants.' This name will prevent the confusion of these stars with the white dwarfs proper which are very much fainter."[22] Due to sparse data about their true nature, subdwarfs were not designated luminosity class VI in the original MKK/Yerkes system in 1943. Astronomer Nancy Roman, who contributed to the MKK system, first designated them luminosity class VI in 1955, but the designation is not universally used and they are more often designated "sd."[23]

---

[21] Walter S. Adams, "Some Radial Velocity Results," PASP, 25 (1913), 259-260.
[22] Gerard Kuiper, "Two New White Dwarfs; Notes on Proper Motion Stars," ApJ, 89 (1939), 548-551.
[23] Nancy Roman, "A Catalogue of High-Velocity Stars," Ap. J. Supplement, vol 2 (1955), p. 195. Roman used "sd" for subdwarfs earlier than G0, and VI for stars later than G0.

   Thus began what science writer Ken Crosswell has called "the rise of the subdwarfs," the discovery of a relatively small class of stars destined to have an outsized effect on astronomy. It is now known that the relative faintness of subdwarfs, amounting to one or two magnitudes less than normal main sequence stars, is due to their composition. In particular, it is due to their low "metallicity," meaning they have a lower content of what astronomers refer to as "metals," which—strangely and contrary to common usage of the term "metal"—are any elements other than hydrogen and helium. This causes them to be hotter and bluer than most main sequence stars of the same luminosity.

   This effect for subdwarfs was first demonstrated in a famous and controversial paper in 1951 by Lawrence Aller and Joseph W. Chamberlain. They showed that the two high velocity stars first detected by Adams in 1913 were actually F spectral type, not A spectral type. This in turn had major implications for the stars' composition, indicating that their iron abundances, for example, were only 1/100th that of the Sun, what would today be defined as a metallicity of minus two (two orders of magnitude less than the Sun). "This was the beginning—the breaking down of this prejudice that all stars had the same composition," Aller noted.[24]

   Aller and Chamberlain's paper did indeed open the way to numerous studies of metal-poor stars. We now know that metallicity is in general a difference between the two stellar populations (S 36) that Walter Baade detected in the 1940s. In this terminology, subdwarfs are population II stars that have high velocities and highly elliptical orbits because they are not confined to the galactic disk but orbit in the bulge and halo of the Galaxy (G 12). Their lower metallicity indicates they are earlier generation stars that have not been greatly enriched by heavier elements. They play an important role in our current understanding that stars in different parts of the Galaxy have different ages, that the Galaxy is becoming more metal-rich with age as heavier elements are generated inside stars, and that we are in fact "starstuff," as Carl Sagan liked to say.

   As more subdwarfs were discovered over the next half century, astronomers attempted to classify them as a means to enhance understanding. In 1997 they first distinguished subdwarfs and

---

[24] Ken Crosswell, "The Rise of the Subdwarfs," chapter 8 in *Alchemy of the Heavens: Searching for Meaning in the Milky Way*, (New York: Doubleday, 1995), 85-100: 89; Joseph W. Chamberlain and Lawrence H. Aller, "The Atmospheres of A-Type Subdwarfs and 95 Leonis," ApJ, 114 (1951), 52-72.

extreme subdwarfs based on their metallicity, and ten years later added the class of ultrasubwarfs. These three classes (or Types we would say), have metallicities of -0.5, -1.0, and -2.0 respectively. They are the classes still used today.

We now know that cool subdwarfs, which span spectral types G to M, represent a late stage in main sequence stellar evolution, just before helium fusion begins. Kapteyn's star, a high velocity M1 V star seen on Fig. 8.2, is an example of a cool subdwarf. Discovered by Dutch astronomer J. C. Kapteyn in 1898, it was at the time the star with the highest known proper motion and still holds second place in that category today after Barnard's star. Because ultracool subdwarfs have very high velocities and unusual orbits in the galaxy, they are susceptible to search methods using these criteria. A 2016 survey over several thousand square degrees using the Sloan Digital Sky Survey and the Two Micron All-Sky Survey found 49 subdwarfs, 25 extreme subdwarfs, and six ultrasubwarfs, doubling the number of known cool subdwarfs.[25] They are classified as spectral type M7 and later, and can be 10,000 times fainter than the Sun. These low-mass subdwarfs are believed to be relics of the early galaxy with ages exceeding 10 billion years. They are therefore important for tracing the structure of our Galaxy.

More recently, helium-burning O and B stars have also been referred to as subdwarfs, illustrating the inconsistent and sometimes confusing nomenclature of astronomy.[26] These so-called "hot subdwarfs" have nothing to do with the traditional cool subdwarfs found just below the lower main sequence. They are also known as extreme "horizontal branch" stars from their position on the HR diagram, to the left and below the hot end of the main sequence but above the white dwarfs. B subdwarfs are designated sdB and represent the next stage for low-mass stars after the red giant stage. They are believed to be pure helium stars that are converting helium to carbon in their cores. They have extremely thin hydrogen envelopes, and their total mass is about half that of the Sun. They are often found in globular clusters, in close binaries with white dwarf or low-mass main sequence stars as companions, and are often variables.

---

[25] N. Lodieu et al., "New ultracool subdwarfs identified in large-scale surveys using Virtual Observatory tools," A&A, 598, A92 (2017), https://arxiv.org/abs/1609.08323;

[26] Gray and Corbally, *Stellar Spectral Classification*, p. 553

By contrast, sdO stars are hotter and more luminous, and are believed to be the final stage before the star end as a white dwarf. The central star of most planetary nebulae are sdO stars, with electron-degenerate cores composed of carbon and oxygen. By contrast to cool subdwarfs, more than 5,000 hot subdwarfs have been catalogued.

Subluminous hot subdwarfs were discovered in the 1950s using Milton Humason and Fritz Zwicky's 1947 survey of the North Galactic Pole region. Many more were discovered after the Palomar-Green survey of the northern Galactic hemisphere, published in 1986. B subdwarfs outnumber white dwarfs at all apparent magnitudes brighter than 18.[27] They have recently been studied by NASA's Galaxy Evolution Explorer (GALEX). Hot subdwarfs are so different from "normal" cool subdwarfs, as originally discovered and defined by Adams, Joy, and Kuiper, that one might well argue they are in a class of their own, between giants and white dwarfs in terms of their evolution. But "hot subdwarfs" seems to be sufficient for practical purposes.

---

[27] Ulrich Heber, "Hot Subdwarf Stars," ARAA, 47 (2009), 211-251; Ulrich Heber, Jeffery Simon and Ralf Napiwitzki, *Hot Subdwarf Stars and Related Objects*, PASP Conference Series volume 392 (San Francisco, 2008); S. Geier et al. "The population of hot subdwarf stars studied with Gaia. I. The catalog of known hot subdwarf star," A&A, 600, A50 (2017), https://www.aanda.org/articles/aa/pdf/2017/04/aa30135-16.pdf

Class S 6: Subgiant

If dwarf stars on the main sequence are burning hydrogen, and (as we will see in the next entry) giant stars have evolved off the main sequence and are burning helium, it makes sense that there should be a transitional category between the two. Indeed there is: the subgiants, a class of stars that represents an intermediate stage in low-mass stellar evolution, between main sequence dwarfs and giant stars in the HR diagram. With the "Post-Main Sequence" Subfamily, we enter the realm of stars nearing the end of their lives. They mostly have spectral types from G0 to K4 and absolute magnitudes between 2.5 and 4, slightly brighter than their main sequence counterparts of the same spectral type, but fainter than the giants.

Subgiants, up to ten times the size of the Sun, have begun their expansion into red giants, which can grow up to 100 times the size of the Sun. Fusion in their hydrogen cores has ceased, but the point at which helium fusion begins (the "helium flash") has not yet begun, at least in the early subgiant phase. Their energy derives from a shell of hydrogen that burns outside the inert helium core, called "hydrogen shell fusion." Quite logically, subgiants are luminosity class IV in the MK system, lodged between class V main sequence stars and class III giants. The length of the subgiant phase varies depending on mass but may last up to a billion years—only 10% of its main sequence life as a dwarf star.

The separation of subgiants as a class distinct from giants and main sequence stars was difficult and accomplished only several decades after the HR diagram was first devised, when stellar absolute magnitudes were more accurately determined. Absolute magnitude is defined as the brightness of stars if they were all observed from the same distance of 10 parsecs (32.6 light years). It is thus a measure of luminosity and requires knowing stellar distances. Although there were inklings of a separate subgiant sequence as early as 1917, Gustaf Stromberg first used the term "subgiants" only in 1930.[28] A catalog of the entire Mt. Wilson pro-

---

[28] Gustaf Stromberg, "The Distribution of Absolute Magnitudes among K and M Stars Brighter than the Sixth Apparent Magnitude as Determined from Peculiar Velocities," ApJ 71 (1930), 175-190; This paragraph and those that follow are based on Dick (2013), pp. 99-101. For a more detailed description of the separation of the subgiants from the main sequence and giant stars, see Allan Sandage, *The Mount Wilson Observatory*, (Cambridge: Cambridge University Press, 2004), pp. 249-250, 363-367; 539-554.

gram of spectroscopic parallaxes (and thus distances) published in 1935 included an HR diagram that clearly showed a separate sequence between the main sequence and giant stars, though populated by only 90 stars out of the 4,179 plotted. The authors cautiously wrote that "the existence of a group of stars of types G and K somewhat fainter than normal giants has been indicated by the statistical studies of Stromberg ... Although these stars may not be entirely separated from the giants in absolute magnitude, there is some spectroscopic evidence to support the suggestion."[29]

Because a class of subgiants did not fit into the Henry Norris Russell's picture of stellar evolution, doubts were expressed about the reality of their existence. Doubt was removed in 1936 when reliable trigonometric parallaxes (and thus distances and luminosities) of more subgiants became available. W.W. Morgan at Yerkes also saw these spectral differences independently, and the first edition of what became the canonical MKK catalog of luminosity spectral types published in 1943 included five stars as the defining examples of subgiants: Beta Aquilae, Eta Cephei, Gamma Cephei, delta Eridani, and Mu Herculis. As defining members of their class, all were designated luminosity class IV, with spectral types today known to vary from G5 to K0.

The place of subgiants in stellar evolution was not resolved until the early 1950s, when astronomers used older population II stars in the globular clusters M3 and M92, and younger population I stars in the open cluster M67, to infer the relation between subgiants and main sequence stars (see S 36 for more on stellar populations). Allan Sandage later characterized the discovery of the relation of subgiants to main sequence stars as "a serendipitous discovery" that "arose from the solution of the independent problem posed in 1948 by [Walter] Baade in his outline of stellar population programs for the 200-inch Palomar telescope." In particular, by making observations to fainter magnitudes in globular clusters, astronomers found that "the main sequence was attached to a stubby subgiant sequence, which in turn merged continuously into the giant branch .... It seemed clear that the subgiant and giant-branch stars had once been main-sequence stars; they had

---

[29] Walter S. Adams, Alfred H. Joy, Milton L. Humason and Ada Margaret Brayton, "The Spectroscopic Absolute Magnitudes and Parallaxes of 4179 Stars," 187-291; Sandage (2004), p. 540.

evolved those two branches as they aged." Sandage characterized this discovery as a landmark in the study of stellar evolution.[30]

Theory had indicated that once hydrogen burning ceased, no stable configuration existed for the star. Theorist Martin Schwarzschild soon showed that when a star with a core about 10% of its total mass converts all its hydrogen to helium, the core contracts and releases gravitational energy so that hydrogen begins burning in a shell around the core. The star expands in radius and leaves the main sequence. The existence of subgiants was thus explained, and stars were shown to expand as they age rather than contract, as Russell thought. "By mid-1952," Sandage wrote, "it was clear that the evolution of a star off the main sequence and into the subgiant sequence is secure, both from the Mount Wilson/ Palomar observational side and now from the theoretical perspective as well."[31] A similar result in 1955 from the open cluster M67 explained the subgiant sequence in population I stars. Subgiants thus played a major role uncovering the process of stellar evolution as we understand it today.

At least six subgiants or proto-subgiants are located within 30 light years of our Sun. These include Altair, Procyon A, Delta Pavonis, Beta Hydri, Mu Herculis, and Rana. The bluish-white Altair (A 7 IV-V), part of a multiple star system with three companions, is one of those transitional stars that may be just beginning its subgiant phase. Located about 16.8 light years distant, it is the brightest star in the constellation Aquila and one of the brightest and closest in the sky. Altair is characterized by an extremely rapid rotation rate of about 200 miles per second, completing one rotation in 9 hours compared to 25.4 days for the Sun at its equator. As a result, interferometry has shown it has the shape of a flattened ellipsoid. The brightness of Altair varies by one-thousandth of a magnitude, and in 2005 it was identified as a type of pulsating variable known as delta Scuti after its prototype.

The yellowish-white binary star Procyon A (F5 IV-V), the Little Dog Star in the constellation Canis Minoris, is also believed to be a star at the beginning of its subgiant stage (see Fig. 8.2). It is about 1.5 solar masses and 1.7 billion years old, only one third of the Sun's age, located at 11.5 light years. It has a white dwarf

---

[30] Sandage (2004), pp. 539-554; quotations on pp. 541-542 and 547.

[31] Sandage (2004), pp. 548-549; A. Sandage and M. Schwarzschild, "Inhomogeneous Stellar Models. II. Models with Exhausted Cores in Gravitational Contraction," ApJ, 116 (1952), 463-476.

companion and will likely expand to the red giant phase in less than 100 million years. The yellow-orange stars Delta Pavonis, Beta Hydri, and Mu Herculis A are spectral types G5-8 V-IV, G2-IV, and G5 IV respectively, while Rana (delta Eridani) has moved to the orange-red subgiant stage as a K0 IVe star.

More than 200 subgiants are known to exist within 100 light years of the Sun, including some of the brightest in the night sky. They are mostly in the mid- to lower- temperature sequence, since any O, B, and most A stars born close to the Sun evolved much faster due to their high mass and therefore have already moved through the subgiant, giant, and supergiant stage to one of the stellar evolutionary endpoints (S 11–14). A recent study using data from the Gaia and Kepler spacecraft found that about 23% of the 186,000 stars observed by Kepler are subgiants. Since Kepler's targets range to a distance of about 3,000 light years, this means that about 40,500 stars are subgiants in the small section of the sky (in the constellations Cygnus and Lyra) observed by Kepler out to this distance. 65% are main sequence stars, and 12% red giants. Eight confirmed planets and 34 planet candidates have been found around some of these stars, including hot super Earths around some of the subgiants.[32]

---

[32] Travis Berger et al., "Revised Radii of Kepler Stars and Planets using Gaia Data Release 2," preprint, August 14, 2018, https://arxiv.org/abs/1805.00231.

Class S 7: Giant

Giant stars are those stars that have ceased their core hydrogen burning and left the main sequence on the HR diagram. They are thus part of the "post-main sequence" Subfamily that includes even brighter giants (S 8), supergiants (S 9), and hypergiants (S 10). They constitute luminosity class III in the MK classification system of stars. Giants have masses up to five times, and luminosities hundreds of times, that of the Sun. And whereas a subgiant (S 6) may be ten times bigger than the Sun, a giant star can be 100 times larger than the Sun.

Red giants are especially important because they are the next stage after hydrogen burning and subgiants in the evolution of low-mass stars, those (including the Sun) with less than 8 solar masses. Because they span the temperature sequence from blue-white B stars through red giants, some high-mass hotter giants may still be near the end of their main-sequence lives while cooler giants are fully fusing helium. But like all post-main sequence stars, most giants have exhausted their core hydrogen fuel. Unlike subgiants, whose energy derives from a shell of hydrogen that burns outside the inert helium core, fully developed giant stars have advanced to another stage: they are undergoing helium fusion in their cores, even though they may also be slightly pre- or post-helium burners (Fig. 8.5). Blue, yellow, and orange giants are a transitory phase, because they represent hotter giants before cooling and expanding to become a red giant. Loosely speaking, yellow and orange giants are sometimes included in the red giant class.

Our Sun is midway through its main sequence lifetime and will become a red giant in about five billion years, at which time it will be 100 times larger and hundreds of times brighter than it is today. In the process it may swallow the Earth, extinguishing all life. Because red giants have lower surface gravity than typical main sequence stars, large amounts of mass escape in the form of stellar winds (S 29), much more than in the solar wind (P 16) now emanating from the Sun. The red giant phase for the Sun will only last about 100 million years, and, after a few more last gasps, the Sun will shed it outer layers and become a planetary nebula (S 17) with a central white dwarf (S12).

Some of the best-known stars in the sky are red or orange giants, including Arcturus (K1.5 III), Aldebaran (K5 III), and Pollux (K0 IIIb). Capella is part of a multiple star system consisting of two giant stars, a yellow-orange G8 III and G1 III, and two small red

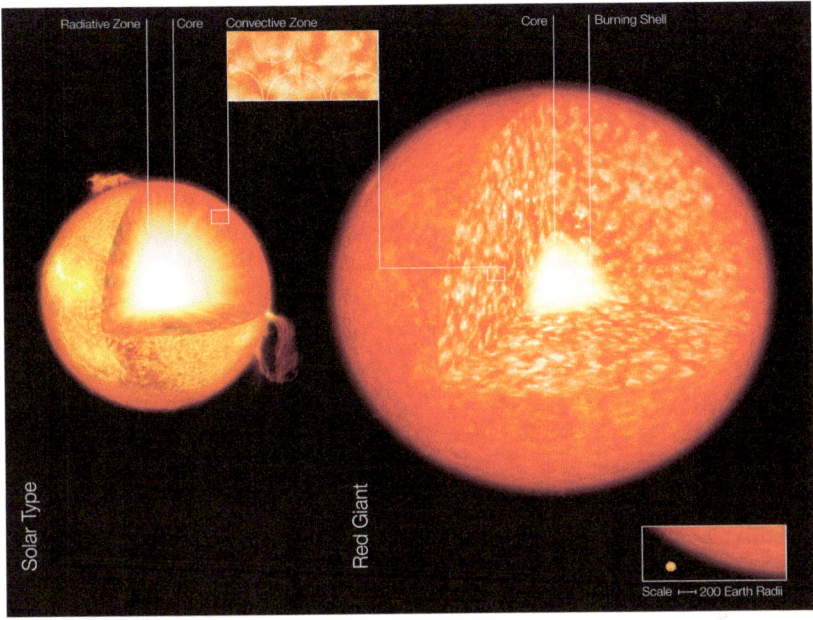

Fig. 8.5. Artist's impression of the structure of a Sun-like star and a red giant, comparing the core, radiative, and conductive zones. Unlike the Sun converting hydrogen to helium, the red giant is undergoing helium fusion in its core. The scale is given in the bottom right insert. Credit: ESO

dwarfs. At the hotter end of the temperature sequence, Alcyone, the brightest star in the Pleiades open cluster, is an example of a blue B giant. It is in a transitory stage before it becomes a bright giant or a red or blue supergiant. Its surface temperature is 13,000 K and its luminosity is 750 times the Sun's. Unlike its red giant brethren, hydrogen fusion is slowing and it is not yet fusing helium, analogous to subgiants for low-mass stars. Thuban, the brightest star in the constellation Draco, is a white A giant.

Although stellar spectroscopy had begun in 1859, it took almost 50 years for the first solid evidence that stars existed of great dimensions. Stellar spectra could be observed for any star that was bright enough; these were observed at an increasing pace as telescopes improved and spectroscopic and photographic techniques advanced in the late 19th and early 20th century. By 1924 the Henry Draper Catalogue emanating from Harvard College Observatory had classified 225,000 stars to 8.5 magnitude. But in order to determine the true nature of a star, one needed to know its absolute magnitude and luminosity, and this required a knowledge of stellar distances. Such distances had to be estimated through stellar proper motions (on the assumption that smaller proper motions implied greater distance) or determined through trigonometric parallaxes, which existed for only a few hundred stars at the turn of the 20th century.

Based on proper motions, the amateur astronomer William Monck first suspected a luminosity difference among stars in a paper published in 1895, where he stated "I suspect ... that two distinct classes of stars are at present ranked as Capellan, one being dull and near us, and the other bright and remote like the Sirians." Monck gave alpha Centauri and Procyon as examples of the first class, and Canopus as the second class; they are now classed as main sequence, subgiant, and bright giant respectively. Monck's paper had no effect, perhaps because, as Allan Sandage has conjectured, it had only one intelligible paragraph![33]

By contrast, a decade later the work of two widely separated astronomers would have a great effect on the idea of a giant/dwarf dichotomy among the stars. In 1905 and 1907, the Danish chemist and amateur astronomer Ejnar Hertzsprung, building on Antonia Maury's observation that stars of the same spectral type could have different spectral line widths, wrote two papers in an obscure photographic journal that in effect revealed the existence of what he called "Riesen" (giant) and Zwerge (dwarf) stars of the same spectral temperature. It turned out that Maury's narrow-lined stars (designated "c") had smaller proper motions and thus were more distant, suggesting that they were much more luminous than the nearer ones. Maury's distinction had theretofore been ignored, but Hertzsprung wrote Edward C. Pickering, Maury's boss at Harvard, that "To neglect the c-properties in classifying stellar spectra is nearly the same thing as if the zoologist, who has detected the deciding differences between a whale and a fish, would continue in classifying them together." Pickering was unconvinced. Karl Schwarzschild called attention to Hertzsprung's "Giganten" stars, but to little avail at first.[34]

By 1910, Henry Norris Russell was independently convinced that giant stars existed, based on direct and inferred parallaxes. By 1913 Russell proposed "two great classes of stars"—giants and dwarfs, now following the terminology of Hertzsprung. And, like Hertzsprung, he plotted a diagram of stellar luminosities ver-

---

[33] William Monck, "The Spectra and Colours of Stars," JBAA, 5 (1895), 416-419: 418; Sandage, 239. This paragraph and the following are based on Dick (2013), pp. 95-99.

[34] Ejnar Hertzsprung, "Zur Stralung der Sterne [On the Radiation of Stars]," Zeitschrift fur Wissenschaftliche Photographie, 3 (1905), 449, in part translated into English in Bartusiak, 243. DeVorkin gives the context in his biography of Russell, *Henry Norris Russell: Dean of American Astronomers* (Princeton:Princeton University Press, 2000), pp. 84-86; Karl Schwarzschild, *Astronomische Jahresbericht*, 1909, 179.

sus temperatures, which after 1933 (at the urging of the Danish astronomer Bengt Stromgren) became known as the Hertzsprung-Russell diagram. Also in 1913, Russell and Shapley made direct distance determinations of giants belonging to eclipsing binaries, indicating radii 300 times the Sun's and 20 solar masses, and thus densities a million times less than the Sun's.[35]

The determination of luminosities received an enormous boost when Walter S. Adams and Arnold Kohlschütter found a method for determining stellar distances using spectral lines, first published in 1914. Unaware of the work of Monck, Hertzsprung, or Russell, they concluded that "Certain lines are strong in the spectra of small proper motion stars. The use of the relative intensities of these lines results for absolute magnitudes in satisfactory agreement with those derived from parallaxes and proper motions."[36] The monumental volume *Spectroscopic Absolute Magnitudes and Distances of 4179 Stars*, published in 1935 and mentioned in the previous entry, increased the number of stellar distances a hundredfold and gave clear indications not only of giant and dwarf classes but also of subgiants.[37] With the further development of spectroscopic techniques, luminosity effects could be detected by the width and intensity of stellar spectra lines, and the more formal two dimensional temperature and luminosity MKK system (1943) was developed, which included bright giants and supergiants, for a total of five luminosity classes. The MKK system (later the MK system with further developments) incorporates the Harvard spectral classification based on temperature.[38]

The so-called RR Lyrae stars, named after the prototype bright giant R R Lyrae, are among the most important Types of giant stars,

---

[35] Henry Norris Russell, "Giant and Dwarf Stars," *The Observatory*, 36 (1913): 325; and Russell, "Relations between Spectra and Other Characteristics of Stars," *Popular Astronomy*, 22(1914?), 275-94. The *Popular Astronomy* article contains the first appearance of what later became known as the H-R diagram. See also Sandage (2004), p. 402. Sandage has written about the irony of Stromgren's naming of the H-R diagram, Sandage (2004), pp. 240 ff.

[36] W. S. Adams and A. Kohlschütter, "Some spectral criteria for the determination of absolute stellar magnitudes," ApJ, 40 (1914), 385-398; MWC, 89; Sandage (2004), pp. 239-244.

[37] Walter S. Adams, Alfred H. Joy, Milton L. Humason and Ada Margaret Brayton, "The Spectroscopic Absolute Magnitudes and Parallaxes of 4179 Stars," 187-291; Sandage 540.

[38] W. W. Morgan, P. C. Keenan and E. Kellman, *An Atlas of Stellar Spectra, with an Outline of Spectral Classification* (Chicago: The University of Chicago Press, 1943). Online at https://ned.ipac.caltech.edu/level5/ASS_Atlas/paper.pdf

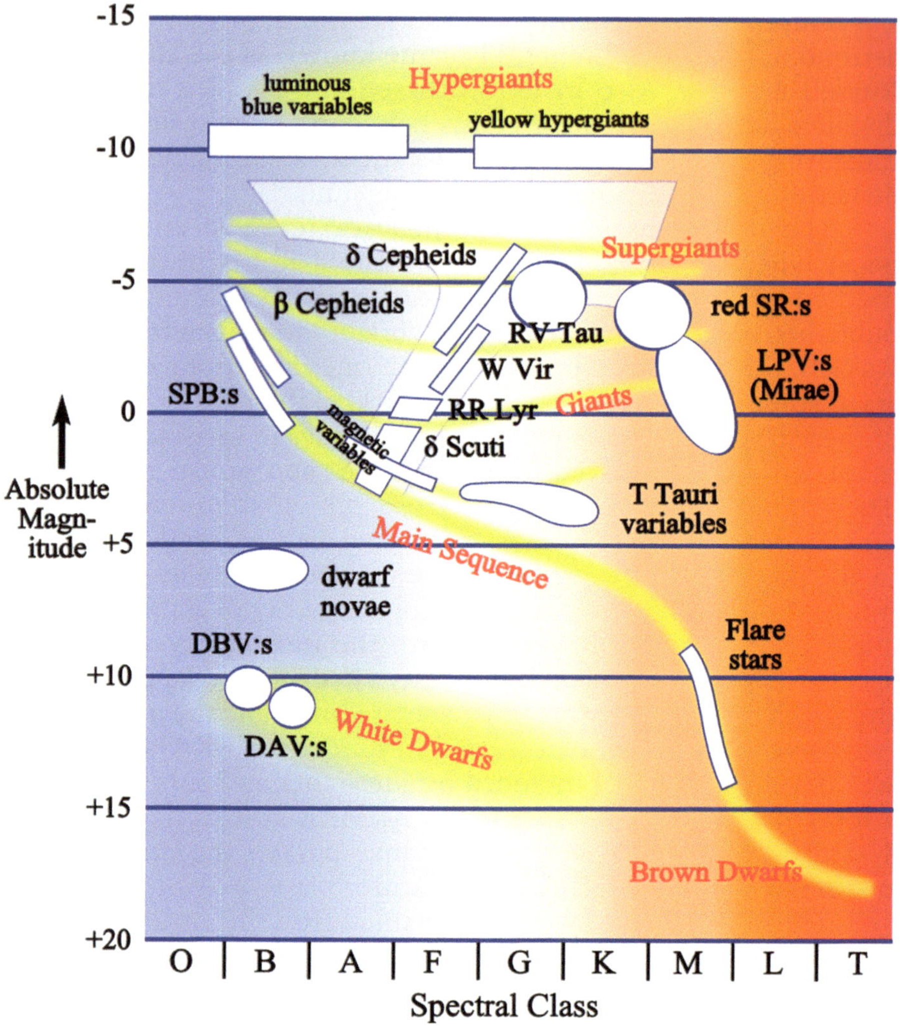

Fig. 8.6. Variable stars on the HR diagram, including RR Lyrae giants and classical Cepheids that range over the giant and supergiant classes. Credit: Rursus via Wikipedia Creative Commons 3.0

usually of spectral type A or F. R R Lyrae itself is classified as an A7 III–F8 III giant (Fig. 8.6). Harvard astronomers E.C. Pickering and Williamina Fleming, as well as Dutch astronomer J.C. Kapteyn, first discovered this Type of star in the 1890s. They are unstable helium-burning yellow or white giants. Their instability causes them to be pulsating variables with periods of a few hours, as evidenced in their light curves, a much shorter period than their brighter supergiant Cepheid variable counterparts (see S 9). Like Cepheids, they are "standard candles" used for distance scale determinations, but since they are not as bright, they are used within and near the Milky Way galaxy. They are found mostly in globular clusters and have been seen as far away as the Andromeda galaxy, some 2.5 million light years from Earth. Several thousand stars of this Type have now been identified. RR Lyrae stars can be further divided into at least two subtypes, depending on their light variations as indicated in a plot of their light curves.

Videos about relative star sizes are common but impressive: https://www.youtube.com/watch?v=vfXHZJ3zN-A; https://www.youtube.com/watch?v=Bcz4vGvoxQA;https://www.youtube.com/watch?v=YVMK1ugQdhM; https://www.youtube.com/watch?v=mjyxTTJWOAE. An ESO video on giant red stars is at https://www.youtube.com/watch?v=L0E8moAlEDk

See also http://hubblesite.org/news_release/news/2016-13/107-illustrations.

Class S 8: Bright Giant

Bright giants are luminosity class II stars intermediate between giant (S7) and supergiant (S9) stars, according to the MK Yerkes classification. Of the five classes delineated in that system, bright giants are in some ways the least distinct, since they are brighter than the giants of class III but in a similar state of evolution, having exhausted their core hydrogen and started or nearly started to undergo helium fusion.

Whereas main sequence "dwarfs" were separated from giant and supergiant stars in the HR diagram around 1910, and subgiants and subdwarfs by the 1930s, bright giants did not gain class status until the original MKK system defined them in 1943. The MKK originators distinguished such a class because five or six magnitudes separate the giants from the supergiants at some points in the spectral sequence. In the introduction to the MKK atlas, the authors simply stated "Stars of class II are intermediate in luminosity between the supergiants and ordinary giants," and Plate 47 in that Atlas made the star theta Herculis the standard star defining the spectral type K1 for class II. In other words, theta Herculis was the star whose spectrum was to be compared when classifying any other stars as bright giants of that spectral type. What counted for the classifier was that the technical definition of a bright giant was determined by the "intensity ratios" of specific spectral lines of various wavelengths. Standards have since been established for bright giants of other spectral types, including Gamma Lyrae (B9 II) and Beta Canis Majoris (B1 II-III) for the early stars, and Iota Aurigae (K3 II) and 37 Leo Minoris (G0 II) for later stars.[39]

The fact that the physical differences between giants and bright giants are not as great as the differences between other classes is less surprising if one considers that a defining characteristic of the MK system is the "MK process," according to which stars are classified by only their spectral characteristics—no physical characteristic is allowed to influence the classification. This has been an important characteristic in keeping the MK system

---

[39] See http://stellar.phys.appstate.edu/Standards/std3_6.html for MK standard stars online. See Gray and Corbally, 562 for late-type standards. More MK standard stars are listed at http://articles.adsabs.harvard.edu//full/1989B ICDS..36...27G/0000027.000.html

in use on a permanent basis. As astronomer James Kaler and others have pointed out, "In any given range of spectral class there are criteria that depend on luminosity, not through temperature changes, but through density changes. While the mass range across the stellar landscape is large, it is nothing compared to the range in radius, hence volume." Thus, a class K giant is more massive than a class K dwarf, but it is also much larger, so that the photosphere that forms the spectrum is much more tenuous.[40] To put it another way, the surface gravities of stars, which are correlated with luminosities, vary with the different classes, with main sequence stars (class V) having high surface gravities and supergiants (class I) having low surface gravities. The effect would be even less for the difference between giants and bright giants, but the difference is detectable in their spectra when compared to the standard stars of each class.

Among the bright giants already mentioned, Beta Canis Majoris is the fourth brightest star (despite its "beta" designation) after Sirius in the Large Dog constellation (Fig. 8.7). It is a hot variable star of the Beta Cephei type (see placement on Fig. 8.6), varying in brightness between 1.95 and 2.00 in six hours. The variation is due to pulsations on the star's surface, likely because of instability caused by the end of hydrogen burning and the beginning of helium burning. Its traditional Arabic name Murzim ("The Herald") probably refers to its heralding the rising of Sirius, the "Dog star," in the same constellation. Sirius (an A1 V main sequence star) is the brightest star in the night sky, but if Beta Canis Majoris (60 times more distant than Sirius) were brought to the distance of Sirius (8.6 light years), it would be 15 times brighter than Venus. On the much cooler side, Theta Herculis, the original standard for the bright giant class, is a K1 type that has evolved from a hot B star over the last 75 years. It is an irregular variable that ranges between 3.7 and 4.1 magnitude over 9 days. Similarly, Iota Aurigae is a cool orange K3 type bright giant that has likely begun core helium fusion. It is of particular recent interest because it may harbor two brown dwarf companions.

---

[40] James Kaler, *The Cambridge Encyclopedia of Stars* (Cambridge: Cambridge University Press, 2006), p. 111.

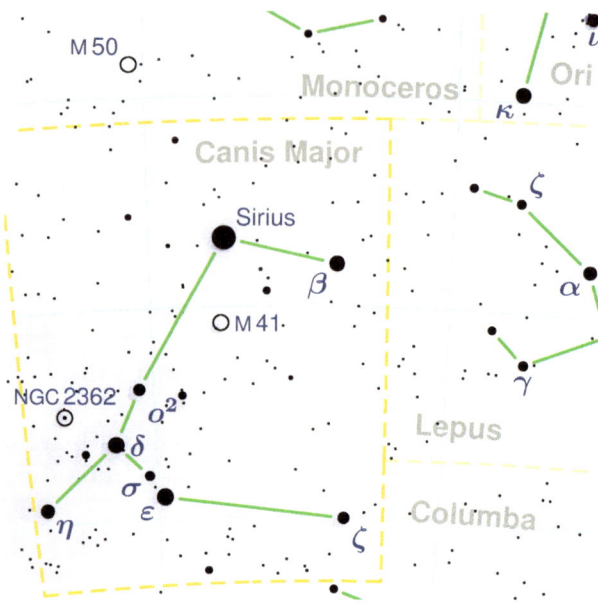

Fig. 8.7. Star map showing Beta Canis Majoris, heralding the bright star Sirius. While Sirius is the brightest star in our night sky, Beta Canis Majoris would dominate by far if brought to the distance of Sirius. Every dot on such a star map has its own dramatic story, in this case a bright giant pulsating variable. Credit: Torsten Bronger

Canopus (Alpha Carina) is an example of the ambiguities of star classification. In the past, it has been classified both as an F0 Ia supergiant and as an F0 II bright giant (in the Simbad database, for example). It is currently thought to be an A9 bright giant. The second brightest star after Sirius, a highlight of the Southern Hemisphere summer, it is 13,000 times more luminous than the Sun and 309 light years distant. It is therefore much more intrinsically luminous than Sirius at 8.6 light years, and the corona of Canopus is a prodigious producer of X-rays. It is not certain if Canopus is evolving to or away from red giant status, but in any case it is fusing helium. Similarly, Alpha Hydrae is classified as a K3 II-III on the border between giants and bright giants. While classification is essential, no one said it would be easy or rigidly unambiguous, especially in the case of hot stars that are continuously and rapidly evolving.

Class S 9: Supergiant

If giants like Arcturus and Aldebaran seem big, supergiants are even bigger, with 10 to 70 solar masses, radii up to 1,000 times the Sun's, and luminosities hundreds of thousands of times the Sun's (Fig. 8.8). They are the result of the evolution of high-mass stars—those greater than eight solar masses. As a member of the Subfamily of post-main sequence stars (luminosity class I in the MK system) they too have exhausted their core hydrogen supply and are burning hydrogen around the inert helium core, or (at a later stage) are burning the helium core, or (at a still later stage) have multiple shell burning, producing the heavier elements, including those necessary for life. Supergiants are essential for life not only because they produce the heavy elements, but also because once they have fused elements as high as iron, most of them explode as supernovae (S 11), distributing these elements into the interstellar medium, becoming available for protostars and planets in the continuous cosmic cycle. The most luminous supergiants are sometimes called hypergiants (S 10).

Supergiants provide a good illustration of how a classification system can expand with new knowledge. The term "supergiant" was used already in 1917 by William H. Pickering in a paper on the stars in Orion, when he noted that a certain star must be a "giant among giants," and that "perhaps we may properly describe

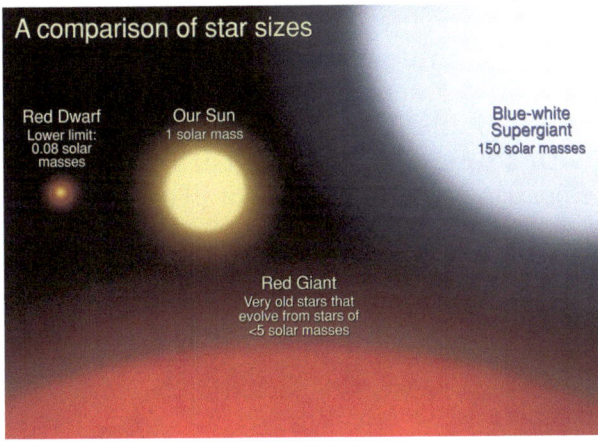

Fig. 8.8. Star masses and sizes compared, from red dwarfs to supergiants. Credit: NASA Goddard Spaceflight Center

these stars as consisting of many giants and a few supergiants." The term was first used in a paper title by Cecelia Payne and Carl T. Chase in two articles in 1927. In 1943, "Supergiant" became luminosity class I in the MKK system. Because of the large range of luminosities, in the revised 1953 MK system Luminosity class I was subdivided into Ia and Ib.[41]

Supergiants in general span the entire temperature spectrum: they can be hot blue O stars where the core nuclear reactions are burning slightly slower (making the star smaller and hotter), relatively cool, highly evolved red M stars, and everything in between. Supergiants are relatively rare, particularly red supergiants. Only about 200 red supergiants are known in our Galaxy, in part because they last in that phase for less than a million years. Cool yellow hypergiants (S 10) are believed to be post-red supergiants evolving toward blue supergiants.

Because of their high masses, blue supergiants have short lifetimes, on the order of a few million years. They are often found in young objects like open clusters (S 34). Rigel, the brightest star in the constellation Orion, is the closest blue supergiant, a B8 star located at 800 light years. With 40,000 times the luminosity of the Sun, it is the most luminous object in our part of the Galaxy. Many blue supergiants are actually hypergiants, and many are still mysterious objects. Blue supergiants are also found outside the Galaxy. The blue luminous variable known as S Doradus, 169,000 light years distant, is the brightest star in the large Magellanic Cloud, a satellite galaxy to the Milky Way Galaxy. Supernova 1987A resulted from the explosion of a B3 blue supergiant, which may in fact be a hypergiant.

To the right of the blue supergiants on the HR diagram (Fig. 8.2) are the cooler yellow supergiants, represented by Canopus (classified as both A9 II and F0 Ia), the second brightest star in the sky after Sirius with a luminosity more than 10,000 times the Sun. If placed in the position of the Sun, it would extend beyond the orbit

---

[41] William H. Pickering, "The Distance of the Great Nebula in Orion," HCO Circular 205 (1917), 1-8: 5; he refers to supergiants again in connection with the Pleiades in HCO 206, 1-3. Cecelia Payne and Carl T. Chase, "The Spectra of Supergiant Stars of Class F8," Harvard College Observatory Circular, 300 (Jan 1927), 1-10; Cecelia Payne, "Photometric Line Intensities for Normal and Supergiant Stars," HCO Circular 306 (1927). H. L. Johnson and W. W. Morgan, "Fundamental stellar photometry for standards of spectral type on the revised system of the Yerkes spectral atlas," ApJ, 117 (1953), 313-352; pp. 318-319 lists changes from MKK to MK.

of Mercury. Epsilon Aurigae (F0 Iab) is another yellow supergiant, part of a multiple star system in which a possible eclipsing binary makes the brightness of the supergiant drop by about one magnitude for 640–730 days every 27 years. The system remains rather mysterious, with some theories postulating a massive dust disk to account for the observations. The supergiant itself also appears to be pulsating, and its mass has even been called into question.

The Earth's Pole star, Polaris, is also a yellow supergiant or bright giant (F7 Ib-II), and a classical Cepheid with a mass 5.4 times that of the Sun and a radius 37.5 times the Sun's. The Hipparcos satellite determined a distance of about 433 light years from Earth, but more recent estimates indicate it could be 100 light years closer. It is 2,000 times the brightness of the Sun, the closest Cepheid but not the brightest—an honor reserved for supergiant Cepheids like its namesake delta Cephei. Although you couldn't tell with just the naked eye, Polaris is part of a multiple star system (S 32) whose triple components were imaged in 2006 by the Hubble Space Telescope, resolving the third and closest companion for the first time (see Fig. 12.1).

Yellow supergiants also exist in other galaxies; in 2018 astronomers reported a high-speed and highly evolved one in the Small Magellanic Cloud, the first "runaway" yellow supergiant ever found outside our own Galaxy. It is a rare find, because yellow supergiants are believed to be a phase that lasts only 10,000 to 100,000 years before becoming a red supergiant. The authors suggested it may have originated in a binary system where the primary star exploded as a supernova and flung the yellow supergiant out of the system from which it is now fleeing at a speed of 300 km/second.[42]

Moving left to right across the HR diagram and down the temperature scale, Betelgeuse in Orion and Antares in Scorpio are the nearest red supergiants. Both are M stars, the former at 640 light years distant with a luminosity more than 100,000 times that of the Sun, and the latter at 600 light years. In 1995, the Hubble Space Telescope imaged Betelgeuse and found a mysterious hot spot the diameter of the Earth's orbit (remember, if Betelgeuse were put in the place of our Sun, its outer atmosphere would extend beyond the orbit of Jupiter). Hubble also detected 14 red supergiants in a massive cluster in our Milky Way, some 18,900 light years distant (Fig. 8.9).

---

[42] Kathryn F. Neugent et al., "A Runaway Yellow Supergiant Star in the Small Magellanic Cloud," AJ, 155 (2018), 7 pp.

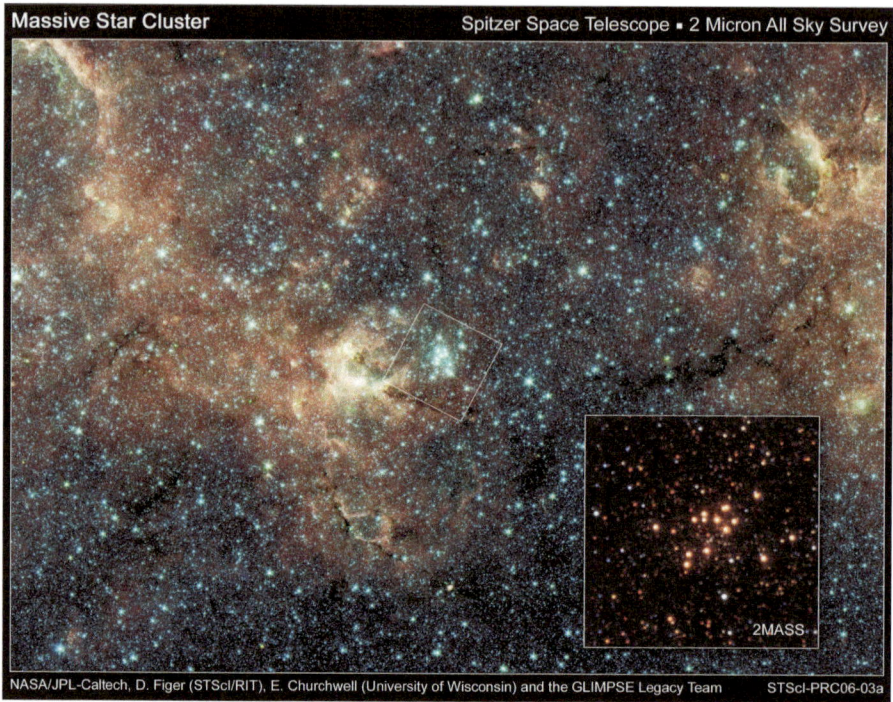

Fig. 8.9. The Spitzer Space Telescope captured 14 massive red supergiants on the verge of exploding as supernovae in a massive star cluster known as RSGC1 in the constellation Scutum. The image is taken in the infrared rather than the optical part of the spectrum. The inset image is from the Two Micron All-Sky Survey (2MASS). Credit for Spitzer Image: NASA/JPL-Caltech, D. Figer (Space Telescope Science Institute/Rochester Institute of Technology), E. Churchwell (University of Wisconsin, Madison), and the GLIMPSE Legacy Team; Credit for 2MASS Image: NASA/NSF/2MASS/UMass/IPAC-Caltech and D. Figer (Space Telescope Science Institute/Rochester Institute of Technology)

Supergiants also include the exceedingly rare Wolf-Rayet stars, highly evolved dying stars between 10 and 25 solar masses that are losing their mass at an enormous rate through stellar winds (S 29). They have luminosities between 100,000 and one million times the Sun. These stars will likely end in as a supernova, enriching the interstellar medium with their chemical remnants. The French astronomers Charles Wolf and Georges Rayet first discovered them in 1867, when they detected three stars with faint emission lines.[43] About 500 are now known in the Milky Way Galaxy and 150 in the Magellanic Cloud galaxies.

---

[43] Charles Wolf and Georges Rayet, *Comptes Rendus*, 65 (1867), 292. Gray and Corbally, *Stellar Spectral Classification*, 441-464.

Because of their brightness, supergiants can be seen at great distances and are extremely important for determining extragalactic distances, in particular in the form of those pulsating bright giants and supergiants known as Cepheids. For this reason, Cepheid variables (so called because the prototype, delta Cephei, was discovered in 1784 in the constellation Cepheus) are among the most important types of yellow giant and supergiant stars. In 1912 Henrietta Leavitt at Harvard College Observatory, observing numerous variables in the Magellanic Cloud galaxies, found a relationship between the observed period of variability of Cepheids and their luminosity: the longer the period, the greater the luminosity. We now know these are pulsating variables, expanding and contracting in unsteady equilibrium as the star tries to match its internal energy to its energy radiating into space. Cepheids have periods ranging from one to 60 days, and rarely, even longer. It is this period-luminosity relationship that has become an essential tool for distance determination, because once the intrinsic luminosity of a star is known, the distance can be calculated based on its apparent magnitude. Because of their brightness, they are useful for this purpose to much greater distances than the RR Lyrae giant stars mentioned in class S 7. See Fig. 8.6 for variable star classes.

Because of this period-luminosity relation, Cepheids have become spectacularly useful in astronomy. In 1915, Harlow Shapley published the first distance scale using Cepheids in our Galaxy. In 1924, Edwin Hubble famously found Cepheids in M31 (the Andromeda Galaxy), proving it to be an extragalactic object, not a "nebula" in our own Galaxy. But Cepheids also proved spectacularly misleading for a few decades in the history of astronomy because astronomers did not at first recognize the existence of different Types. In 1942, Mt. Wilson Observatory astronomer Walter Baade found from his study of stellar populations (S 28) in the Andromeda Galaxy that there were actually two types of Cepheids, young metal-rich Population I stars now known as classical Cepheids, and older metal-poor Population II Cepheids. The prototypes for the latter are W Virginis, BL Herculis, and RV Tauri stars. Classical Cepheids are F and G bright giants and supergiants (luminosity classes I and II), while Population II Cepheids are the same spectral types, but are giants and bright giants (luminosity classes II and III). The net effect, for example, is that the luminosity of classical Cepheids is 1.5 times more than W Virginis Cepheids, and thus they have a different period-luminosity relationship. Because of the failure to separate these two types of Cepheids, after Baade's discovery in the 1940s the size of the universe doubled overnight.

The Hubble Space Telescope still uses Cepheids for improving the distance scale, affirming the continuing utility of these giant and supergiant stars. One of the brightest classical Cepheids in our own Galaxy, known as RS Puppis (Fig. 8.10), is a southern hemisphere star 10 times more massive than our Sun, 200 times larger, and 15,000 times greater luminosity. It pulsates with a period of 41.4 days. Numerous Cepheids of varying periods continue to be found in other galaxies. One such Hubble image of the spiral galaxy NGC 5584, at http://hubblesite.org/image/2826/category/112-hubble-heritage, distinguishes those with periods less than 30 days, less than 60 days, and a few rare specimens at more than 30 days. An image of Polaris and its companions may be seen at http://hubblesite.org/newscenter/archive/releases/2006/02/image/a. Hubble's image of Betelgeuse is at http://hubblesite.org/image/394/category/107-illustrations.

Fig. 8.10. The classical Cepheid RS Puppis is at a distance of 6,500 light years and is surrounded by reflective dust that changes in brightness as the stars pulsates. The image was released in December, 2013. Credit: NASA, ESA, and the Hubble Heritage Team (STScI/AURA), Hubble/Europe Collaboration; Acknowledgment: H. Bond (STScI and Pennsylvania State University)

Class S 10: Hypergiant

Believe it or not, there are stars even larger than supergiants. In 1956, South African astronomers Michael W. Feast and Andrew (David) Thackeray, working at the Radcliffe Observatory, suggested that the brightest supergiants (greater than magnitude minus 7) be called super-supergiants.[44] In 1971, the indefatigable classifier of the stars Philip Keenan suggested a new class, designated 0, to define hypergiants, and in the next revision of MK system in 1973, Morgan and Keenan wrote "In the new and tighter network of supergiants the highest luminosity class, 0, is defined (Keenan, 1971) by the four reddest of the brightest stars in the Large Magellanic Cloud. These are the stars originally called 'super supergiants' by Feast & Thackeray (1956)."[45] Class 0 hypergiants are distinct from what are called Ia-0, supergiants/hypergiants, the latter being an example of the common practice of classifying intermediate luminosity classes. Philip Kennan, who devoted most of his 71-year career to classifying stars, died in the year 2000 at the age of 92.

The subdivision of class I into subclasses Ia and Ib on the one hand and the declaration of a new class 0 of hypergiants on the other seems somewhat arbitrary. But Keenan justified the new class 0 by saying the situation was confused for stars of luminosity class Ia and brighter, and that the super-supergiants discovered in the Large Magellanic Cloud "provide the means of anchoring the upper end of the luminosity sequence for these types," for which he placed the absolute magnitudes at around -9, 500,000 times solar luminosity.[46] Nevertheless, the class has been only sporadically and inconsistently used over the last four decades, and one can argue whether hypergiants should be given separate class status. However classified, they are some of the most stupendous objects in the universe, their 100 solar masses on the verge of instability with large mass loss and extended circumstellar envelopes. In the end, like all supergiants, they will become supernovae and black holes.

---

[44] Feast and Thackeray, "Red supergiants in the Large Magellanic Cloud," MNRAS, 116 (1956), 587-590. Dick (2013), pp. 256-257.

[45] Philip Keenan, "Classification of supergiants of types G, K, and M," in G. Wesley Lockwood and H. Melvin Dyck, eds., *Late-Type Stars* (Kitt Peak National Observatory, 1971), pp. 35-39. Morgan and Keenan, "Spectral Classification," ARAA, 11 (1973), 29-50: 44. The latter is a major article on classification.

[46] Keenan (1971), 35.

As with other classes of stars, astronomers typically classify hypergiants according to their temperatures and thus colors. As we move left to right across the very top of the HR diagram from hot to cool, they are known as luminous blue variables, blue hypergiants, yellow hypergiants, and red hypergiants. These are what we might call in the Three Kingdom system four Types of hypergiants.

The star with the enigmatic name P Cygni is the prototype for a Luminous Blue Variable (LBV), one of only about a dozen known. It is about 6,000 light years from Earth, visible in the constellation Cygnus the Swan. It is known to have undergone a series of eruptions, most recently in the 17th century, but also 2,400 and 20,000 years ago, as evidenced from the faint shells that of emitted matter than surround it. As an LBV, it is an extremely massive star, perhaps 50 times the Sun's mass, and therefore will shine for only a few million years before exhausting its nuclear fuel and exploding as a supernova (S 11).

Even P Cygni is outdone by the double star system known as Eta Carina, which has been studied extensively from ground-based telescopes, the Hubble Space Telescope, and other instruments. It is some 7,800 light years distant and is actually composed of two massive stars orbiting each other; at their closest, they are separated only by the average distance between Mars and the Sun! The brighter star, with about 90 solar masses, is one of the most massive stars in the Galaxy, with a luminosity five million times that of the Sun. The "lesser" star has about 30 masses and is "only" one million times the luminosity of the Sun. When these two mammoths are at their closest to each other, known as periastron, they emit spectacular X-ray flares.[47] Like P Cygni and other luminous blue variables, Eta Carina has undergone massive eruptions. It is surrounded by the homunculus nebula (Fig. 8.11), most of which was formed during the 1840s and 1850s, when Eta Carina became the second brightest star in the sky after Sirius. The spectacular form of this gas and dust emission and reflection nebula (S 24 and S 28) is shaped by these eruptions and the two stars in

---

[47]NASA Release, January 5, 2015, "NASA Observatories Take an Unprecedented Look into Superstar Eta Carinae," https://www.nasa.gov/content/goddard/nasa-observatories-take-an-unprecedented-look-into-superstar-eta-carinae; Lee Billings, "Fact or Fiction?: The Explosive Death of Eta Carinae Will Cause a Mass Extinction," *Scientific American*, December 16, 2014, https://www.scientificamerican.com/article/fact-or-fiction-the-explosive-death-of-eta-carinae-will-cause-a-mass-extinction/

Fig. 8.11. Gas and dust surround the massive hypergiant Eta Carinae in this Hubble Space Telescope image. **Credit:** J. Hester/Arizona State University, NASA/ESA

the system. Eta Carina is a massive powder keg on the verge of exploding as a supernova.

Moving toward the right at the top of the HR diagram (Fig. 8.2), the yellow hypergiant Rho Cassiopeiae, located about 8,200 light years from Earth, is 500,000 times brighter and 500 times larger in diameter than the Sun. Only about a dozen yellow hypergiants are known in the Milky Way Galaxy, but they are so bright that a few have also been detected in other galaxies such as the Large and Small Magellanic Clouds.

The red hypergiant VY Canis Majoris (Fig. 8.12) features as the location of extraterrestrial intelligence in Douglas Phillips' highly original science fiction novel *Quantum Space*. It is 3,900 light years distant, up to 25 times as massive as the Sun, and if placed where our Sun is would extend to the orbit of Jupiter or Saturn. Its luminosity is at least 300,000 times greater than the Sun's. It is surrounded by an extensive nebula that has been imaged by the Hubble Space Telescope.[48]

---

[48] HST Release, January 8, 2007, "Astronomers Map a Hypergiant Star's Massive Outbursts, http://hubblesite.org/news_release/news/2007-03

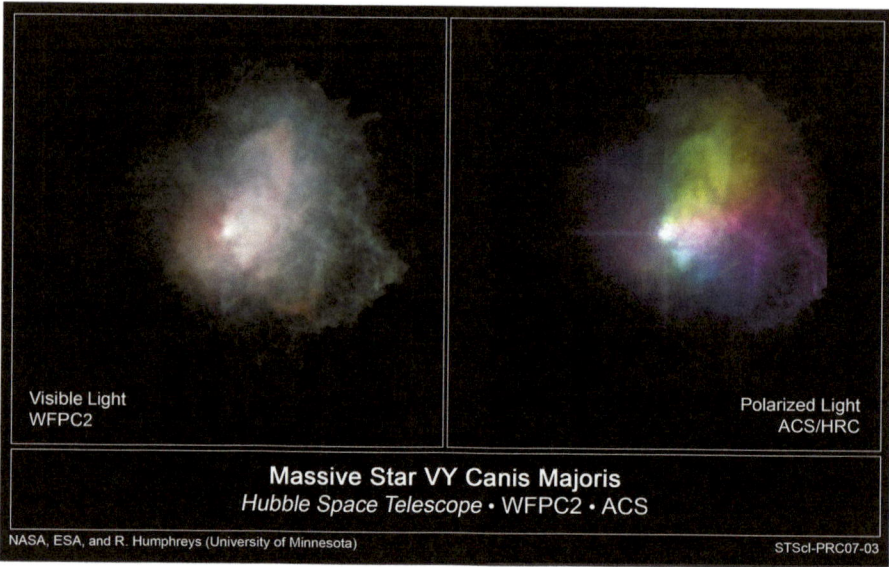

Fig. 8.12. The massive outflows from the red hypergiant VY Canis Majoris, believed to be produced by localized eruptions from active regions on the star's surface. Credit: NASA, ESA, and R. Humphreys (University of Minnesota)

With the hypergiants, we come to the end of the stellar classes, at least as intact stars before they race toward their mass-dependent evolutionary endpoints as supernovae, white dwarfs, neutron stars, or black holes (S 11 through S 14). Summarizing our stellar classes from dwarfs to hypergiants, we can say that most stars were distinguished as separate classes over a period of some 30 years beginning with giants and dwarfs in the early 20th century (Hertzsprung and Russell), supergiants in 1917 (Pickering), subgiants and subdwarfs in the 1930s (Stromberg and Kuiper), and bright giants in 1943, as defined in the MKK system. Despite hints earlier, it was almost three decades later, in 1971, that Keenan declared the rare hypergiants in a class of their own.

A spectacular video from NASA's Goddard Spaceflight Center explains more about Eta Carina: https://www.nasa.gov/content/goddard/nasa-observatories-take-an-unprecedented-look-into-superstar-eta-carinae.

Class S 11: Supernova

A supernova is a catastrophic explosion triggered either when a massive star exhausts its nuclear fuel and collapses under its own weight, or when a white dwarf (S 12) in a binary system explodes after accreting 1.4 solar masses. The latter is known as Chandrasekhar limit. We can therefore speak of two main Types of supernovae: core collapse supernovae that produce neutron stars (S 13) or black holes (S 14), and white dwarf supernovae, believed to leave behind only material for the interstellar medium or the companion to the exploding white dwarf. Core collapse supernovae produce spectacular circumstellar remnants (S 19), while white dwarf supernovae produce a different kind of remnant (S 26). In both cases the complex elements of life are dispersed into the unending life cycles of stars and planetary systems. Supernovae also release galactic cosmic rays (S 30) and neutrinos.

Despite these two main Types of physical mechanisms, supernova nomenclature can be confusing both because of historical circumstance and because we now know many subtypes of these two main categories. In the lingo of astronomers, Type II supernovae are those that result when a progenitor star between about eight and 15 solar masses collapses to become a neutron star. More precisely, it is the core of the progenitor star that collapses after the star's massive outer shells crash down onto the burnt-out core and material is carried away by the supernova explosion. For a neutron star, this core will have a mass from 1.4 to two or three solar masses. It is thus very important to distinguish the core mass from the original "progenitor" star mass. Supernovae that result from the collapse of stars greater than 15 to 20 solar masses (with core remnants greater than three solar masses) result in a black hole. But despite their similar collapse mechanism, for historical reasons they are not labeled subtypes of Type II, but Types Ib and Ic, massive Wolf-Rayet stars that have lost their outer hydrogen layer. Objects resulting from the collapse of extremely massive stars greater than 30 solar masses are sometimes referred to as hypernovae, superluminous supernovae, or collapsars. If massive enough, they may leave behind no remnant at all.

On the other hand, those supernovae that occur in binary systems where material from a normal star accretes onto a white dwarf until it explodes are labeled Type Ia (Fig. 8.13). The companion star is believed to be often ejected. Type Ia supernovae may

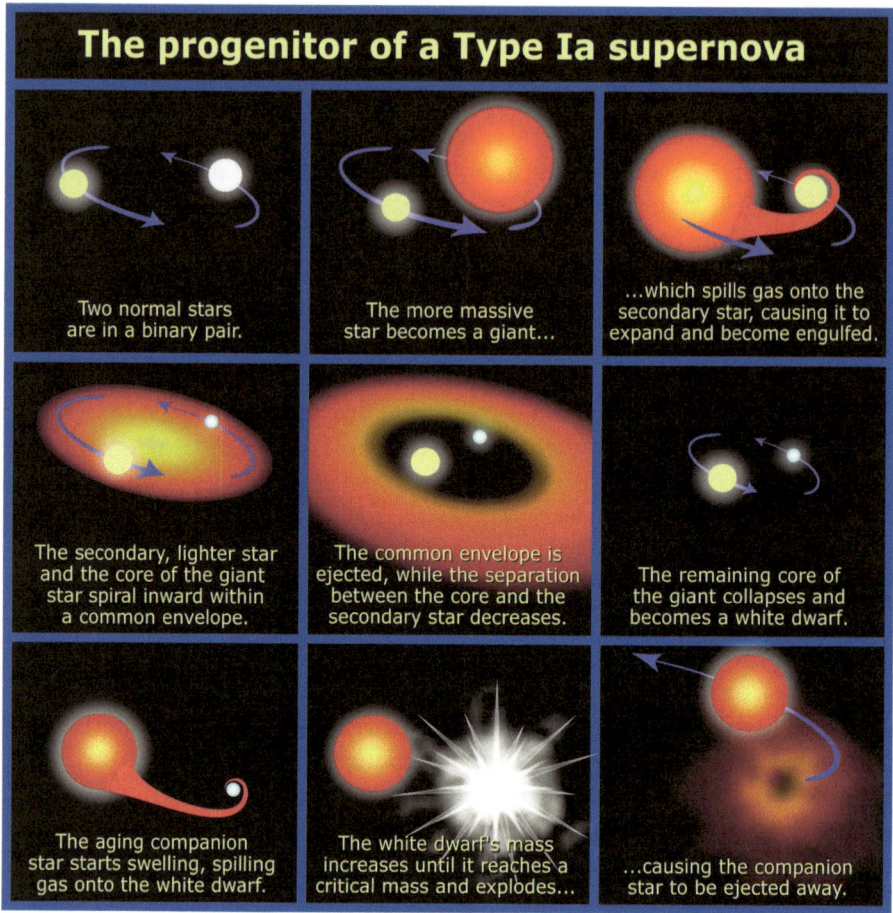

The progenitor of a Type Ia supernova

Two normal stars are in a binary pair.

The more massive star becomes a giant...

...which spills gas onto the secondary star, causing it to expand and become engulfed.

The secondary, lighter star and the core of the giant star spiral inward within a common envelope.

The common envelope is ejected, while the separation between the core and the secondary star decreases.

The remaining core of the giant collapses and becomes a white dwarf.

The aging companion star starts swelling, spilling gas onto the white dwarf.

The white dwarf's mass increases until it reaches a critical mass and explodes...

...causing the companion star to be ejected away.

Fig. 8.13. The Type Ia supernova process. Credit: NASA, ESAm and A. Feild (STScI)

also occur when two white dwarfs collide. One rare breed of supernova occurs in a binary system consisting of two normal stars where the more massive star (the supernova progenitor) evolves into a red giant or supergiant, and the companion star siphons off most of the hydrogen from the outer shell, resulting in an instability causing the more massive star to explode. Although this is different from a core collapse, it is still classified as Type IIb for spectroscopic reasons. The Hubble Space Telescope has imaged the surviving companion of a supernovae believed to be of this type, also known as "stripped-envelope supernovae." The companion star, about 10 solar masses, appears to have stripped away the more massive 13 solar mass star's outer hydrogen shell before the latter exploded as Supernova 2001ig in the galaxy NGC 7424.[49]

Core collapse supernovae are distinguished from other types by the presence of hydrogen in their spectrum. Aside from their luminosities, which are typically three to ten times brighter than core collapse supernovae, white dwarf supernovae are distinguished from other supernovae types by the lack of hydrogen lines and the presence of silicon in their spectra. White dwarf supernovae are similar to novae in that both involve white dwarfs accreting material from a normal star. But in the case of the nova, the white dwarf undergoes increased nuclear fusion without exploding, resulting in much less brightness than a supernova and a circumstellar nova remnant (S 18). Supernovae can release a million times more energy than novae.

Astronomers can perhaps be forgiven for the illogical nomenclature of supernovae because of their limited knowledge at the time it was instituted. Until the 20th century, absolute magnitudes and luminosities were not known well enough to distinguish novae from supernovae. Already in 1920, Knut Lundmark spoke of "giant novae," then in 1923 "much more luminous novae," and in 1927 "upper class novae." In 1929 Walter Baade called them "Hauptnovae" [chief novae], and in 1929 Hubble called them "exceptional novae." Walter Baade and Fritz Zwicky coined the term "supernovae" in print in 1934 and theorized that "supernovae represent the transitions from ordinary stars into neutron stars."[50] From 1934 to 1939, they found many more supernovae with telescopes at Mt. Wilson, and Zwicky theorized in more detail that a supernova represented a massive star explosion that resulted in a high density star only 12 miles in diameter, 100 million times the

---

[49] HST Release, April 26, 2018, "Stellar Thief Is the Surviving Companion to a Supernova," http://hubblesite.org/news_release/news/2018-20

[50] Baade and Zwicky, "On Supernovae," PNAS, 20 (May 15, 1934), 254-259, reprinted in Bartusiak, 341-343; Baade and Zwicky, "Supernovae and Cosmic Rays," Physical Review, 45 (1934), 138. Donald E. Osterbrock, "Who Really Coined the Word Supernova/Who First Predicted Neutron Stars," BAAS, 33: 1330-1331. Osterbrock's answer to both questions is Baade and Zwicky, who used the term supernovae in seminars and courses at Caltech in 1931. See also Hilmar Duerbeck, "Novae: An Historical Perspective," in Michael F. Bode and Adeurin Evans, eds., *Classical Novae* (Cambridge: Cambridge University Press, 2nd edition, 2008), p. 2.

density in a white dwarf.[51] These proved later to be the neutron stars observed as pulsars.

This history is also a lesson in the difficulties of classification with incomplete knowledge. By 1941, enough supernovae had been observed that Rudolph Minkowski could separate them into two types based on their spectra. "Spectroscopic observations indicate at least two types of supernovae. Nine objects...form an extremely homogeneous group provisionally called 'type I.' The remaining five objects...are distinctly different; they are provisionally designated as 'type II'," he wrote. He wrote further that "Supernovae of type II differ from those of type I in the presence of a continuous spectrum at maximum and in the subsequent transformation to an emission spectrum whose main constituents can be readily identified. This suggests that the supernovae of type I have still higher surface temperature and higher level of excitation than either ordinary novae or supernovae of type II."[52]

These designations were soon adopted by astronomers such as Zwicky and Baade, and the latter found that supernovae light variations could also be classified by the same two types. What we now know as white dwarf supernovae (Minkowski and Baade's Type I, our Type Ia) exhibit a rapid rise to maximum brightness over two weeks, then a steady decline over a few weeks, with a "half-life" of about 50 days during which they successively fade to half their previous brightness. Type II core collapse supernovae remain at maximum brightness for several weeks and dim more slowly than Type I.

Furthermore, Baade found that Type I white dwarf supernovae occurred in all types of galaxies from elliptical to spirals and were therefore Population II objects, while Type II were Population I objects (S 36) occurring only in spirals.[53] We now know that this is because Type II core collapse supernovae occur where massive stars are found, such as in the arms of spiral galaxies. They are not seen in elliptical galaxies, which are dominated by older low-mass stars. It was only later that Type I supernovae were divided

---

[51] Fritz Zwicky, "On the Theory and Observation of Highly Collapsed Stars," *Physical Review*, 55 (1939), 726-43. This and the following paragraphs based on Dick (2013), pp. 207 ff.

[52] R. Minkowski, "Spectra of Supernovae," PASP, 53 (1941), 224-225; abstract of paper given at the Pasadena meeting of the ASP.

[53] Donald Osterbrock, *Walter Baade: A Life in Astrophysics* (Princeton: Princeton University Press, 2001), 117.

into Type Ia, Ib, and Ic.[54] The mechanism for Type Ia supernovae involving exploding white dwarfs was not theorized to involve binary stars until the work of Fred Hoyle and William Fowler in 1960, following the realization in the early 1950s that novae also involved binary stars.[55] Astronomers did the best they could at the time with nomenclature and classification, because it was based on spectra and an incomplete understanding of the underlying physical nature. Once such a nomenclature is established, however, it is difficult to change.

Because supernovae are such energetic explosions, they have been observed throughout history, as described in David Clark and Richard Stephenson's classic volume *The Historical Supernovae*.[56] Among the most famous are Tycho's supernova in 1572 and Kepler's in 1604, both now known to be Type Ia white dwarf supernovae. Neither Tycho nor Kepler was the first to observe the supernovae that bear their names, but they did study them carefully as "new stars" (*stella nova*). Their observations were very dramatic at a time when the heavens were believed to be immutable. Tycho first observed his supernova on November 11, 1572, as bright as the planet Jupiter. It soon equaled the brightness of Venus in the sky, at -4.5 magnitude, and was observed in daylight for two weeks. It began to fade at the end of the month, vanishing completely by March 1574.

While Tycho's supernova remnant has been long observed (Fig. 11.5), in 2004 an international team of astronomers announced they had discovered the probable surviving companion star. This provided important evidence of the long-held belief that Type Ia supernovae indeed arise from a binary star system in which one star has evolved more quickly to the white dwarf stage, while a normal star like our Sun is expanding to the red giant stage, spilling hydrogen onto the white dwarf until it explodes. The surviving

---

[54] Supernovae classification types and methods are discussed in Alexei V. Filippchenko, "Optical Spectra of Supernovae," ARAA, 35 (1997), 309-355. For example Type Ia show strong silicon II, Ib show prominent helium I, Ic show neither.

[55] Fred Hoyle and William A. Fowler, "Nucleosynthesis in Supernovae," AJ, 132 (1960), 565-590.

[56] David H. Clark and F. Richard Stephenson, *The Historical Supernovae* (Oxford: Pergamon Press, 1977); F. R. Stephenson and D. A. Green, *Historical Supernovae and their Remnants* (Oxford: Clarendon Press, 2002).

companion in this case is traveling at high speed and appears to be in the process of being ejected.[57]

Only three decades after Tycho's supernova, observers spotted another one on October 9, 1604. When the clouds parted in Prague, on October 17 Kepler himself saw it, studied it in detail, and wrote about it in his volume *De Stella Nova*. Three of NASA's Great Observatories have observed the remnants of this supernovae (Fig. 8.14) in different wavelengths, a bubble of gas and dust now 14 light years wide and expanding at four million miles per hour (2,000 km per second).

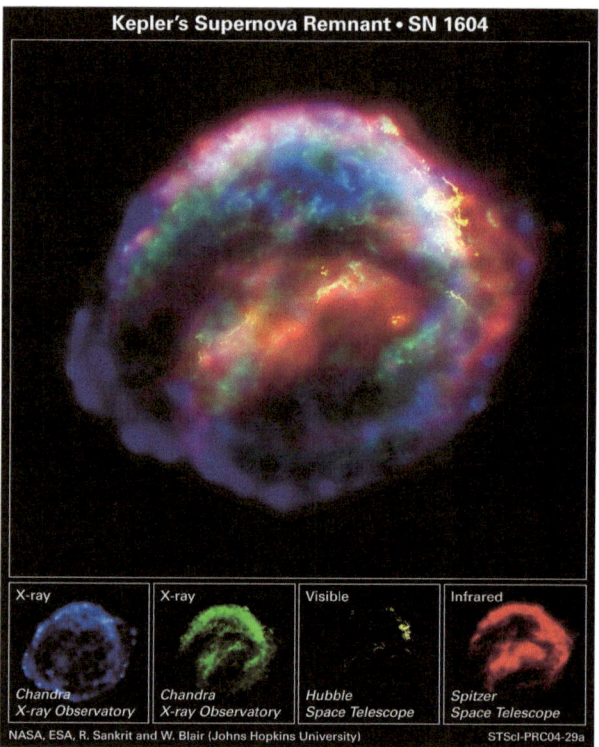

Fig. 8.14. The Hubble, Spitzer and Chandra Telescopes combined to produce this image of Kepler's supernova remnant, now known to be a white dwarf supernova. Credit: NASA, ESA, R. Sankrirt, and W. Blair (Johns Hopkins University)

---

[57] HST Release, October 27, 2004, "Stellar Survivor from 1572 A.D. Explosion Supports Supernova Theory," http://hubblesite.org/news_release/news/2004-34

Kepler's was the last naked eye supernova observed in our Galaxy, although at least two others not visible to the naked eye are known to have occurred since: the Cassiopeia A supernova about 300 years ago (the brightest radio source in the sky), and an object now known as G1.9+0.3 about 140 years ago, discovered in 2005 by the Chandra X-ray Observatory. Cassiopeia A is now known to be a massive star supernova of Type II, and may have been unknowingly observed by British Astronomer Royal John Flamsteed as a 6th magnitude star. G1.9+0.3, at about 28,000 light years distant, has expanded 16% in 22 years. It is not yet known whether it is a core collapse or Type Ia supernova.[58] Although these two currently hold the record as the youngest supernovae in the Milky Way Galaxy, astronomers estimate that they occur at a rate of about three per century in our Galaxy, so a few younger supernovae probably remain undiscovered in the Galaxy.

Supernovae are more often found outside our Galaxy, where they may not be so obscured by gas and dust. Supernova 1987A (Fig. 8.15), first observed February 24, 1987 in the Large Magellanic Cloud, was the first naked-eye supernova since Kepler's. It was a Type II core collapse of a blue supergiant with about 20 solar masses, located some 168,000 light years from Earth. This event was observable only from the Southern Hemisphere (I had the pleasure of observing it serendipitously while stationed in New Zealand). It was easily visible to the naked eye at magnitude 3 at its peak. Astronomers have been following it closely since the explosion occurred. The neutron star that should in theory form from this collapse has not yet been observed, but the ever-expanding shells of material from the explosion have been observed in detail.

The pace of supernovae discovery has accelerated with improved telescopes and detectors. As of 1984, more than 600 supernovae were known outside our galaxy, and several thousand have now been identified.

---

[58] On the 2005 Chandra discovery see Chandra Release, May 14, 2008, "Chandra Uncovers Youngest Supernova in Our Galaxy," http://www.nasa. gov/mission_pages/chandra/news/08-062.html; William B. Ashworth, "A Probable Flamsteed Observation of the Cassiopeia A Supernova," JHA, 11 (1980), 1-9.

Supernova 1987A • December 6, 2006
Hubble Space Telescope • Advanced Camera for Surveys
NASA, ESA, P. Challis, and R. Kirshner (Harvard-Smithsonian Center for Astrophysics)   STScI-PRC07-10a

Fig. 8.15. Supernova 1987A, photographed almost two decades after the explosion was first seen on Earth. The glowing ring is about one light year across, and was probably shed by the star 20,000 years before it exploded. The material in the center is from the supernova explosion itself and the outer edge of that material is slamming into the ring of material released earlier, causing it to glow. Credit: NASA, ESA, P. Challis, and R. Kirshner (Harvard-Smithsonian Center for Astrophysics)

White dwarf supernovae have proven extremely important as objects for probing the distances of galaxies. Because all exploding white dwarfs have exactly the same mass (the Chandrasekhar limit), their explosion always results in the same luminosity, as opposed to other types of supernovae, which result from exploding stars of different masses, and thus different luminosities. White dwarf supernovae thus serve as "standard candles," in a similar way to the RR Lyrae and Cepheid variables mentioned earlier. White dwarf supernovae are most famous for being used in the late 1990s as proof that the universe is not only expanding but also accelerating its expansion.[59] An accelerating universe leads to the

---

[59] Adam G. Riess, Alexei Filipchenko et al, "Observational Evidence from Supernovae for an Accelerating Universe and a Cosmological Constant," ApJ, 116 (1998), 1009-1038. This article also provides some of the history of Type Ia supernova studies. Proof of the accelerating universe was the "breakthrough of the year" for *Science* magazine in 1998.

idea of "dark energy," which must be a repulsive force counteracting gravity and dark matter. This renews the idea of a "cosmological constant," which Einstein considered his greatest blunder after Hubble showed the universe was expanding. Far from being a blunder, dark energy is now believed to represent more than 70% of the mass-energy of the universe.

White dwarf supernovae are relatively rare, occurring at a rate of only one per galaxy per hundred years. High redshift ("high Z") supernovae searches needed for distance determination therefore must observe hundreds of galaxies to find enough supernovae for their surveys. Because it takes a long time to produce a white dwarf (the endpoint of stars of less than 1.4 solar masses that last billions of years), there can be no white dwarfs in the early universe. As telescopes look back beyond 2 or 3 Z redshifts, there should be no white dwarfs, and thus no Type Ia supernovae.

A comprehensive list of supernovae reported since 1885, sponsored by Harvard University and the International Astronomical Union, is found at http://www.cfa.harvard.edu/iau/lists/Supernovae.html. A much shorter list of supernova of historical significance, with links to a description of each, can be found at https://en.wikipedia.org/wiki/List_of_supernovae. More on Kepler and Tycho's supernova remnants can be found at http://hubblesite.org/newscenter/archive/releases/2004/34/ and http://hubblesite.org/newscenter/archive/releases/2004/29/. The Chandra X-Ray Observatory has a gallery of supernovae and supernovae remnants at http://chandra.harvard.edu/photo/category/snr.html, including an image of the supernovae of 1006 http://chandra.harvard.edu/photo/2005/sn1006/. Images of Supernova 1987A: are at http://hubblesite.org/newscenter/archive/releases/2007/10/image/a/ and at https://www.nasa.gov/feature/goddard/2017/the-dawn-of-a-new-era-for-supernova-1987a. The press release about the probable surviving companion to Tycho's supernova is at http://hubblesite.org/newscenter/archive/releases/2004/34/. A more recent example is at http://hubblesite.org/news_release/news/2018-20/4-galaxies.

Class S 12: White Dwarf

A white dwarf is the final stage in the evolution of a star whose core is less than 1.4 solar masses, known as the Chandrasekhar limit. The original progenitor star may have been as much as 10 solar masses before shedding most of its mass. After burning its core hydrogen through nuclear fusion, a lower main sequence star such as our Sun will become a helium-burning red giant (S 7, S 8), producing carbon and oxygen. Eventually it will slough off its outer layers and produce a planetary nebula (S 17). It then settles down as a white dwarf star in which fusion no longer takes place. The star, incredibly small at about the size of the Earth, has lost its constant battle between outward pressure due to fusion and the inward gravitational force. Its faint luminosity derives only from leftover thermal energy that flows to the surface by conduction. The white dwarf is thus an extremely dense object with the mass of a Sun-sized star squeezed into an area the size of the Earth.

Not all white dwarfs are white; their surface temperatures may range from 150,000 K down to 4,000 K as they cool and redden over a long period of time. In theory, over extremely long periods, they would become cold black dwarfs. These temperature differences lead to different Types of white dwarfs as revealed by their spectra. Most white dwarfs have atmospheres rich in hydrogen, but about one third are helium white dwarfs, and a tiny percentage have carbon atmospheres. In a system dating back to the 1980s, white dwarfs are commonly classified according to their spectra as DA (hydrogen dominated), DB (helium), and DQ (carbon), with a variety of subtypes also recognized. Hot white dwarfs with temperatures exceeding 25,000 K are also recognizable because they emit X-rays.

A white dwarf consists primarily of a degenerate electron gas, in which the electrons roam free from their carbon and oxygen nuclei, the products of previous helium fusion. The carbon and oxygen nuclei produced from helium fusion are embedded in the electron gas. In such a degenerate gas, the lower electron energy levels are filled, and electrons in higher energy levels cannot move to lower energy levels due to the Pauli Exclusion Principle, which specifies that "two is a crowd." The free-roaming electrons exert an outward pressure because they move with high speed. The only thing supporting the star against total gravitational collapse is this electron degeneracy pressure.

Because the vast majority of main sequence stars are low-mass stars, up to 98% of the stars in our galaxy will end as white dwarfs. Those with higher masses will produce neutron stars (S 13) or black holes (S 14), as the electron degeneracy pressure in turn can no longer hold back the force of gravity. A white dwarf that exceeds the Chandrasekhar limit due to accretion from a companion star or collision with another star will explode as a supernova (S 11), known as Type Ia.

Henry Norris Russell at Princeton, and E. C. Pickering and Williamina Fleming at Harvard College Observatory, first recognized the existence of white dwarfs in 1910. Russell wrote:

> The first person who knew of the existence of white dwarfs was Mrs. Fleming; the next two, an hour or two later, Professor E. C. Pickering and I. With characteristic generosity, Pickering had volunteered to have the spectra of the stars which I had observed for parallax looked up on the Harvard plates. All those of faint absolute magnitude turned out to be class G or later. Moved with curiosity I asked him about the companion of 40 Eridani. Characteristically, again, he telephoned to Mrs. Fleming who reported within an hour or so, that it was of Class A. I saw enough of the physical implications of this to be puzzled, and expressed some concern. Pickering smiled and said, "It is just such discrepancies which lead to the increase of our knowledge." Never was the soundness of his judgment better illustrated.[60]

In other words, the companion to the star 40 Eridani (now known as 40 Eridani B) did not follow the pattern of other known stars—it was a faint, relatively hot star in contrast to the many faint cool stars known. It appears well to the bottom left of the main sequence stars on what is now known as the Hertzsprung-Russell diagram (Fig. 8.2). Russell classified it as a dwarf star, and Hertzsprung called it a "dark white star."

The second object that turned out to be a white dwarf was Sirius B. In 1844, Wilhelm Friedrich Bessel suggested that the perturbed motion of the bright star Sirius was caused by an unseen companion. In 1862, telescope maker Alvan Clark first observed this companion to Sirius A while testing one of his telescopes;

---

[60] Henry Norris Russell, "Notes on white dwarfs and small companions," AJ, 51 (1944), 13-17: 13. *White Dwarfs*, E. Schatzman, ed, (Amsterdam: North-Holland, 1958), p. 1. This and the following paragraphs are from Dick (2013), pp. 105-107.

it was ten magnitudes fainter than its primary star, a factor of 10,000 in luminosity. It was so faint that some astronomers suggested it was actually a large planet.[61] In 1914 Walter S. Adams obtained the first spectrum, showing that it was spectral class A, similar to 40 Eridani B. Sirius B has a mass equal to the Sun packed into a diameter 90% of the Earth, yielding a surface gravity some 400,000 times that of Earth.[62]

The third of the classical white dwarfs is van Maanen 2, located in the constellation Pisces, discovered in 1917 by the Dutch astronomer Adrian van Maanen, working at Mt. Wilson on high proper motion stars. Now known to be at a distance of only 14 light years, it was the first white dwarf discovered that is not part of a multiple star system.[63] The fact that all three white dwarfs were relatively close to the Sun hinted that they were common objects throughout the Galaxy.

Astronomer Willem J. Luyten coined the term "white dwarf" in 1922 as part of his work at Lick Observatory studying the high proper motion of nearby stars. In a series of papers that year, Luyten first spoke of "faint white stars" of high proper motion, and then "white dwarfs," extending the contemporary terminology of "yellow dwarfs" and "red dwarfs." The term was popularized by the theoretician Arthur S. Eddington in 1924.[64] By 1950 more than 100 were known; 50 years later over 2000, and by now more than 10,000 white dwarfs have been catalogued, thanks both to large surveys and especially the Sloan Digital Sky Survey. Because they are the evolutionary endpoint of low-mass stars, many more undoubtedly exist but are too faint to be seen.

---

[61] J. B. Holberg and F. Wesemael, "The discovery of the companion of Sirius and its aftermath," JHA, 38 (2007), 161-174. U. J. J. Leverrier made the suggestion of a planet in 1855 even before the observation was made; see Holberg (2009) below.

[62] Walter S. Adams, "An A-type Star of Very Low Luminosity," PASP, 26 (1914), 198, and "The Spectrum of the Companion of Sirius," PASP, 27 (1915), 236-237, both in Bartusiak, 329-330.

[63] A. van Maanen, "Two Faint Stars with Large Proper Motions," *Publications of the Astronomical Society of the Pacific,* 29, (1917), pp. 258–259.

[64] J. B. Holberg, "How Degenerate Stars Came to be Known as White Dwarfs," *Bulletin of the American Astronomical Society* 37 (2005), p. 150; and J. B. Holberg, "The Discovery of the Existence of White Dwarf Stars: 1860 to 1930," JHA 30 (2009), 137-154.

The true nature of white dwarfs dawned only slowly. Eddington captured their incredible nature when he wrote that "The message of the Companion of Sirius, when decoded, ran: 'I am composed of material 3,000 times denser than anything you have come across; a ton of my material would be a little nugget you could put in a matchbox.' What reply can one make to such a message? The reply which most of us made in 1914 was—'Shut up. Don't talk nonsense.'"[65]

How to explain this in terms of physical theory? The astronomer J.B. Holberg has detailed the long road to the discovery of white dwarfs, including the interplay between theory and observation, and the understanding of their remarkable nature. In 1926, the British mathematician Ralph Fowler applied the new quantum mechanics to white dwarfs and concluded that their electrons and nuclei were packed into the smallest volume possible, resisted only by the degeneracy pressure produced by electrons, as required by Fermi-Dirac statistics.[66] Five years later, reasoning that white dwarf electrons would be moving near the speed of light, Subrahmanyan Chandrasekhar applied relativistic degeneracy to show that the compaction of white dwarfs could not continue forever. He calculated the upper mass of a white dwarf, now known as the Chandrasekhar limit, to be slightly less than one solar mass; it is now known to be 1.4 solar masses. Beyond that, it would collapse.[67]

Not everyone agreed, least of all the great British theorist Arthur Eddington. In his book *Empire of the Stars*, Arthur Miller details the heated controversy between Eddington and Chandrasekhar regarding white dwarfs and the fate of more massive stars. Chandra embraced relativistic degeneracy as the basis of his work, while Eddington believed it was a mathematical result with no astrophysical meaning. It was Chandra who began

---

[65] A. S. Eddington, *Stars and Atoms* (Oxford: Clarendon Press, 1927), p. 50.

[66] Ralph Fowler, "On Dense Matter," MNRAS, 87 (Dec, 1926), 114-122, in Bartusiak, 331-333.

[67] S. Chandrasekhar, "The Maximum Mass of Ideal White Dwarfs," ApJ, 74 (1931), 81-82. Edmund Stoner had come to a similar conclusion several years earlier; see Michael Nauenberg, "Edmund C. Stoner and the discovery of the maximum mass of white dwarfs," JHA, 39 (2008), 297-312.

to glimpse the truth about white dwarfs and more massive stars as endpoints of stellar evolution when he wrote in 1935, "for a star of small mass the natural white dwarf stage is an initial step towards complete extinction. A star of large mass [greater than the upper limit for white dwarfs] cannot pass into the white dwarf stage, and one is left speculating on other possibilities."[68]

White dwarfs have proven increasingly important to astronomical research over the last few decades. As recounted in entry S 11, white dwarf supernovae (Type Ia) were used in the late 1990s to show that the expansion of the universe is accelerating, giving rise to the concept of a repulsive force known as dark energy, a revival of Einstein's idea of a cosmological constant. Faint white dwarfs are also one of the candidates for baryonic dark matter known to exist from the anomalous rotation rates of galaxies. And because of their strong gravitational attraction, white dwarfs are known to shred planets surrounding them, allowing astronomers to determine the composition of those planets before their destruction, one of the few ways to determine the internal composition of exoplanets. Along with the iron-rich elements, water is also detected.

The latest news on white dwarfs from the Hubble Space Telescope is at http://hubblesite.org/news/90-white-dwarfs. An X-ray image of Sirius B from the Chandra Observatory is at http://chandra.harvard.edu/photo/2000/0065/.

---

[68] J. B. Holberg, "The Discovery of the Existence of White Dwarf Stars: 1860 to 1930," JHA, 30 (2009), 137-154; Arthur Miller, *Empire of the Stars* (Houghton-Mifflin: Boston, 2005), especially pp. 102, 132-133.

Class S 13: Neutron Star

Like white dwarfs (S 12) and stellar black holes (S 14), neutron stars are one of the endpoints of a star's life. A neutron star results when a stellar core between about 1.4 and 2 or 3 solar masses, having consumed its nuclear fuel at the end of its life, explodes as a supernova (S 11), leaving behind a dense core squeezed so tightly it consists only of neutrons. The neutrons form when an electron and a proton combine under tremendous pressure to form a neutron and a neutrino. The neutrons are held together by gravity, and only neutron degeneracy pressure keeps the object from collapsing altogether, in the same way electron degeneracy pressure prevents a white dwarf from collapsing. A neutron star is estimated to be only about 10 to 15 miles in diameter, making it so dense that one teaspoon of its matter would weigh about two billion tons.

Neutron stars are normally detected as pulsars, rapidly rotating neutron stars with extremely precise periods ranging from a few milliseconds to several tens of seconds. Pulsars emit intense bipolar radiation along their magnetic field axis, giving the effect of a rotating lighthouse beam; they may radiate in this way for as much as 20 million years. Because some neutrons stars are oriented such that their lighthouse beams do not sweep past Earth, all pulsars are neutron stars, but not all neutron stars are pulsars, at least from our point of view. But from a more universal point of view, virtually all neutron stars are probably pulsars. More than 2,600 pulsars have been detected, but they are far rarer than the neutron stars we cannot see; astronomers estimate that in excess of 100 million neutron stars may exist in our own Milky Way Galaxy. Astronomers using the Chandra X-ray telescope have also discovered neutron stars in globular clusters. Although first detected in the radio region, some pulsars have also been detected by optical, X-ray, and gamma-ray telescopes. They have proven very important to astronomy, for reasons ranging from precise timekeeping to stringent tests of Einstein's general theory of relativity.

Neutron stars were first theorized by Walter Baade and Fritz Zwicky in 1934 as objects formed from supernovae. Jocelyn Bell (later Burnell) and Anthony Hewish first discovered pulsars, which turned out to be the theorized neutron stars, in 1967 at the Mullard Radio Astronomy Observatory in Cambridge, England. It was a serendipitous discovery in the sense that Hewish was actually looking for quasars (G 8) using the technique of interplanetary scintillation, a twinkling effect due to diffraction of radio waves when they pass through the solar wind (P 16). And it was an increasingly controversial discovery, because Hewish received the 1974 Nobel Prize in physics for finding pulsars, even though Bell was largely responsible for the discovery as his graduate student. This was somewhat remedied only in 2018 when Bell-Burnell received the $3 million Breakthrough Prize. She donated the money to scholarships for women and minorities to study physics.[69]

In any case, the extremely precise periods of pulsars were very unusual, so unusual that they led to the "Little Green Men" hypothesis. As Hewish recalled, "the short duration of each pulse suggested that the radiator could not be larger than a small planet. We had to face the possibility that the signals were, indeed, generated on a planet circling some distant star, and that they were artificial. I knew that timing measurements, if continued for a few weeks, would reveal any orbital motion of the source as a Doppler shift, and I felt compelled to maintain a curtain of silence until this result was known with some certainty. Without doubt, those weeks in December 1967 were the most exciting in my life."[70] The hypothesis was rejected when Doppler shifts in the signal showed only the orbital motion of the Earth, not of a planet with extraterrestrials. The discovery of similar signals coming from Cassiopeia A also mitigated this possibility, since two civilizations would not likely be signaling at the same frequency. In February 1968 the data was reported in *Nature*, where the authors speculated that the signals could be due to radial pulsations of white dwarfs or neutron stars.[71]

---

[69] Walter Baade and Fritz Zwicky, "Remarks on Super-Novae and Cosmic Rays," *Phys. Rev.* 46 (1934): 76–77; Anthony Hewish, "Pulsars and High Density Physics," Nobel Lecture, December 12, 1974, online at https://www. nobelprize.org/prizes/physics/1974/hewish/lecture/. A more extensive first-person account is Jocelyn Bell Burnell, "The Discovery of Pulsars," in *Serendipitous Discoveries in Radio Astronomy*, Ken Kellermann and B. Sheets, eds. (NRAO, 1983), 160-170. For more details of the discovery see Dick (2013), pp. 107-109, from which parts of this account are taken. And for a recent popular review of neutron stars see Francis Reddy, "Fantastic Stars: A Field Guide," *Astronomy*, 45 (2017), 26-32.

[70] Hewish, 1974 Nobel Lecture, p. 178.

[71] Anthony Hewish, S. Jocelyn Bell, John D. H. Pilkington, Paul Frederick Scott, Robin Ashley Collins, "Observation of a Rapidly Pulsating Radio Source," *Nature*, 217 (February 24, 1968), 709-713; reprinted in Bartusiak, 515-518.

Theorist Thomas Gold at Cornell quickly developed the model that explained pulsars as the rotating neutron stars that had been predicted by Baade and Zwicky; other papers in *Nature* immediately following Gold's discussed vibrating white dwarfs or a satellite orbiting a neutron star as the explanation. Gold had predicted the pulse period should increase with time, and as Hewish put it, this prediction "soon received dramatic confirmation with the discovery of the pulsar in the Crab Nebula. Further impressive support for the neutron star hypothesis was the detection of pulsed light from the star which had previously been identified as the remnant of the original explosion. This, according to theories of stellar evolution, is precisely where a young neutron star should be created. Gold also showed that the loss of rotational energy, calculated from the increase of period for a neutron star model, was exactly that needed to power the observed synchrotron light from the nebula." By 1969, astronomers at the Steward Observatory in Tucson, Arizona made the first optical identification of a pulsar—the central star of the Crab Nebula. It has since been imaged many times at multiple wavelengths, even showing the motion of the surrounding nebulosity (Fig. 8.16)[72]

Within a year after the first discovery, 27 more radio pulsars were found. But pulsars had more surprises in store than just their number. In 1982 Don Backer, Shri Kulkarni and colleagues reported a pulsar with the astoundingly fast rotation period of 1.558 seconds, 21 times faster than the previously known record, the Crab pulsar. That rotation period corresponds to 640 hertz, a high-pitch "concert A" tone. The current record is held by a pair of pulsars that rotate more than 700 times per second, including one in the Terzan 5 globular cluster in our own Milky Way galaxy.

These so-called "millisecond pulsars" were soon explained by the phenomenon of "pulsar recycling," in which a red giant star transfers mass to the pulsar, spinning it up to enormous rates, leaving the remnant white dwarf in orbit with the pulsar.[73] Despite the

---

[72] Thomas Gold, "Rotating Neutron Stars as the Origin of the Pulsating Radio Sources," *Nature*, 218 (1968), 731-732; reprinted in Bartusiak, 518-521; Hewish Nobel Lecture, op. cit., pp. 178-179; David H. Staelin and Edward C. Reifenstein III, "Pulsating Radio Sources near the Crab Nebula," *Science*, 162 (1968), 1481-1483; W. J. Cocke , M. J. Disney, and D. J. Taylor, "Discovery of Optical Signals from Pulsar NP 0532," *Nature*, 221 (Feb 8, 1969), 525-527.

[73] D. C. Backer, S R. Kulkarni et al., "A millisecond pulsar," *Nature*, 300 (1982), 615-618; M. A. Alpar et al., "A new class of radio pulsars," *Nature*, 300 (1982), 728.

Fig. 8.16. Composite image of the Crab Nebula in X-ray (blue) and visible (red) light, from the Chandra and Hubble Space Telescopes. Matter and anti-matter are being propelled near the speed of light by the Crab pulsar, a rapidly rotating neutron star the size of Manhattan. A detailed description of this complex environment is given in the press release at http://hubblesite.org/news_release/news/2002-24. Credit X-ray image: NASA/CXC/ASU/J. Hester et al.; optical image: NASA/HST/ASU/J. Hester et al. Compare with composite image released in 2018 at http://chandra.harvard.edu/photo/2018/crab/

extreme difficulty of detecting them, millisecond pulsars are being discovered at an increasing pace. From 2000 to 2010 their numbers quadrupled, so that as of 2010 about 250 were known, 140 in globular clusters. Terzan 5 is now known to contain at least 37 of them, more than any other globular. All millisecond pulsars are gamma-ray emitters (38 were discovered in the Fermi spacecraft's first four years of operation), and, through ingenious timing techniques, they held the promise of direct detection of gravitational wave radiation. That promise was achieved in 2017.

Yet another surprise was the discovery of a rare type of neutron star now known as a magnetar—a neutron star with an extremely strong magnetic field 100 times that of normal radio pulsars, a thousand trillion times stronger than the Earth's. The story of magnetars began when objects emitting brief intense outbursts of low-energy gamma rays, known as soft gamma-ray repeaters, were first discovered in 1979. Robert Duncan and Christopher Thompson published a theory in 1992 interpreting these soft gamma-ray repeaters as young magnetars, formed when conditions for dynamo action are met during the first few seconds after gravitational collapse, but quickly losing their rotational energy via magnetic braking. In 1998 Chryssa Kouvelioutou and colleagues reported observations with the Rossi X-ray Timing Observatory of an X-ray pulsar with characteristics that "demonstrate the existence of magnetars."[74] Magnetars have magnetic field strength of about 800 trillion Gauss, the most intense known magnetic fields in the universe. By comparison, Earth's magnetic field is a mere 0.6 Gauss at the poles, and the best we can sustain in laboratories on the ground is 1 million Gauss. Normal radio pulsars reach about 1 trillion to five trillion Gauss, strong but still short of a magnetar. Magnetars may represent a violent stellar afterlife, lasting about 10,000 years. As of 2017, 29 magnetars were known in the Milky Way Galaxy and one in the Small Magellanic Cloud.[75]

Pulsar astronomy today is a robust area of research. About 2,000 were known as of 2016. We now know that pulsars and magnetars give rise to a variety of even more exotic phenomena. In 1974, Russell Hulse and Joseph Taylor discovered the first binary neutron star system, PSR 1913 + 16 (PSR stands for pulsar, and 1913 + 16 specifies the pulsar's position in the sky).[76] Neutron stars in binary systems (S 31) are important for testing one of the last predictions of the general theory of relativity, and for their work in this area, Hulse and Taylor received the Nobel Prize in physics for 1993.

That was not the end of the binary neutron star story. On August 17, 2017, an even more amazing discovery was made: the detection of gravitational waves from the merger of two neutron stars. Only the second time gravity waves were observed (The first detection was only 27 months earlier when binary black holes spiraled together (S 14)). The detection of the neutron star collision was made with the Laser Interferometer Gravitational-Wave Observatory (LIGO). The binary system was located some 130 mil-

---

[74] Robert Duncan and Christopher Thompson, "Formation of very strongly magnetized neutron stars - Implications for gamma-ray bursts, ApJ Letters, 392 (1992), L9-L13; Chryssa Kouveliotou, "An X-ray pulsar with a super-strong magnetic field in the soft γ-ray repeater SGR1806 – 20," *Nature*, 393 (1998), 235-237.

[75] A magnetar catalog is at "McGill Online Magnetar Catalog," http://www.physics.mcgill.ca/~pulsar/magnetar/main.html

[76] Russell A. Hulse and Joseph H. Taylor, "Discovery of a Pulsar in a Binary System," ApJ, 195 (Jan 15, 1975), pp :51-:53; reprinted in Bartusiak, 548-551. See Hulse's 1993 Nobel lecture at http://nobelprize.org/nobel_prizes/physics/laureates/1993/hulse-lecture.html and Taylor's at http://nobelprize.org/nobel_prizes/physics/laureates/1993/taylor-lecture.html

lion light years away. The event was observed at 70 observatories over the spectrum ranging from gamma-ray detectors to radio telescopes. It was this detection of gravitational waves that fulfilled one of the predictions of Einstein's general theory of relativity.[77]

When they are not colliding, neutron stars can still be spectacular. For example, some neutron stars are X-ray binaries. A low-mass X-ray binary has a companion with a star less than one solar mass, from which the neutron star pulls mass into an accretion disk, the analogue to white dwarf novae. A high-mass X-ray binary has a neutron star paired with a star greater than five solar masses. A dozen spacecraft detected short gamma-ray bursts (lasting .2 second) on December 27, 2004 from magnetar SGR 1806-20. Six spacecraft picked up bursts from the direction of M81 on November 3, 2005. The Rossi X-Ray Timing Explorer Satellite, which operated from 1996 to 2012, was one important observer of such phenomena. Much research today concentrates on millisecond pulsars, which are notoriously difficult to detect and comprise about 1% of all pulsars.

Despite the exotic nature of neutron stars—about as far as one can imagine from a star like our Sun—in 1980 physicist Robert Forward published a famous novel, *Dragon's Egg*, about life on a neutron star. The star was inhabited by sesame-seed-sized intelligent cheela, who live out their lives with 67 billion times the surface gravity of Earth. Their biology utilizes nuclear rather than chemical reactions, so they think and develop a million times faster; one minute for humans is many years for a cheela. Their civilization therefore develops very rapidly as the humans watch with astonishment. While this may seem far-fetched, in one of the great surprises of pulsar research, the first pulsar planets (P 5) were detected in 1992.

The latest news on neutron stars from Hubble observations is at http://hubblesite.org/news/70-neutron-stars. A 2002 time lapse video of the Crab Nebula over several months is at http://hubblesite.org/videos/news/release/2002-24 and http://hubblesite.org/news_release/news/2002-24/79-pulsars. A 2016 video and other images are at http://hubblesite.org/news_release/news/2016-37/79-pulsars.

---

[77] Adrian Cho, "Cosmic Convergence," *Science*, 358 (2017), 1520-1521. The role of many observatories in this detection is described at HST Release, October 16, 2017, "NASA Missions Catch First Light From a Gravitational-Wave Event" http://hubblesite.org/news_release/news/2017-41/4-galaxies

Class S 14: Black Hole

A black hole is an infinitely dense region of space with a gravitational field so intense that even light cannot escape. Active Galactic Nuclei (G 6—G 9), those galactic centers such as quasars that emit fantastic quantities of radiation, are powered by black holes. But once galaxies become quiescent, black holes are revealed only by the glow of the surrounding gases heated by the black hole's gravitational energy, or by the motions of nearby stars or gas clouds as they swirl around it at high velocities. Astronomers have indirectly observed or inferred three Types of black hole based on their masses: those that are supermassive and located at the centers of galaxies (millions to billions of times the mass of the Sun), those of intermediate mass (a few thousand solar masses, found at the centers of globular clusters, S 35), and those of stellar mass (up to tens of times more massive than the Sun, found, for example, in X-ray binaries, S 31). The merger of two black holes in a binary system gave rise to one of the great discoveries of 21st century astronomy: the detection of gravitational waves.

The matter of a black hole is compressed into an infinitely small volume known as the singularity. The so-called "event horizon," the distance at which it is impossible to escape the pull of the black hole, depends on the mass of the black hole, and may vary from a few miles to the size of the Solar System. Aside from mass, black holes are characterized by only two other properties: spin and charge.

Stellar black holes are the final stage in the evolution of a stellar core greater than about three solar masses. Thus they are classified here in the Subfamily of stellar evolutionary endpoints, along with supernovae (S 11), white dwarfs (S12), and neutron stars (S13). The collapse of a massive star may often be observed as a supernova explosion emitting gamma rays before a neutron star or black hole is formed. It is important to distinguish the collapsing stellar remnant from the original mass of the star before the supernova ejected large amounts of material; a three-solar mass remnant core may have originated as a 20 solar mass star. One might argue that supermassive black holes should be classified under galaxies, but their origin remains unclear and may also be related to stars.

The idea of black holes has a long history dating back to the 18th century British natural philosopher John Michell, the

French mathematician Pierre-Simon de Laplace, and the German astronomer Karl Schwarzschild.[78] Schwarzschild described the gravitational field of a point mass in mathematical terms in 1916, the year of his death at the young age of 42 from a painful disease contracted on the Russian front during World War II. Schwarzschild discovered the solutions of the equations of general relativity that describe non-rotating black holes. In 1939 J. Robert Oppenheimer, soon to be famous for his work on the Manhattan Project, gave the first modern description of a black hole with his graduate student, Hartland Snyder. With the discovery of quasars and other active galaxies in the early 1960s, black holes became one explanation among many for their observed energies but remained unconfirmed observationally. John Archibald Wheeler undertook further research on such "gravitationally collapsed objects" and coined the term "black hole" in 1967 during a talk at NASA's Goddard Institute for Space Studies.[79]

It is not easy to identify the discoverer of the first black hole. As early as 1971, the X-ray binary Cygnus X-1 was suspected to harbor a stellar mass black hole after measurements of the orbiting blue giant or supergiant star (S 9) indicated an object of at least ten solar masses. By 1974, after a massive effort, one astronomer put the odds at 80% that Cyg X-1 harbored a black hole. But there was no unequivocal black hole signature, and physicist Stephen Hawking bet physicist Kip Thorne that it was not a black hole. Hawking conceded to Thorne only in 1990.[80] Cyg X-1 is now known to be at a distance of about 6,000 light years, making the black hole's mass about 14.8 solar masses, placing it firmly in the stellar Type black hole mass. The mass of the blue giant companion is about 19 solar masses. The black hole is believed to spin at 800 revolutions per second, orbiting its companion in 5.6 days.

---

[78] Simon Schaffer, "John Michell and black holes", *JHA*, 10, 42-43 (1979). See Dick (2013), pp. 148-154.

[79] John Archibald Wheeler, *Geons, Black Holes, and Quantum Foam: A Life in Physics* (New York: W W Norton, 1998), p. 296

[80] Kip Thorne, *Black Holes & Time Warps* (New York: W W Norton, 1994), pp. 314-321: 317. Thorne's confidence estimate 20 years later had risen to 95%. As he pointed out there was no unequivocal signal of a black hole, unlike neutron star pulsars with their regular pulses.

In the realm of the galaxies, in the early 1990s the Hubble Space Telescope found circumstantial evidence for a supermassive black hole in the giant elliptical galaxy M87 (Fig. 15.2) 50 million light years distant, as well as in other galaxies. These were revealed because of an increase in the number of stars around galaxy centers, possibly due to the gravitational pull of black holes. With the installation of the corrective optics known as COSTAR during its first servicing mission in 1993, astronomers using Hubble were able to measure spectroscopically the motion of the gas around the putative black hole using the Faint Object Spectrograph and thereby determine its mass. By 1994, Hubble astronomers were able to claim "seemingly conclusive evidence" for an astonishing black hole of about at three billion solar masses at center of M87, based on velocity measurements of orbiting gas.[81] Later in 1994, radio astronomers confirmed a 40-million solar mass black hole in the spiral galaxy NGC 4258, and numerous other supermassive black holes have been "observed" in this way since then.

By 1997, Hubble astronomers were suggesting that most galaxies contain black holes, some in more quiescent form. In 2008 astronomers confirmed a supermassive black hole of four million solar masses in the center of our own Milky Way Galaxy; at a distance of 28,900 light years, it is the nearest supermassive black hole. Surprisingly, more than a million stars exist within four light years of this black hole, known as Sgr A*, and they are still forming; in 2017 the Alma Large Millimeter Array (ALMA) detected 11 low-mass protostars (S 1) forming within three light years of this violent environment, and numerous stars are known to exist even closer.

It is now clear that large black holes formed even in the early universe. The record for the farthest supermassive black hole, announced in 2017, is associated with quasar J1342+0928, 13.1 billion light years away, formed only 690 million light years after the Big Bang. Its mass is 800 million solar masses. Because most galaxies are believed to harbor a supermassive black hole, and because 100 billion galaxies exist that can be seen from Earth, 100 billion supermassive black holes may exist in the observable universe.

---

[81] HST Release, May 25, 1994, "Hubble Confirms Existence of Massive Black Hole at Heart of Active Galaxy," http://hubblesite.org/newscenter/archive/releases/exotic/black-hole/1994/23/text/

Intermediate black holes long remained the most elusive. Although in 2000 the ROSAT and Chandra X-ray Observatories identified extremely bright X-ray sources that could be interpreted as intermediate black holes in starburst galaxies such as M82, the Hubble Space Telescope seems to have clinched the discovery in 2002 with the announcement of intermediate black holes in the globular cluster M15 at a distance of 32,000 light years in Pegasus, and in the giant globular cluster G1, located at a distance of 2.2 million light years away in the neighboring Andromeda Galaxy.[82] Those black holes were 4,000 and 20,000 solar masses, respectively. In both cases, the black hole masses were identified by the speed of stars orbiting them. This announcement culminated a 30-year search for intermediate black holes. Since then, astronomers using galaxy spectra from the Sloan Digital Sky Survey (SDSS) have reported ten candidates for middleweight black holes in other galaxies.[83]

As for stellar black holes, it has been estimated that one out of every thousand stars in our galaxy is massive enough to end its life as a black hole, in which case the Milky Way galaxy may contain 100 million objects of this class. Because of difficulties observing them, only a few have been identified, the nearest at a distance of 1,600 light years. Just as supermassive black holes power active galactic nuclei, stellar-mass black holes power the most luminous X-ray binaries. At least one such black hole has been identified in a stellar-mass binary system in the galaxy M83 known as MQ1. MQ1 is known as a microquasar because it is a black hole binary system that spews large amounts of matter and energy into the surrounding interstellar gas, in the form of stellar jets (S 20). The prototype of this kind of system is SS433, located in our galaxy about 18,000 light years from Earth.[84]

---

[82] Roeland P. van der Marel, Joris Gerssen et al., "Hubble Space Telescope Evidence for an Intermediate-Mass Black Hole in The Globular Cluster M15, Parts I and II, AJ, 124: 3255–3269 and 3270–3288; NASA Press Release, September 17, 2002, "Hubble Discovers Black Holes in Unexpected Places," http://hubblesite.org/newscenter/archive/releases/2002/18/text/

[83] Daniel Clery, "Middleweight Black Holes Found at Last," *Science* 360 (June 2018), p. 1057.

[84] Andrew King, "Testing the Limits of Accretion," *Science*, 343 (2014), 1318-1319.

In 2010, astronomers announced that they had been following the birth of a likely stellar black hole, formed by the implosion of supernova 1979C in the Galaxy M100 some 50 million light years distant in the Virgo Cluster. That explosion was initially observed by amateur astronomer Gus Johnson, and was followed by astronomers using the Chandra, XMM, Swift and ROSAT X-Ray spacecraft, allowing them to witness the birth of a black hole during its first 30 years. The newfound black hole is about five times the mass of the Sun and likely originated from a 20-solar mass object (Fig. 8.17). Much closer to home, in 2018 astronomers detected a dozen potential stellar-mass black holes within three light years of the supermassive black hole at the center of our Milky Way Galaxy. They conjectured that up to 20,000 stellar black holes could exist around our galactic center.[85]

Fig. 8.17. Supernova 1979C in the galaxy M100 gave birth to a stellar black hole. Credit: X-ray: NASA/CXC/SAO/D.Patnaude et al, Optical: ESO/VLT, Infrared: NASA/JPL/Caltech

[85] Chandra Release, November 15, 2010, "SN 1970C: NASA's Chandra Finds Youngest Nearby Black Hole," http://chandra.harvard.edu/photo/2010/sn1979c/; Marc Kaufman, "Scientists Witness Apparent Black Hole Birth," *Washington Post*, November 16, 2010, page A6; Charles J. Hailey et al., "A density cusp of quiescent X-ray binaries in the central parsec of the Galaxy," *Nature*, 556 (2018), 70-73.Read more at: https://phys.org/news/2018-04-tens-thousands-black-holes-milky.html#jCp

In one of the greatest discoveries in the history of science, in 2015 astronomers made the first detection of gravitational waves from the merger of two stellar black holes, each of about 30 solar masses. The binary system, known as GW150914, is located about 1.3 billion light years from Earth. The discovery was made with the LIGO collaboration, an interferometer with detectors in Hanford, Washington and Livingston, Louisiana. It was announced on February 11, 2016, and confirmed one of the main predictions of Einstein's general theory of relativity. In 2017 the Nobel Prize in Physics was awarded to Ranier Weiss, Kip Thorne, and Barry Barish for their roles in the detection.[86]

The formation of black holes remains a subject of intense research. Of the three Types of black holes, the formation of the stellar type is the best understood as the collapse of a stellar core greater than three solar masses. The formation of the other two types of black holes remains somewhat mysterious. It is known that black holes formed early and often in the universe, since quasars with black holes of a billion solar masses are found less than a billion years after the Big Bang, at redshifts (z) greater than six.

Two schools of thought theorize how these first supermassive black holes were formed: by the collapse of massive early Population III stars that formed small black holes and then grew, or by the direct collapse of a massive gas cloud. The Population III scenario starts with smaller seeds earlier in the universe and therefore provides more time for black holes to grow, whether by mergers or accretion. The direct collapse scenario has larger seeds and less growth time, mainly by accretion. Mitch Begelman and colleagues have worked out the latter scenario in some detail: first a massive gas cloud collapses, prevented from fragmenting into normal stars by turbulence and low angular momentum. This collapsed cloud then forms a supermassive star, larger than any now known. That star in turn collapses to form a small black hole less than 1,000 solar masses, but which rapidly grows. The black hole then swallows the remaining "quasistar," a theorized object that resembles the structure of a red giant star but may be 100 astronomical units in diameter with the luminosity of a Seyfert

---

[86]Benjamin P. Abbott et al. (LIGO Scientific Collaboration and Virgo Collaboration) "Observation of Gravitational Waves from a Binary Black Hole Merger". *Phys. Rev. Lett.* **116** (6), 061102, preprint at arXiv:1602.03837, popular summary with imagery at https://www.ligo.caltech.edu/system/media_files/binaries/301/original/detection-science-summary.pdf

galaxy. In this scenario, a quasistar consists of a massive hydrostatic gaseous envelope from which the black hole accretes more matter. Neither the supermassive stars nor quasistars have been observed, but both might be in the future with the James Webb Space Telescope.[87] Both the Population III and the direct collapse scenarios are still in play as supermassive black hole formation theories. Many astronomers believe the recent discovery of intermediate black holes tilts in favor of the theory that a stellar black hole acts as a seed for the intermediate mass black hole, which in turn acts as a seed for the supermassive black holes found at the centers of some galaxies.

An important clue to black hole formation was the discovery that the mass of a black hole is proportional to the size of its host galaxy, in particular showing a correlation between black hole mass and galactic bulge mass (Fig. 8.18). Supermassive black holes

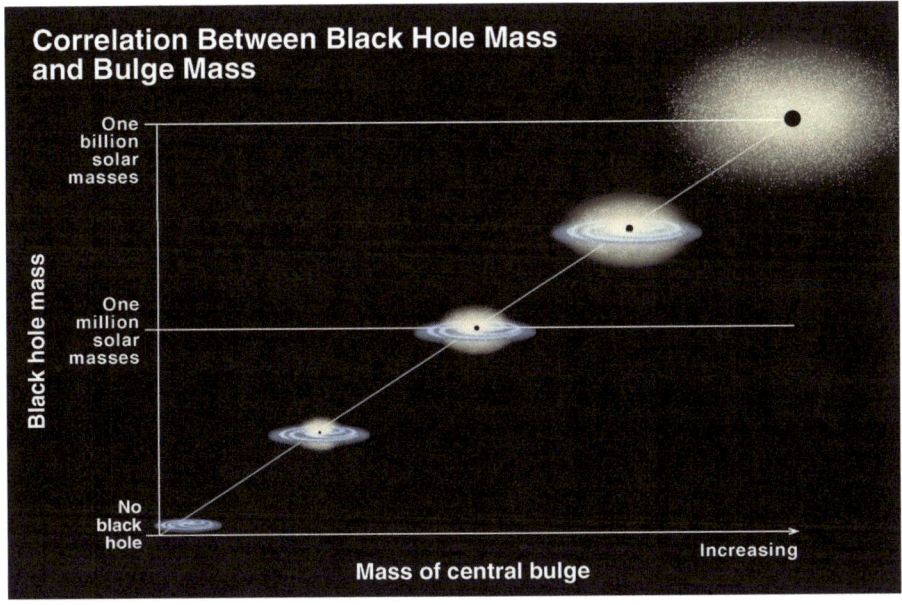

Fig. 8.18. Correlation of black hole mass with galaxy bulge mass. Credit: K. Cordes, S. Brown (STScI). Since this graph from 2000, data shows that this correlation holds for black holes in galaxies

---

[87] Mitchell Begelman, E. Rossi et al., "Quasi-stars: accreting black holes inside massive envelopes," MNRAS, 387 (2008), 1649-1659; Mitch Begelman and Martin Rees, *Gravity's Fatal Attraction: Black Holes in the Universe* (W. H. Freeman, 1998, 2nd edition, 2007).

are about one-thousandth of the mass of their host galaxy, and the same has been found true of intermediate black holes in their host globular clusters.

Whether or not a supermassive black hole should be classified as a distinct class of object in the galactic kingdom remains an open question; what is certain is that they power the Subfamily of objects known as active galactic nuclei and those that are now quiescent form the central core of many normal galaxies as well, including our Milky Way Galaxy. We should not become too complacent, however. Although a black hole may exhaust the material that feeds it and become quiescent, it may also be rejuvenated by interacting galaxies (G 20). Recent observations of the early universe indicate that galaxy mergers may produce quasars that are hidden by gas until radiation blows away the obscuring material, making them visible. In 4.5 billion years, the Andromeda Galaxy is expected to collide with ours, perhaps reactivating its central black hole.[88]

Hubble and Chandra Observatory news on black holes can be found at http://hubblesite.org/news/10-black-holes and http://chandra.harvard.edu/photo/category/blackholes.html. The Chandra claim for an intermediate black hole in M82 is at http://chandra.harvard.edu/photo/2000/m82bh/.

---

[88] Joel Primack, "Hidden Growth of Supermassive Black Holes in Galaxy Mergers," *Science* (30 April, 2010), 576-578, and pp. 600 ff.

9. The Circumstellar Family

Class S 15: Debris Disk

Debris disks are residual materials observed around stars in the form of a flattened disk after planet formation has taken place. As such, this class is distinguished from protoplanetary disks, such as proplyds that exist before planet formation (P 1), and from circumstellar shells (S 16) formed by mass loss from stars later in stellar evolution.

Because of the planetary formation process, debris disks normally lack large amounts of gas and are likely similar to the Kuiper belt (P 21) in our own Solar System. They are believed to be continually replenished through collisions between planetesimals or protoplanets.[1] Debris disks may have a long lifetime and can therefore be found around young or older stars; they tend to be thickest and most robust around younger stars a few hundred million years old. Protoplanetary disks are by definition found only around very young stars. Many debris disks, such as those around Vega, Beta Pictoris, Alpha Lyrae, Alpha Piscis Austrini, and Epsilon Eridani, were believed to be protoplanetary at the time of their discovery beginning in the 1980s but are now classified as debris disks.

The first direct evidence of debris disks came from the Infrared Astronomical Satellite (IRAS), a joint project of NASA, the Netherlands, and the United Kingdom. During its opening calibration tests in 1983, a team of astronomers centered at NASA's Jet Propulsion Laboratory serendipitously found an "infrared excess" around the star Vega, a main sequence dwarf star 2.5 times the mass of the Sun but only about 350 million years old. This infrared excess was the thermal signature of the presence of a relatively cool cloud of millimeter-sized solid particles, in this case at a distance of about 85 astronomical units from the star, twice the distance of Pluto from our Sun. At the time, the material was interpreted to be the remnant of the cloud out of which Vega had formed. The results, the authors of the discovery paper wrote, "provide the

---

[1] Mark C. Wyatt, "Evolution of Debris Disks," ARAA, 46 (2008), 339–383.

© Springer Nature Switzerland AG 2019
S. J. Dick, *Classifying the Cosmos*, Astronomers' Universe,
https://doi.org/10.1007/978-3-030-10380-4_9

first direct evidence outside of the Solar System for the growth of large particles from the residual of the prenatal cloud of gas and dust."[2] Although the scientific paper stopped short of calling the discovery a Solar System, a news release from JPL noted that the material could be a Solar System at a different stage of development than our own, an interpretation highlighted by the press.

By early 1984, IRAS astronomers had found similar "circumstellar shells" or "protoplanetary disks" around six more stars, including Beta Pictoris, Fomalhaut, and Epsilon Eridani. By summer of that year, with a total of 40 such stars observed, the shells were being reported as a widespread phenomenon. The discoverers in general were careful to emphasize that planets had not been found; instead, "the presumption is that these rings will eventually condense into Solar Systems like our own; if so, that makes the Vega phenomenon the first semi-direct evidence that planets are indeed common in the universe."[3] Though that interpretation proved erroneous, the IRAS finding of debris disks, rather than protoplanetary disks, constituted the discovery of a new class of objects. IRAS found that in general, 15% of main sequence stars it observed in the solar neighborhood had an infrared excess, and undoubtedly this phenomenon is found throughout the Galaxy.

The environments of some of the IRAS stars have since been observed in much more detail and illustrate how complex and multifaceted the process of discovery really is. Following up on the IRAS observations, in 1984 astronomers Brad Smith and Richard Terrile used new ground-based CCD technology to image the first debris disk around Beta Pictoris, producing what is now a famous image in the history of 20th century astronomy. In a statement that highlights the difficulty of separating protoplanetary disks from debris disks at the time, the authors wrote, "Because the circumstellar material is in the form of a highly flattened disk rather than a spherical shell, it is presumed to be associated with planet formation. It seems likely that the system is relatively young and that planet formation either is occurring now around Beta Pictoris or has recently been completed."[4]

---

[2]H. H. Aumann, F. C. Gillett et al., "Discovery of a Shell around Alpha Lyrae," ApJ Letters, 278 (1984), L23-L27: L23; front page of *Washington Post* for August 10, 1983, "Satellite Discovers Possible Second Solar System."

[3]"Protoplanetary Systems," *Science*, 225 (July 6, 1984), 39; "Infrared Evidence for Protoplanetary Rings around Seven Stars," *Physics Today* (May, 1984), 17–20.

[4]Smith, B. A. and Terrile, R. J., "A Circumstellar Disk around Beta Pictoris," Science, 226 (1984), 1421–1424.

Beta Pic, as it is known to astronomers, is now known to be a 3.8 magnitude spectral type A main sequence star 63 light years distant and only 10 million years old. Since the first image, evidence has been published indicating planetesimal belts, cometary activity, and one or more planets. The Hubble Space Telescope has imaged the disk many times, as has the European Southern Observatory in Chile and other observatories. A planet was inferred already in 1997 based on an inner warp in the disk plane. Three years later, a planet was inferred from observed comet infall, possibly due to gravitational perturbations from a giant planet. Near-infrared images in 2003 revealed an object that appeared to be within the dust disk, but the observation was ambiguous because it could have been a background star. In 2008, A.M. Lagrange and colleagues used adaptive optics with the ESO's Very Large Telescope in Chile to image a giant planet eight to 15 astronomical units from its star, the closest extrasolar planet to its parent star yet imaged. They confirmed the discovery in 2009, but only in 2018 did precise data from the Hipparcos and Gaia missions peg the planet's mass at nine to 13 times that of Jupiter. Following conventional notation, it is called Beta Pictoris b.[5]

In 2003 the Spitzer Space Telescope was launched, a successor to IRAS capable of imaging the infrared spectrum from three to 200 microns. It promptly imaged debris disks around a number of stars. By 2004, Spitzer had imaged six stars the same age and size as the Sun, and found the disk is ten to 100 times thinner than around younger stars and therefore much more difficult to observe. In 2005 it imaged the disk of Vega, the prototype debris disk observed by IRAS, which in addition to being relatively young and massive is only 25 light years from Earth and thus offers excellent opportunities for detailed observation. The authors of that discovery paper concluded that a ring 86 to 200 AU from the star may well be analogous to the Kuiper belt, but that much of the debris

---

[5] D. Mouillet et al, "A planet on an inclined orbit as an explanation of the warp in the Beta Pictoris disc," MNRAS, 292 (1997), 896–904; H. Beust and A. Morbidelli, "Falling Evaporating Bodies as a Clue to Outline the Structure of the β Pictoris Young Planetary System," Icarus, 143 (2000), 170–188; A.M. Lagrange et al, "A Giant Planet Imaged in the Disk of the Young Star Beta Pictoris," Science, 329 (2010), 57–59.

disk further from Vega out to 800 AU "is ephemeral, the conse-
quence of a large and relatively recent collisional event" from one
or more of the asteroidal bodies in the ring.[6] Unconfirmed claims
have also been made for a planet around Vega. As it neared the end
of its mission, Spitzer had detected 25 debris disk systems around
Sun-like stars over a range of ages, as well as debris disks around
brown dwarfs, pulsars and hypergiant stars, and binaries.[7]

In 2006, Paul Kalas and his colleagues released Hubble Space
Telescope images of two debris disks around stars about 60 light
years from Earth, bringing to nine the number imaged at visible
wavelengths.[8] An even more spectacular debris disk surrounds the
star Fomalhaut, like Vega about 25 light years from Earth and relatively young at about 200 million years. In 2003, more than two
decades after IRAS observations of an infrared excess around Vega,
Karl Staplefeldt and colleagues obtained the first infrared images
of the debris disk with the Spitzer Space Telescope. Kalas and colleagues observed the debris disk in the optical region with Hubble
in 2004, clearly showing a structure about 21.5 billion miles across
with a sharp inner edge. Kalas and others proposed that the ring
was being gravitationally affected by a planet lying between the
star and the ring's inner edge, and in 2008 the team announced a
planet lying 1.8 billion miles inside the ring's inner edge—the first
such disk to have a planet imaged at visible wavelengths (Fig. 9.1).[9]
The planet, dubbed Fomalhaut b, lies some 119 AU (10.7 billion

---

[6]K. Y. L. Su, G. H. Rieke et al, "The Vega Debris Disk: A Surprise from
Spitzer," ApJ, 628 (2005), 487–500.

[7]Lynne A. Hillenbrand et al, "The Complete Census of 70-micron Bright
Debris Disks Within 'The Formation and Evolution of Planetary Systems'
Spitzer Legacy Survey of Sun-Like Stars," ApJ, 677 (2008), 630–656.

[8]HST Release, January 19, 2006, "Dusty Planetary Disks around two Nearby
Stars Resemble Our Kuiper Belt," http://hubblesite.org/newscenter/archive/
releases/2006/05/image/a/; Robert Sanders, UC Berkeley Release, January
19, 2006, "Two new dusty planetary disks may be astrophysical mirrors of
ourKuiperBelt,"http://www.berkeley.edu/news/media/releases/2006/01/19_
kuiper.shtml

[9]Spitzer Release, December 18, 2003, "Fomalhaut Circumstellar Disk,"
http://www.spitzer.caltech.edu/images/1100-ssc2003-06i%20-Fomalhaut-
Circumstellar-Disk; Paul Kalas et al., "Optical Images of an Exosolar Planet
25 Light-Years from Earth," *Science*, 322 (2008), 1345–1348; E. Chiang et al.,
"Fomalhaut's Debris Disk and Planet: Constraining the Mass of Fomalhaut b
from disk Morphology," ApJ, 693 (2009), 734 ; HST Release November 13,
2008, "Hubble Directly Observes Planet Orbiting Fomalhaut," http://hub-
blesite.org/newscenter/archive/releases/2008/39/full/

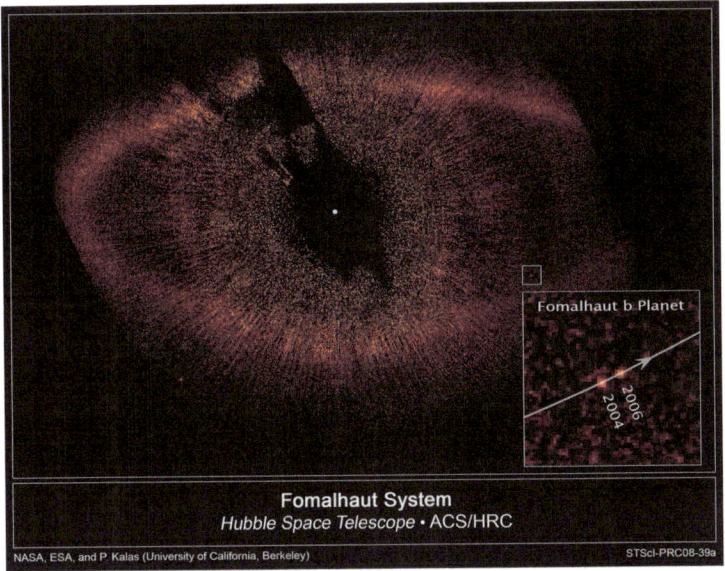

Fomalhaut System
Hubble Space Telescope · ACS/HRC

NASA, ESA, and P. Kalas (University of California, Berkeley)   STScI-PRC08-39a

Fig. 9.1. The Fomalhaut debris disk and the embedded planet Fomalhaut b, showing its motion from 2004 to 2006. The planet orbits Fomalhaut in 872 years. Credit: NASA, ESA, P. Kalas, J. Graham, E. Chiang, E. Kite (University of California, Berkeley), M. Clampin (NASA Goddard Space Flight Center), M. Fitzgerald (Lawrence Livermore National Laboratory), and K. Stapelfeldt and J. Krist (NASA Jet Propulsion Laboratory)

miles) from its star, ten times the distance of Saturn from the Sun. It is estimated to have a mass no more than three times that of Jupiter.

In 2017, a team of astronomers using the Atacama Large Millimeter/submillimeter Array (ALMA) detected a dusty debris disk around Proxima Centauri, the nearest star to Earth. The debris disk lies outside the orbit of the planet Proxima Centauri b circling at .05 astronomical units (compared to Mercury's .39 AU), and extends from one to four astronomical units from the star.[10] There are also hints of another dusty colder ring at 30 AU. Hundreds of stars are now known to have the infrared excess associated with debris disks. Based on the small sample observed

---

[10] Guillem Anglada et al., "ALMA Discovery of Dust Belts around Proxima Centauri," ApJ, 850, 5 pp., preprint at https://arxiv.org/abs/1711.00578

thus far, debris disks seem to come in two Types: narrow ones with a width between 20 and 30 astronomical units and broader ones wider than 50 AU. The Kuiper belt of our own Solar System extends from about 30 to 50 AU. The Fomalhaut debris disk is obviously of the second type, and Proxima Centauri's hot inner ring may be "the dust component of a small-scale analog to our Solar System Kuiper belt." It might also be analogous to our aster-oid belt or zodiacal dust. Debris disks are thus important not only in their own right, but also because they allow us to observe from the outside what we can only observe from the inside for our own Solar System.

A catalogue resolved circumstellar disks, including 22 debris disks, can be found at http://circumstellardisks.org/; another use-ful list with links is at http://en.wikipedia.org/wiki/Debris_disk. Images of the Beta Pictoris disk can be seen by entering those keywords at http://hubblesite.org/images/gallery. The 2003 Spitzer image of Fomalhaut is at http://www.spitzer.caltech.edu/images/1100-ssc2003-06i%20-Fomalhaut-Circumstellar-Disk, and Hubble's first image of Fomalhaut in 2005 is at http://hub-blesite.org/newscenter/archive/releases/2005/10/image/a. A size comparison of the Fomalhaut debris disk with the Solar System is at http://hubblesite.org/image/1665/news_release/2005-10.

Class S 16: Circumstellar Shell

In addition to circumstellar proplyds (P 1) and debris disks (S 15), material may exist around a star totally unrelated to planetary formation. Whereas proplyds by definition surround young stars, and debris disks such as our own Kuiper belt (P 21) may linger well past stellar middle age, circumstellar shells (also called circumstellar envelopes) are related to mass loss late in stellar evolution, via stellar winds (S 29). One mechanism occurs when low and intermediate mass stars such as the Sun reach the cool red giant stage and exhaust their helium fuel, eventually ejecting their outer layers. Because they ascend the Hertzsprung-Russell diagram a second time to become cooler and more luminous, paralleling and asymptotically approaching the earlier red giant branch ascent, astronomers have given such stars the unwieldy name "asymptotic giant branch" (AGB) stars. The AGB stage lasts about a million years, and the circumstellar shells evolve to become planetary nebulae (S 17), while the star itself becomes a white dwarf (S 12). AGB stars are cool (<2000 K), very luminous (about 10,000 solar luminosities), and shed mass copiously up to 1/10,000 solar mass per year. The mass is in the form of molecules and dust grains (perhaps the major contribution to interstellar dust) and so is studied at radio, millimeter, and infrared wavelengths. Because AGB stars are an evolutionary stage between red giants and planetary nebulae, studying this mass is important for understanding stellar evolution and the chemical evolution of the interstellar medium.

Two well-studied types of AGB stars are Mira variables and the carbon stars. Mira variables, known for their dramatic changes of brightness from two to 11 magnitudes, are long-period variables, with pulsations ranging from 80 to 1,000 days. Mira (Omicron Ceti), 420 light years distant, is the prototype, discovered by David Fabricius in 1596, the first variable star to be discovered aside from a nova or supernova. It is 500 times larger than the Sun and 15,000 times more luminous, making all the more remarkable the fact that it physically expands and contracts over a period of 332 days, the reason it was named "Mira" (wonderful). Several thousand Mira variables are known, including the well-known R Leonis, discovered in 1782 by J. A. Koch in Danzig. The 5th variable star, and 4th long period variable, to be discovered, it varies from a dim 12th magnitude to naked-eye 4th magnitude over a period of 312 days. One of the best-studied circumstellar envelopes of an AGB star is that of IRC+10216, now mapped in considerable detail.[11]

Carbon stars, those in which carbon rather than oxygen predominates, are another type of AGB star with even thicker circumstellar envelopes. The pioneering spectroscopist Angelo Secchi found spectral anomalies already in the 1860s in this type of star, but even when William Bidelman wrote his classic book *The Carbon Stars—An Astrophysical Enigma* in 1956, their nature and evolutionary status remained unknown. Their real nature was not determined until the late 1970s. Of 512 red giants and planetary nebulae observed with the Infrared Astronomical Satellite in the 1980s, 76 had circumstellar shells, and 40% of these were found to be carbon stars.[12]

Despite their numerical dominance, the first stars to be observed with evolved circumstellar shells were not AGB stars, but red (M) supergiants, which also lose large amounts of mass late in their evolution. Walter S. Adams and Elizabeth MacCormack were the first to observe these evolved circumstellar shells indirectly in 1935, though their observations were not understood at the time. Using the coude spectrograph on the 100-inch reflector at Mt. Wilson, they found M supergiant spectra that showed blue-shifted resonance lines. In a classic paper in 1956 Armin Deutsch, also using high-resolution visual spectroscopy, explained similar blue-shifted lines of the M supergiant alpha Herculis, part of a multiple system, as a circumstellar shell. In the 1960s and 1970s, new tools were devised to probe circumstellar shells other than high resolution visual spectroscopy. Polarimetry and infrared photometry yielded information on dust in the shell, while microwave spectroscopy employed molecular rotational transitions to probe the gas.[13] Only by the 1970s were a reasonable number of stars examined to draw quantitative conclusions, including mass-loss rates.

---

[11] C. Leao et al, "The circumstellar envelope of IRC+10216 from milli-arcsecond to arc-minute scales, A & A, 455 (2006), 187–194.

[12] K. Young et al, "Circumstellar Shells resolved in IRAS Survey Data. II Analysis. ApJ, 409 (1993), 725–738. See also M. Rowan-Robinson et al., "Models for IRAS observations of circumstellar dust shells around late-type stars," MNRAS, 222 (1986), 273–286. William Bidelman, *The Carbon Stars – An Astrophysical Enigma* (1956); Robert D. McClure, "The Carbon and Related Stars," JRASC, 79 (1985), 277–293.

[13] A. J. Deutsch, "The Circumstellar Envelope of Alpha Herculis," ApJ, 123 (1956), 210; Kenneth Hinkle, "High-Resolution Spectroscopy of Late-Type Circumstellar Shells," PASP, 95 (1983), 550–555.

Fig. 9.2. V838 Monocerotis, in the constellation Monoceros, located about 20,000 light years from Earth at the outer edges of the Milky Way. The red supergiant star at the middle illuminates the surrounding dust shell. Credit: NASA and the Hubble Heritage Team (AURA/STScI)

Evolved circumstellar shells may exist under circumstances as yet little understood. V838 Monocerotis is an example of a star with a circumstellar shell that is still mysterious (Fig. 9.2). It is possibly a very massive supergiant star going through its helium flash, where helium ignites to burn carbon, resulting in heavy mass loss, before settling down as a Wolf-Rayet star. Or it may be a post-asymptotic giant branch star. Evolved shells remain an active area of research both from the ground and space. In addition to IRAS, the Hubble Space Telescope, Infrared Space Observatory, and Spitzer Space Telescope have observed circumstellar dust shells.[14]

---

[14] One set of Hubble Space Telescope observations is reported in Brian Ferguson and Toshiya Ueta, "Differential Proper-Motion Study of the Circumstellar Dust Shell of the Enigmatic Object HD 179821," ApJ, 711 (2010), 613. Infrared Space Observatory observations are reported in B. F. M. Waters et al., "Mineralogy of Oxygen-rich dust shells," A &A, 315 (1996), 361–364.

Class S 17: Planetary Nebula

A planetary nebula represents one of the steps on the way to the evolutionary endpoint of a low-mass star like the Sun. It occurs when the dying star, having expanded to a red giant, ejects its outer envelope. The star itself eventually becomes a white dwarf (S 12) while the expanding ejected envelope is what we see as a planetary nebula, some 500 to 1,000 times the size of the Solar System at the orbit of Pluto. The temperature of the gas is about 10,000 K, and the shell is expanding at about 20 km/second. Like H II regions (S 24), planetary nebulae are emission nebulae, but their strongest emission line arises from extremely rarefied ionized oxygen (O III) rather than hydrogen. Unlike H II regions, they are ephemeral objects, lasting only a few tens of thousands of years before dissipating into the interstellar medium, seeding it with some of the heavier elements. For this reason, they are observed relatively nearby, within a few thousand light years. Though ephemeral, planetary nebulae provide some of the most beautiful images in astronomy.

Planetary nebulae have nothing to do with planets but received their name because William Herschel thought they were reminiscent of a fuzzy planet when seen through his telescope. To add further confusion, to this day the stage between intense mass ejection by the red giant and the formation of the planetary nebula is sometimes referred to as a "protoplanetary nebula," but this also has nothing to do with planets or protoplanets.

Charles Messier discovered the first planetary nebula in 1764, now known as the Dumbbell Nebula (M27) in Vulpecula. 15 years later, Antoine Darquier discovered the Ring Nebula (M57) in Lyra, followed by Pierre Mechain's discovery of the Little Dumbbell Nebula (M76) in Perseus and the Owl Nebula (M97) in Ursa Major. Of the 103 objects in Messier's famous catalogue of nebulae and star clusters of 1781, only these four are planetary nebulae. William Herschel discovered many more with his sweeps of the heavens in the late 18th century. These objects yielded the first evidence that nebulae exist as separate classes of objects from star clusters. It was William Herschel's observation in 1790 of the central star surrounded by nebulosity in the planetary nebula now known as NGC 1514 that convinced him of the existence of true nebulosity.[15] Proof of this came on August 29, 1864 when William Huggins,

---

[15] Michael Hoskin, "William Herschel's Early Investigations of Nebulae: A Reassessment," JHA, vol. X (1979), 165–176. Herschel's discovery article "On Nebulous Stars, properly so called," appears in Hoskin, *William Herschel and the Construction of the Heavens* (London: Oldbourne, 1963), pp. 118–129.

using the newly invented spectroscope, observed the bright planetary nebula in Draco.[16] The spectrum consisted of bright lines without the continuous spectrum characteristic of stars, proving it was gaseous rather than stellar in nature.

Identifying the nature of the gas proved more difficult. Early spectroscopy showed that the strongest emission line in planetary nebulae was at a wavelength that did not correspond with any known element. Astronomers postulated the cloud was mostly composed of an element they dubbed nebulium. Unlike helium, first seen in the Sun and then discovered on Earth, nebulium was never found on Earth. Because of this in his classic 1927 textbook, Henry Norris Russell proposed such nebula were composed of a known element in unfamiliar conditions. In 1928 the physicist Ira Bowen showed that gas at extremely low density produces "forbidden lines," and that doubly ionized oxygen gives rise to the 500.7 spectral line astronomers had seen. [17]

Although as early as 1956 the Soviet astronomer Joseph Shklovskii suggested that planetary nebulae are formed from the outer layers of some red giants, the American astronomers George Abell and Peter Goldreich presented the first coherent evidence that planetary nebulae indeed evolve from red giants. In their 1966 paper, they gave due credit to Joseph Shklovskii, Robert O'Dell and others who had found pieces of the puzzle, and wrote that "In our opinion there are some rather compelling arguments that most, or more probably all, population II red giants ultimately become planetary nebulae and further that the nebulae are formed directly from giant stars. To our knowledge, all of these arguments have not been stated in the literature or at least have not been given in one place. In any case it is our experience that many investigators working in the field are not familiar with the arguments..."[18] While planetary nebulae had been observed for centuries, this may be taken as the beginning of modern studies of the subject, surpassing a mere natural history collection of objects.

Early observations showed most planetary nebulae to be spherically symmetrical, as one might expect when a star sloughs off it

---

[16] William Huggins, "On the Spectra of Some of the Nebulae," *Philosophical Transactions of the Royal Society of London*, 154 (1864), pp. 437–444.

[17] I. S. Bowen, "The origin of the nebular lines and the structure of the planetary nebulae," *ApJ*, 67 (1928), 1–15.

[18] G. O. Abell and P. Goldreich, "On the Origin of Planetary Nebulae," PASP, 78 (1966), 232–241; Abell and Goldreich cite I. S. Shklovskii, Astr. Zhurnal USSR, 33 (1956), 315, and C.R. O'Dell, "The Evolution of the Central Stars of Planetary Nebulae," ApJ, 138 (1963), 67–78.

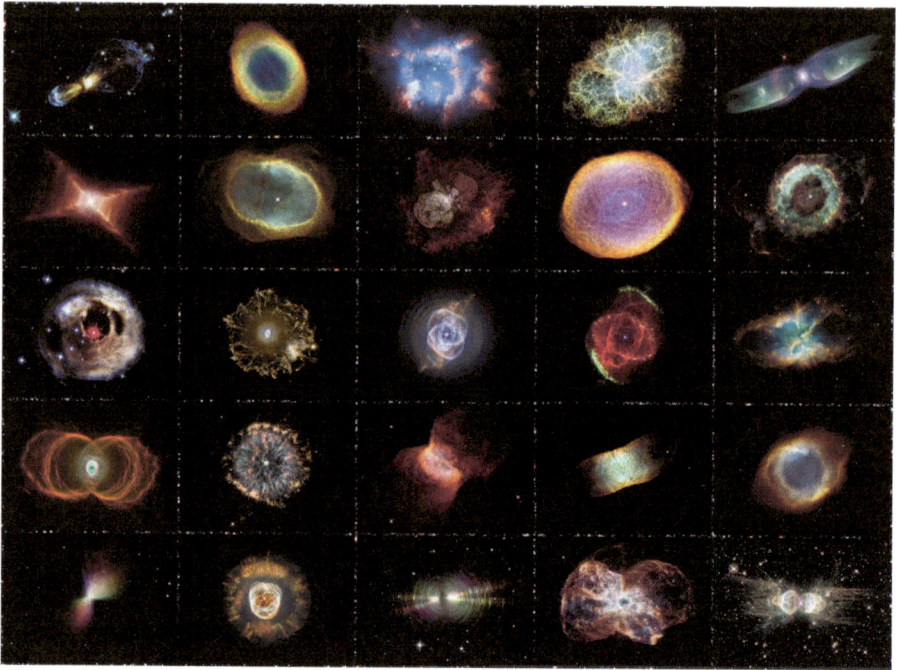

Fig. 9.3. A sample of the great variety of planetary nebula shapes, which depends both on our viewing angle and on the stellar environment. Credit: NASA/ESA

layers. The Ring Nebula in the constellation Lyra and Abell 39 are examples. But with the rise of larger telescopes, both on the ground and in space, planetary nebulae are now known to exhibit a great variety of shapes, depending on our viewing angle, multiple episodes of mass loss, the interstellar gas and dust encountered as the envelope is ejected, and whether the star is part of a binary system (Fig. 9.3). The ejected dust and gas is shaped both by slow winds and by million-mile-per hour "superwinds." Such variable winds were unknown until discovered in the late 1970s by NASA's International Ultraviolet Explorer (IUE) spacecraft. Such faster and slower stellar winds are believed to interact to produce the broad variety of shapes observed as the stellar mass loss proceeds. Attempts have been made to place some of these shapes in an evolutionary sequence. For example, the Egg nebula (Fig. 9.4) may represent an incipient planetary nebula, while those showing more complex features are more highly evolved.[19] In general they are often classified as spherical, elliptical, and bipolar, but not all fit into these categories.

---

[19] Bruce Balick, "The Shapes of Planetary Nebulae," *American Scientist* (July-August, 1996), 342–351; Sun Kwok, *The origin and evolution of planetary nebulae* (Cambridge: Cambridge University Press, 2000).

Fig. 9.4. The Egg Nebula is a "bipolar" planetary nebula about 3,000 light years from Earth. It is sometimes misleadingly termed a "protoplanetary nebula," not because it will form planets but rather because it is believed to be an early-stage planetary nebula. Credit: Raghvendra Sahai and John Trauger (JPL), the WFPC2 science team, and NASA

Catalogs of planetary nebulae have grown by leaps and bounds since the four listed in Messier's catalogue of nebulous objects and star clusters of 1781. By 1992 some 1,143 "planetaries" were listed in the 1992 Strasbourg-ESO Catalogue of Galactic Planetary Nebulae published by the European Southern Observatory. More than 3,000 are known today in the Milky Way Galaxy. By some estimates more than 100,000 planetary nebulae may exist in our Galaxy, since 99% of its stars will go through the red giant stage. They have also been observed in other galaxies and (rarely) in globular clusters. Only four planetary nebulae were known in globular clusters in 2000, including M15 in Pegasus.[20]

Images and news releases of planetary nebulae observed by the Hubble Space Telescope over the last 25 years are available at http://hubblesite.org/news/34-planetary-nebulae. An observer's guide to planetaries is found at http://www.reinervogel.net/pdf/Large_PN.pdf. More details about the Egg Nebula are in the press release at http://hubblesite.org/image/392/category/34-planetary-nebulae.

---

[20] AAS Nova, "The Curious Case of Planetary Nebulae in Globular Clusters," https://aasnova.org/2017/01/24/the-curious-case-of-planetary-nebulae-in-globular-clusters/

Class S 18: Nova Remnant

A nova is a close binary star (S 31) usually consisting of a white dwarf (S 12) and a main sequence normal star, in which hydrogen from the normal star is accreted onto the white dwarf. When sufficient material accumulates on the surface of the white dwarf, the temperature rises enough for the carbon cycle of nuclear fusion to ignite, causing the white dwarf to brighten by about 10 magnitudes and producing as much energy as the Sun does in 1,000 years. A similar process happens for a so-called "Type Ia" white dwarf supernovae (S 11), but in that case the accreted material accumulates to the point that the entire white dwarf explodes, releasing even more energy, generally obliterating the white dwarf and its companion star, and leaving a supernova remnant (S 26) as part of the interstellar medium. The companion may also be ejected from the system, but in either case no star remains.[21] A nova remnant, on the other hand, still surrounds its parent stars and is thus a circumstellar object. While it might seem that a nova and a supernova should both have class status at the same taxon level, they are in fact two very different phenomena, with a supernova white dwarf exploding and being destroyed while a nova white dwarf repeatedly erupts over a period of time.

Most novae take 10,000 to 100,000 years to accumulate enough hydrogen from their companion to ignite the eruption. But some, such as T Pyxidis, brighten several times a century. Such "recurrent novae" are believed to have massive white dwarfs, but not so massive that it would explode as a supernova and become a neutron star or black hole. Only nine "recurrent novae" are known. Occasionally, a pair of merging red dwarf stars will create a nova in the absence of a white dwarf.

Because novae and supernovae were not distinguished until the late 1920s when their absolute magnitudes and luminosities could be determined, nova and supernova remnants could also not be distinguished from each other before that time. Nevertheless, what turned out to be the first nova remnant—nebulosity around a star now known as GK Persei (Fig. 9.5)—was photographed by Yerkes astronomer G.W. Ritchey in 1901. Ritchey wrote that "a fairly satisfactory negative of the spiral or annular nebula surrounding the Nova was obtained," and less than two months later another observation showed that "a large change had taken place in

---

[21] HST Release, October 27, 2004, "Stellar Survivor from 1572 A.D. Explosion Supports Supernova Theory," http://hubblesite.org/newscenter/archive/releases/2004/34/

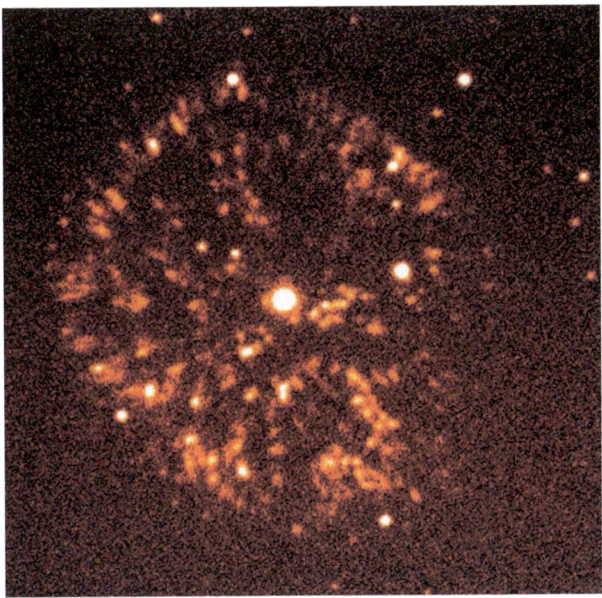

Fig. 9.5. The GK Persei nova remnant, about 1,500 light years away, imaged in 1994. Credit: National Optical Astronomy Observatory, Association of Universities for Research in Astronomy, National Science Foundation

the nebula," which he described in detail. In 1916 E.E. Barnard also photographed the ejecta.[22] Beginning in 1942, Walter Baade began to image the "envelopes," nebular remnants of novae, using the telescopes at Mt. Wilson and Palomar Observatories. DQ Herculis was the first nova remnant he photographed, and its expansion and development is apparent in photographs taken since.

When Cecelia Payne wrote her classic book *The Galactic Novae* in 1927, about 150 novae had been reported, all in our own galaxy. About 200 classical novae were known as of 1990, and 10% of them had confirmed nebular remnants in the optical region and 1% in the radio region. Thanks to systematic surveys and improved technologies such as charge-coupled devices (CCDs) and the Hubble Space Telescope, about twice as many nova remnants are now known in the optical, a half dozen in the radio, plus several in the infrared and at least one (GK Persei) in the X-ray.[23]

---

[22] G. W. Ritchey, "Nebulosity about Nova Persei," ApJ, 14 (1901), 167–168; and "Changes in the Nebulosity about Nova Persei," ApJ, 14 (1901), 293–294.

[23] T. J. O'Brien and M. F. Bode, "Resolved Nebular Remnants," in *Classical Novae*, M. F. Bode and A. Evans, eds. (Cambridge: Cambridge University Press, 2nd edition, 2008), pp. 285–307: 285, including a list of known remnants and an imagery see gallery, pp. 287–289 ff.

Because novae eject large amounts of material up to one ten thousandth of a solar mass, and because the high velocity of about 700 miles per second spreads the material rapidly, nova remnants have been resolved as extended objects in imagery over many wavelengths. It has been estimated that the Hubble Space Telescope can resolve a nova remnant about 1.5 months after its explosion, while ground-based resolution would take about 1.5 years.

The first Hubble image of a nova remnant was of Nova Cygni 1992, one of the brightest novae in 20 years, which erupted February 19 of that year. The image, taken in May of the following year, revealed a ring-like structure of hot gas 37 billion miles across, 10 times the diameter of the Solar System out to Pluto. At the time this was the earliest stage of a nova eruption that had been imaged; it would have taken another five years to resolve the ejecta from a ground-based telescope. At a distance of 10,430 light years, this was no small feat, even before the corrective optics was added to Hubble.[24] In 1997, Hubble found that the recurrent nova T Pyxidis produced thousands of gaseous blobs the size of our Solar System rather than a smooth shell of gas. The blobs were interpreted as collisions among successive eruptions of the nova all the way back to the early 1800s. Astronomers believe T Pyxidis, whose companion is a red dwarf, needs only to siphon hydrogen about the mass of the Moon from its companion in order to trigger each ejection.[25]

The variable star V838 Monocerotis (Fig. 9.2) was originally believed to be a nova with a spectacular remnant, after an outburst was observed in 2002. Though it was given the name Nova Monocerotis 2002, subsequent observations showed erratic activity that would make it a very atypical nova outburst. It is more likely a circumstellar shell (S 16) around a dying star or an event associated with a massive supergiant.

More on the G K Persei nova remnant image is at http://www.noao.edu/image_gallery/html/im0008.html. See the Hubble image of Nova Cygni 1992 before and after its optics were repaired at http://hubblesite.org/image/135/news_release/1994-06.

---

[24] HST Release, September 30, 1993, "Gas Shell Around Nova Cygni," http://hubblesite.org/newscenter/archive/releases/1993/21/image/a/; HST Release, January 13, 1994, "Hubble Sees Changes in Gas Shell around Nova Cygni 1992," http://hubblesite.org/newscenter/archive/releases/1994/06/

[25] HST Release, September 18, 1997, "Blobs in Space: The Legacy of a Nova," http://hubblesite.org/newscenter/archive/releases/1997/29/text/

Class S 19: Core Collapse Supernova Remnant

As the name suggests, a supernova remnant is the remains of material ejected during the explosion of a supernova (S 11), a process that seeds the interstellar medium with heavy elements. Only the core collapse of a massive star, known to astronomers as a Type II or Type Ib or Ic supernova, generates a circumstellar structure; a Type Ia white dwarf supernova normally leaves no star, and thus its remnant (S 26) gradually disperses into the interstellar medium, putting it in the interstellar medium Family (Chapter 11). (But see S 26 for a case where a companion star was left behind, or is being ejected from its system.)

Some 274 supernova remnants of all types are listed in the Catalogue of Galactic Supernova Remnants, not including nova remnants (S 18). Astronomers have divided supernovae remnants into three types based on their appearance: shell-type, as in the Cygnus loop, appearing as a ring-like structure; Crab-like (pulsar wind nebulae, or "plerions"), as in the Crab nebula; and composites, a cross between the two. A 2009 study of 17 supernova remnants (10 core collapse, seven white dwarf) showed that white dwarf supernovae generated very symmetric circular remnants, while core collapse supernovae remnants were distinctly more asymmetric.[26]

In their classic study *The Historical Supernovae,* David Clark and F. Richard Stephenson found 8 probable or certain historically recorded "new stars" associated with supernova remnants observed today. Of these they rated four as certain: the supernova observed in 1006, the Crab Nebula (1054), Tycho's Supernova (1572), and Kepler's Supernova (1604). Of these, only the Crab Nebula resulted from the collapse of a massive star, as evidenced by the neutron star (S 13) at its center. Cassiopeia A, the remains of an explosion 300 years ago, and the Veil Nebula in Cygnus, which exploded about 5,000 years ago, are also now known to be likely Type II core collapse remnants. Supernova remnants have also been found in other galaxies; for example, the Hubble Space Telescope has observed an object known as N132D, an oxygen-rich Supernova remnant in the Large Magellanic Cloud. It is often difficult to determine whether these extragalactic objects are core collapse supernovae or white dwarf supernovae.

---

[26] Chandra Release, December 17, 2009, "Supernova Explosions Stay in Shape," http://chandra.harvard.edu/photo/2009/typingsnrs/

Though we now take for granted the association of certain nebulae with supernova precursors, this was hard-won information. The Crab Nebula itself demonstrates how tortured the identification of that supernova remnant was with the supernova of 1054. Chinese and Japanese astronomers observed the explosion itself in 1054, when it was visible for 23 days during daylight, and at night for more than 22 months.[27] It was not until 1731 that the English amateur astronomer John Bevis discovered the nebulosity; the French astronomer Charles Messier rediscovered it in 1758, and it became M1, the first object in his famous catalogue of fuzzy non-cometary objects.

The "Crab Nebula" name originated in the mid-19th century, perhaps from a drawing of Lord Rosse using his 72-inch reflecting telescope; Rosse himself referred to his drawing as the "crab nebula" in 1848. Isaac Roberts obtained the first photograph of the nebula in 1892, and further photography by James Keeler, H.D Curtis, and G.W. Ritchey, among others, established that its oval form and peculiar filamentary structure were unique among nebulae.

The road to identification with a supernova began in 1921 with the discovery of the ongoing expansion of the Crab Nebula, when John C. Duncan compared the positions of nebulous points on the remnant over a period of 11.5 years and followed up in 1939 over an interval of 29.1 years. Based on the expansion rate, he estimated the supernova date at 1172.[28] However, in the same year Nicholas U. Mayall marshaled the evidence for the Crab Nebula association with the supernova of 1054, following a vague suggestion made by Knut Lundmark in 1921, repeated in an obscure publication in 1938.[29] "The identification of the Crab Nebula as a former supernova possesses a degree of probability sufficiently high to warrant its acceptance as a reasonable working hypothesis," Mayall wrote in 1939.[30]

---

[27] David H. Clarke and F. Richard Stephenson, *The Historical Supernovae* (Pergamon Press: Oxford and New York, 1977) 7, 140–160.

[28] John C. Duncan, "Report on the Expansion of the Crab Nebula," ApJ, 89 (1939), 482–486.

[29] Knut Lundmark, "Suspected New Stars Recorded in Old Chronicles and Among Recent Meridian Observations," PASP, 33 (1921), 225–238; "Was the Crab Nebula formed by a Supernova in 1054 A. D.?," *Professor Östen Bergstrand Vetenskapsmannen Och Läraren. Ägnas Denna Skrift På Hans Sextiofemårsdag*, 1 September, 1938, p. 89.

[30] N. U. Mayall, "The Crab Nebula, a Probable Supernova," ASP Leaflet No. 119 (Jan 1939), 145–154.

By 1942, Walter Baade stated that historical data provided by Leiden Sinologist J.J.L. Duyvendak, and the scientific conclusions in an adjacent paper by Jan Oort and Mayall, "leave hardly any doubt" that the new star of 1054 is the parent of the Crab nebula, and that star was a type I supernovae.[31] The spate of papers in 1942, including those of Baade and Rudolph Minkowski, may be taken as the beginning of the modern era of study of the Crab Nebula as a supernova remnant. The Crab Nebula was the first object identified as a radio source in 1949. It was categorized as arising from a Type I event for many years; by 1966, Minkowski showed that it was more likely Type II core collapse supernova, which we know to be the case today. The nebula holds several solar masses of material ejected by the explosion. It has been studied extensively by both ground based and space telescopes (Fig. 9.6).[32]

Cassiopeia A (known as Cass A) is another massive collapse remnant of a supernova of Type IIb. The discovery of Cass A was reported in 1948, among the first discrete radio sources to be found because it is the brightest in the sky. The optical counterpart was found in 1954, when Baade and Minkowski identified it with a rapidly expanding emission nebulosity "of a new type," which they did not relate to a supernova. "So far as we know, it is for the first time that we encounter the type of nebulosity just described," they wrote; spectra showed it to be a "nebulosity of extraordinary properties." Only in 1970, by studying the expansion rate of the nebulosity, was the time of the supernova explosion dated to 1667

---

[31] Walter Baade, "The Crab Nebula," CMWO (1942), 1–11; J. J. L Duyvendak, "Further Data Bearing on the Identification of the Crab Nebula with the Supernova of 1054 A.D. Part I. The Ancient Oriental Chronicles," PASP, 54 (1942), 91–94; N. U. Mayall and J. H. Oort, "Further Data Bearing on the Identification of the Crab Nebula with the Supernova of 1054 A.D. Part II. The Astronomical Aspects," PASP, 54 (1942), 95–104. R. Minkowski, "The Crab Nebula," ApJ, 96 (1942), 199–211.

[32] Minkowski, "Nonthermal Galactic Radio Sources," in *Nebulae and Interstellar Matter*, Stars and Stellar Systems, vol. 7, pp. 623–666: 629. J. Jeff Hester, "The Crab Nebula: An Astrophysical Chimera, ARAA, 46 (2008), 127–155. The historian of science L. Pearce Williams still shows some skepticism with astronomers' conclusions in his article "The Supernova of 1054: A Medieval Mystery," in *The Analytic Spirit: Essays in the History of Science*, ed. Harry Woolf (Cornell, 1981), pp. 329–349.

Fig. 9.6. Crab Nebula supernova remnant 6,500 light years away and six light years in diameter. The colors in this composite of 24 images indicate elements expelled during the explosion: blue in the filaments of the outer part of the nebula is neutral oxygen, green is ionized sulfur, and red is ionized oxygen. Credit: NASA, ESA, J. Hester, and A. Loll (Arizona State University)

plus or minus 8 years.[33] Both ground- and space-based telescopes have studied it intensely ever since. The Chandra spacecraft not only detected the central neutron star of Cass A, but in 2010 it also for the first time measured the cooling rate of a neutron star at the rate of 4% over ten years. Ground-based telescopes have detected high energy gamma rays emanating from the remnant, and in 2010 the Fermi spacecraft's Large Area Telescope also detected lower energy gamma rays.[34]

---

[33] M. Ryle and F.G. Smith, "A new intense source of radio-frequency radiation in the constellation of Cassiopeia," *Nature*, 162 (1948), 462–463; W. Baade and R. Minkowski, "Identification of the radio sources in Cassiopeia, Cygnus A and Puppis A," ApJ, 119 (1954), 206-21, including Palomar 200-inch plates of the nebulosity; Sidney van den Bergh and W.W. Dodd, "Optical studies of Cassiopeia A. I: Proper motions in the optical remnant," ApJ, 162 (1970), 485–493.

[34] Craig Heinke and Wynn Ho, "Direct Observation of the Cooling of the Cassiopeia A Neutron Star," ApJ Letters, 719 (2010), 167–171; NASA Release, February 16, 2010, "NASA's Fermi Closes on Source of Cosmic Rays," http://www.nasa.gov/mission_pages/GLAST/news/cosmic-rays-source.html#

Fig. 9.7. Supernova remnant Cassiopeia A, a composite of three images in the X-ray (green and blue), optical (yellow), and infrared (red). X-ray Credit: NASA/CXC/SAO; Optical: NASA/STScI; Infrared: NASA/JPL-Caltech/ Steward/O.Krause et al.

The 350-year-old Cass A remnant (Fig. 9.7) is a broken shell with a radius of 100 to 150 arcseconds, making the shell diameter about 1/6th the size of the full Moon. At its distance of about 11,000 light years from Earth, this makes the diameter of the shell about ten light years across. The gaseous shell, expanding at about 11 million miles per hour, has temperatures ranging from 10° C in the infrared to 10 million degrees C in the X-ray. Oxygen rather than hydrogen appears to be the most abundant element in the remnant, which also contains iron, silicon, sulphur, argon, calcium, and nitrogen—all elements produced in the massive progenitor star. The ejecta from the supernova has been estimated at about four solar masses and the swept-up mass about nine solar masses, for a total of 14 solar masses for the remnant. Some have suggested its progenitor must have been a massive Wolf-Rayet star.[35]

For many years after the discovery of Cass A, it was a mystery why such an explosion would not have been observed during the late 17th century, when astronomy was in full swing in Europe. In 1980 historian Bill Ashworth argued convincingly that the Type II supernova causing the Cassiopeia A remnant may have been observed in 1680 by British Astronomer Royal John Flamsteed as

[35] J. Vink et al, "A New mass estimate and puzzling abundances of SNR Cassiopeia A," A&A, 307 (1996), L41–44.

the 6th magnitude star 3 Cassiopeia, at the limits of visibility of the naked eye.[36]

Supernova remnants greatly vary in size depending on their age and the velocity of the ejecta. In general, the older the supernova explosion, the larger and more dissipated the ejecta shell. An extreme example is the Veil Nebula, the remnant of a supernova that exploded 5,000 to 8,000 years ago, first detected in 1784 by William Herschel. It covers about 3 x 3 degrees of the sky, six times the diameter of the full Moon. Oxygen, hydrogen, and sulfur have been detected in its shell. At the other extreme is Supernova 1987A in the Large Magellanic Cloud about 179,000 light years away, the brightest supernova observed since Kepler's in 1604. Unlike Kepler's white dwarf supernova (S 26), which was in our own Milky Way Galaxy, 1987A is believed to be another core collapse Type II supernova, even though its neutron star has not yet been detected. It resulted from the collapse of a blue supergiant rather than the usual red supergiant star. Because of its proximity to Earth, astronomers were able for the first time to witness the birth of a supernova remnant. After 15 years, its inner ring was roughly two arcseconds in diameter, equivalent to almost two light years at that distance (see Fig. 8.15). As much as 5,000 solar masses may have been ejected at speeds exceeding 5,000 kilometers per second.

David Green's *Catalogue of Galactic Supernova Remnants* is available at http://www.mrao.cam.ac.uk/surveys/snrs/. Those observed by the Chandra X-ray telescope are in the Chandra Supernova Catalog, found at http://hea-www.cfa.harvard.edu/ChandraSNR/. An extensive gallery of supernova and supernova remnants is at the Chandra X-ray Observatory site http://chandra.harvard.edu/photo/category/snr.html. A multiwavelength three-dimensional (3-D) reconstruction of a supernova remnant is at http://chandra.harvard.edu/photo/2009/casa2/. For more on the neutron star at the center of Cass A see http://chandra.harvard.edu/photo/2009/cassio/. Numerous images of the Crab Nebula and other supernova remnants are available at http://hubblesite.org/news by entering the specific subject matter such as "Crab Nebula." The most recent is at http://hubblesite.org/news_release/news/2017-21/3-nebulae. More on Fig. 9.6 at http://hubblesite.org/image/1823/news_release/2005-37 and http://hubblesite.org/image/3885/category/35-supernova-remnants.

---

[36] William B. Ashworth, "A Probable Flamsteed Observation of the Cassiopeia A Supernova," JHA, 11 (1980), 1–9.

Class S 20: Stellar Jet

Stellar jets are energetic matter and radiation ejected from stars accreting matter, either at star birth or star death and under certain other exotic circumstances such as cataclysmic variables and X-ray binaries. The smallest known astronomical objects with ejection jets are substellar brown dwarfs (S 22). In addition to star birth phenomena, stellar jets are also produced by stellar black holes, pulsars, and white dwarfs in binary systems where accretion takes place. Most, but not all, of these jet phenomena are the result of accretion disks at very different scales. Jets associated with accretion disks are sustained as long as inflowing matter is available. Depending on the circumstances, jets can consist of hydrogen and helium or a mix of electrons, protons, and positrons, but their origin and composition remain the object of much research. The most energetic stellar jets, involving supernova explosions and stellar black holes (S 14), are believed to be one of the long-sought sources of gamma-ray bursts (GRBs). Impressive as these are, the largest objects with observed jets are not in the Stellar Kingdom but the Galactic Kingdom, where supermassive black holes with more than three million times the mass of the Sun produce galactic jets (G 11) emanating from the cores of active galaxies.

Jets associated with star formation consist of materials blasted away from a protostar (S 1), a disk of gas and dust falling into an embryonic star. The earliest stages of jet formation in this scenario occur in molecular cloud nurseries (S 25) such as found in Orion, which can be observed at infrared wavelengths (Fig. 9.8). Jets have the effect of reducing the angular momentum of the disk, allowing the star, and sometimes planets, to form. A protostellar jet may collide with interstellar gas, causing the gas to glow. Such objects are called Herbig-Haro objects (S 21) after the astronomers George Herbig and Guillermo Haro who first detected them. Our Sun would have gone through such a phase during its formation when it was a T Tauri star (S 2).[37]

---

[37] For a review of the state of pre-Hubble Space Telescope knowledge see C. J. Lada, "Cold outflows, energetic winds, and enigmatic jets around young stellar objects," ARAA, 23 (1985), 267–317. The press release for Fig. 9.8 is at HST Release, December 17, 2015, "Stellar Jets," http://hubblesite.org/image/3656/category/25-stellar-jets

Herbig-Haro Jet HH 24

Hubble
Heritage

NASA and ESA • *Hubble Space Telescope* • WFC3/IR • WFPC2 • STScI–PRC15–42a

Fig. 9.8. Stellar jets from a newborn star in the Orion B molecular cloud complex 1,350 light years distant. Herbig-Haro objects are also evident. Credit: NASA, ESA, the Hubble Heritage (STScI/AURA)/Hubble-Europe (ESA) Collaboration, D. Padgett (GSFC), T. Megeath (University of Toledo), and B. Reipurth (University of Hawaii)

Although galactic jets were known as early as 1918 because of their large scale, Herbig-Haro objects were the first observational evidence that jets occurred at stellar scales. They were first suggested as such in the 1950s by the Soviet-Armenian astronomer Viktor Ambartsumian, who believed they represented the early stages of T Tauri stars. By the mid-1970s it was realized that HH objects had very high proper motions, pointing away from their very young stars, and by the early 1980s some of these objects were known to have a highly collimated jet-like nature.[38] They have been studied intensively ever since. The Hubble Space Telescope provided the first close-up views of stellar jets in 1995, and the

---

[38] S. E. Strom et al., "Infrared and optical observations of Herbig-Haro objects," ApJ, 191 (1974), 111–142.

Spitzer Telescope has provided many more images in the infrared. Progress has also been made in modeling the jet phenomenon; in 2009, a multi-trillion watt laser at the University of Rochester in New York produced simulations dramatically similar to what is observed around young stars.[39]

At the opposite end of the life cycle of massive stars, jets are also associated with stellar black holes that form as the result of star death, in particular the collapse of a massive star to form a supernova. The collapsing star results in a black hole (S 14) with an accretion disk, which produces a jet of high-energy particles. Shockwaves within the jet produce a burst of X-rays and gamma rays of a few minutes duration, and the interaction of the jet with the supernova shell produces an X-ray glow lasting days to months. Gamma-ray bursts had been linked with black holes since 1998, but in 2003 the Chandra Observatory confirmed this scenario in detail. And in 2009 the Fermi gamma-ray telescope observed one of the most energetic gamma-ray blasts ever see, believed to emanate from the jet of a collapsing massive star.[40]

A supernova involving an exploding lower mass star between 1.5 and three solar masses will end as a neutron star (S 13) rather than a black hole, and jets can also be produced by neutron stars. In the particular type of neutron star known as a pulsar, the rotational energy generates an electrical field, creating an electromagnetic beam emanating from the poles. In other pulsars, the accretion of matter is the power source rather than the rotational energy. The Vela pulsar jet, powered by the combined action of the fast rotation of the star and its intense magnetic field, emits a stream of extremely energetic particles (probably electrons or positrons) half a light year (three trillion miles) in length and 200 billion miles in width. The jet width is likely confined by magnetic fields created by the jet itself. The spiraling particles produce X-rays, and between 2000 and 2002 the Chandra Observatory

---

[39] HST Release, June 6, 1995, "Hubble Observes the Fire and Fury of a Stellar Birth,"http://hubblesite.org/newscenter/archive/releases/1995/24/text/; Spitzer Release, September 18, 2008, "Water Hit with Young Star's Best Shot," http://www.spitzer.caltech.edu/news/908-feature08-12-Water-Hit-with-Young-Star-s-Best-Shot

[40] Fermi Release, February 19, 2009, "NASA's Fermi Telescope see Most Extreme Gamma-ray Blast Yet;" http://www.nasa.gov/mission_pages/GLAST/news/high_grb.html See also http://spaceflightnow.com/news/n0303/24chandra/

observed the variability of this jet, possibly caused as the jet moves through the surrounding gas at 200,000 miles per hour. In 2003 Chandra detected the counter jet. Chandra has made similar detailed observations of other pulsar wind nebulae, including the famous Crab Nebula.[41]

Stellar jets are also associated with a type of binary star (S 31) known as X-ray binaries, consisting of either a black hole or neutron star together with another star. In this case, infalling material from the normal star creates an accretion disk around the neutron star or black hole. The first object recognized as such was SS 433, the 433rd object in the Stephenson-Sanduleak catalogue of 1977. In 1978, David H. Clark and Paul Murdin drew attention to this energetic object at radio, optical and X-ray wavelengths. It consists of either a neutron star or black hole with a late A-type star as a companion. Located at a distance of 18,000 light years, it was the first relativistic jet discovered in the Milky Way Galaxy.[42] Because of the similarities of the accretion disk process to quasars, such X-ray binaries are also called microquasars, though the scale and energies are much smaller. SS 433 is associated with a supernova remnant known simply as W50. Cygnus X-1, the first X-ray source, discovered in 1964, is another object that was later recognized as a microquasar. It consists of a blue supergiant and a black hole. In X-ray binaries, the mass of the black hole or neutron star is several solar masses.

Yet another jet phenomenon is associated with the late life stages of low-mass stars. Most solitary stars like the Sun will become red giants, produce a symmetric planetary nebula, and go quietly to the white dwarf stage. However, within the last 25 years it was discovered that binary systems consisting of a red giant and a white dwarf may form jets as material is poured from the giant to the dwarf. This is a short-lived phenomenon, and fewer than 200 such stars are known. These "symbiotic stars" are one

---

[41] G. G. Pavlov et al., "The Variable Jet of the Vela Pulsar," ApJ, 591 (2003), 1157–1171; ChandraRelease, June 30, 2003, "Vela Pulsar Jet: Firehose-Like Jet Observed In Action," http://chandra.harvard.edu/photo/2003/vela_pulsar/
[42] D. H. Clark and P. Murdin, "An unusual emission-line star/X-ray source/radio star, possibly associated with an SNR," Nature, 276 (1978), 44–45.

Fig. 9.9. Artist's rendering of a cataclysmic variable, a binary system consisting of a white dwarf and another star whose material is being poured onto the dwarf. Credit: STScI

form of a class of objects known as cataclysmic variables (Fig. 9.9). The R Aquarii and Z Andromedae systems are examples. In such systems the resulting planetary nebulae will show exotic structure. The first planetary nebula to show the bipolar collimated outflow characteristic of jets was the Eskimo Nebula NGC 2392, detected in 1985.[43] The even more famous Butterfly Nebula, also known as Minkowski 2-9 after its discoverer and sometimes dubbed the Twin Jet Nebula due to its appearance, is also believed to be caused by polar jets with a velocity in excess of 200 miles per second. The red giant-white dwarf pair orbits very closely together, and the ejected material that form the jets extend approximately ten times the diameter of Pluto. The stellar outburst that created them may have occurred as recently as 1,200 years ago.[44]

---

[43] J. A. Lopez, "Emission Lines from Jets in Planetary Nebulae," in *Emission Lines from Jet Flows*, W. J. Henney et al, eds., *Revista Mexicana de Astronomía y Astrofísica*, 13 (2002), 139–144.

[44] Mario Livio and Noam Soker, "The 'Twin Jet' Planetary Nebula M2-9," ApJ, 552 (2001), 685–691.

Moving down considerably in terms of the energies involved, in 2005 and 2007 astronomers using European Southern Observatory's Very Large Telescope (VLT) detected for the first time brown dwarfs spewing jets; they have ungainly names like 2MASS 1207-3932 because they were found during the 2 Micron All-Sky Survey. As one would expect, the scale of the jets is much smaller for these less energetic objects. Their jets stretch less than one billion miles in length, with a speed of a few miles per second, and their existence had to be inferred from high-resolution spectra rather than observed directly. In 2009 another brown dwarf jet was detected, this time around the Herbig-Haro object known as HH 211.[45] The discovery of brown dwarf jets has given rise to speculation that gas giant planets may also produce jets during their accretion phase. If so, this would give rise to a class of planetary jets in the planetary kingdom, but they have not yet been observed.

From star birth with Herbig Haro Objects and brown dwarfs to star death involving black holes, neutron stars, and white dwarfs, jets are a widespread astrophysical phenomena in the Stellar Kingdom, a kind of universal jet set marking places where the action is.

---

[45] Emma T. Whelan et al., "A resolved outflow of matter from a brown dwarf jet," *Nature* 435 (2005), 652–654; E. Whelan et al., "Discovery of a Bipolar Outflow from 2MASSW J1207334-393254, a 24 M_{Jup} Brown Dwarf," ApJ, 659 (2007), L45-L48; SAO Release, August 7, 2009, "Jets from a Possible Young Brown Dwarf," http://www.cfa.harvard.edu/news/2009/su200932.html

Class S 21: Herbig-Haro Object

Herbig-Haro Objects are small bright patches of nebulosity formed when the gas from a stellar jet (S 20) collides with gas in the interstellar medium. Such objects are associated with protostars (S 1) and young T Tauri stars (S 2) and are believed to last only a few thousand years until they dissipate into the interstellar medium. The total mass ejected from the star is typically one to 20 Earth masses, consisting of 75% hydrogen and 25% helium, at temperatures of about 8,000 to 12,000 K after settling down in the nebula. These temperatures are similar to other ionized nebulae such as H II regions and planetary nebulae. HH objects are typically found less than two light years from their parent star. About 600 have been observed, and as many as 150,000 may exist in our Galaxy. They are often found near the molecular clouds (S 25) where stars are born. Because of observational difficulties, historical usage has sometimes been ambiguous as to whether the HH object includes the jet (Fig. 9.8) or the nebula the jet causes, but the latter is currently the most common use.

In 1890 astronomer Sherburne W. Burnham first observed an object of this type near the star T Tauri, using the 36-inch refractor at Lick Observatory. Now called Burnham's nebula (HH 255), Burnham believed it was an emission nebula and did not recognize it as a new class of object.[46] In the late 1940s and early 1950s, Lick Observatory astronomer George Herbig and Mexican astronomer Guillermo Haro independently recognized similar objects in the constellation Orion (now known as HH 1, 2 and 3) as a different kind of emission nebula (S 24). Herbig recalled many years later, "While looking around for new T Tauri stars as part of my thesis, I ran across BD -6 1253 (now V380), which illuminates NGC 1999. . . In 1946-47, I took some direct photographs of the region of NGC 1999 with the Crossley reflector at Lick, and noticed some odd little fuzzy blobs nearby; these later became HH-1, -2 and -3." Herbig recalled that he paid no serious attention to the objects at the time, but in 1949 he met Haro at an AAS meeting in Tucson, where Haro gave a paper on the emission line spectra of the same objects Herbig had observed. This reignited Herbig's interest because he had in the meantime also obtained spectra of

---

[46] S. W. Burnham, "Note on Hind's Variable Nebula in Taurus," MNRAS, 51 (1890), 94–95.

Burnham's nebula, which had the same odd combination of emission lines as Haro showed in the Orion objects. "So at Lick in 1950, I obtained slit spectra of HH-1 and -2, from which came the note in ApJ in 1951, in which attention was drawn to the similarity to Burnham's Nebula. It was probably this connection with T Tauri that gave rise to the conjecture that Herbig-Haro Objects, as they were named by Ambartsumian, had something to do with early stages of star formation."[47]

It was indeed in 1954 that the Soviet/Armenian astronomer Viktor Ambartsumian first called these objects Herbig-Haro Objects and suggested they represented the early stages of T Tauri stars.[48] Herbig later showed that HH objects were variable, and that knots could change brightness on the scale of a few years.

By 1974 Herbig had compiled a total of 43 HH objects, and 25 years later about 400 were known.[49] Research on HH objects picked up in the 1970s when it was realized their emission spectra had some similarities to some supernovae and could be understood in terms of shock physics. The shocked nature of the objects was recognized in the early 1980s as a collision between the jets and gases in the interstellar medium. Detailed optical, infrared, ultraviolet, and radio studies followed, and eventually these objects were understood as an important part of star formation.[50] HH objects have now been well observed by the Hubble Space Telescope and other space observatories. They move away from their parent star at speeds ranging from 100 to 1,000 km per second, and can now be seen evolving over scales of a few years. Infrared HH objects are called molecular hydrogen emission line objects (MHOs).

---

[47] Bo Reipurth and Steve Heathcote, "50 Years of Herbig-Haro Research: From Discovery to HST," in Bo Reipurth and Claude Bertout, *Herbig–Haro Flows and the Birth of Stars* (Kluwer Academic Publishers, 1997), pp. 3–18.

[48] V. A. Ambartsumian, Communications of Byurakan Observatory No. 13 (1954); Ambartsumian, "Stars of T Tauri and UV Ceti types and the phenomenon of continuous emission," in *Non-Stable Stars*, George H. Herbig, ed. (Cambridge: Cambridge University Press, 1957), p. 177 ff.

[49] George H. Herbig, "Draft Catalog of Herbig-Haro Objects," *Lick Observatory Bulletin* No. 658 (1974), Reipurth, 1997.

[50] J. Bally, J. Morse, and B. Reipurth "The Birth of Stars: Herbig–Haro Jets, Accretion and Proto-Planetary Disks," in P. Benvenuti, F. D. Macchetto, and E. J. Schreier. *Science with the Hubble Space Telescope – II* (Space Telescope Science Institute, 1996).

Among the many interesting HH objects, the HH 111 jet is such a perfect representation of its class that theoreticians use it as a benchmark to test their models. It is driven by a young star embedded in a compact molecular core in what is known as the L 1617 cloud complex in Orion. The jet spans about 12 light years. HH 47 (Fig. 9.10), according to astronomer Bo Reipurth, is "the object which got the entire HH jet bandwagon rolling" with a seminal article in 1982. It is located in the Gum Nebula, an active region that has been stirred up by several very luminous stars, which have swept away the thinner parts of the Nebula, leaving more dense regions such as Bok globules behind. The bipolar flow of HH 47 emanates from a small molecular cloud known as a Bok globule, which is also a reflection nebula (S 28). The jet recedes from the source with a velocity of about 200 miles per second. The age of this object has been estimated at about 1,000 years. HH 47 was first observed from the ground with the European Southern Observatory's New Technology Telescope, and then from the Hubble Space Telescope, which, with ten times better resolution, revealed unprecedented detail. At the end of the jet is the characteristic bright bow shock where the jet material has collided with interstellar matter.[51]

Fig. 9.10. Herbig-Haro Object 46/47. Image taken by the Spitzer Space Telescope in 2007. The infant star is the white spot at the center of the image. Credit NASA/JPL-Caltech/T. Velusamy (JPL)

---

[51] Bo Reipurth, "Herbig-Haro Jets and their Role in Star Formation," European Space Observatory *Messenger*, 88 (1997), 20–26.

More details on the Spitzer image of HH 46-47 is at http://spitzer.caltech.edu/images/1090-ssc2003-06f-Embedded-Outflow-in-HH-46-47. More information and images of Herbig-Haro objects may be accessed by entering "Herbig-Haro" in the search field at http://hubblesite.org/news.

10. The Substellar Family

Class S 22: Brown Dwarf

Brown dwarfs are objects intermediate in mass between planets and stars, too large to have formed as planets, too small to sustain hydrogen fusion. Although they have been called "a poor excuse for a star," they are embraced by stellar astronomers and have even found a place in the standard stellar classification system. They range in mass from 13 to 80 times the mass of Jupiter, about 8% of a solar mass, but most are about the size of Jupiter. They are completely boiling, convective objects. Brown dwarfs are difficult to detect due to their very low luminosity, which during the first hundred million years or so derives from gravitational contraction, after which they become even fainter. Their temperature of 1,000 K and less dictates that they radiate primarily in the infrared region of the spectrum and are especially amenable to detection by infrared telescopes. Brown dwarfs can undergo deuterium and lithium fusion during their first ten million years.

Brown dwarfs appear at the far bottom right of the HR diagram (Fig. 8.2), the latest extension of the MK system beyond the M and L dwarfs to what are now called "T dwarfs." (There is, however, some crossing over of brown dwarfs into the L and even M dwarfs). In the discovery process, it is often difficult to distinguish high-mass brown T dwarfs from low-mass L stars, and low-mass brown dwarfs from large planets, though spectral differences are becoming better known with time.[1] Even though they are substellar, the placement of T dwarfs on the HR diagram is perhaps justified because brown dwarfs are believed to have formed through a star-like nebular condensation, rather than through a planetary accretion-type process. In addition to orbiting single stars, they have been found as part of binary systems (S 31) and as free-floating objects.

---

[1] Adam J. Burgasser, "The T-type Dwarfs," in Gray and Corbally, *Stellar Spectral Classification*, pp. 388–440.

© Springer Nature Switzerland AG 2019
S. J. Dick, *Classifying the Cosmos*, Astronomers' Universe,
https://doi.org/10.1007/978-3-030-10380-4_10

The theoretical possibility of substellar objects was a natural result of the discovery of lower and lower mass stars. As early as 1963, astronomer Shiv Kumar described stars that might exist with less than .08 solar mass and termed them "black dwarfs."[2] But because "black dwarf" was already being used by some astronomers to describe hypothetical cooled white dwarfs, astronomer Jill Tarter coined the term brown dwarf in her 1975 dissertation, to denote substellar objects that could not sustain nuclear fusion. Though others proposed names like "planetar" and "substar," the term "brown dwarf" stuck. "It was obvious that we needed a color to describe these dwarfs that was between red and black. I proposed brown and Joe [Silk] objected that brown was not a color," in terms of the primary colors of the spectrum, Tarter recalled.[3]

The discovery of actual brown dwarfs was a long time coming. In 1988 Eric Becklin and Ben Zuckerman discovered a low-mass star known as GD 165 B, cooler than the well-known M dwarfs, the first of what turned out to be a class of low-mass stars now known as L dwarfs, still undergoing hydrogen fusion. Their technique involved imaging of the extremely dim object, and their ambiguity regarding the nature of the object was evident: "We have discovered an infrared object located about 120 AU from the white dwarf GD165," they wrote. "With the exception of the possible brown dwarf companion to Giclas 29–38 which we reported last year, the companion to GD165 is the coolest (2,100 K) dwarf star ever reported and, according to some theoretical models, it should be a sub-stellar brown dwarf with a mass between 0.06 and 0.08 solar masses. These results, together with newly discovered low-mass stellar companions to white dwarfs, change the investigation of very low-mass stars from the study of a few chance objects to that of a statistical distribution. In particular, it appears that very low-mass stars and perhaps even brown dwarfs could be

[2] Shiv S. Kumar, "The Structure of Stars of Very Low Mass," ApJ, 137 (1963), 1121–1125; Kumar describes his work on low-mass stars and "black dwarfs" in Shiv S. Kumar, "The Bottom of the Main Sequence and Beyond: Speculations, Calculations, Observations, and Discoveries (1958-2002)," online at http://arxiv.org/pdf/astro-ph/0208096.

[3] Jill Tarter, "The interaction of gas and galaxies within galaxy clusters," PhD dissertation, University of California Berkeley, 1975; J. C. Tarter, "An Historical Perspective: Brown is Not a Color," *Astrophysics of Brown Dwarfs*, Minas C. Kafatos, Robert S. Harrington, Stephen P. Maran (eds.), (Cambridge: Cambridge University Press, 1986), pp. 121–138.

quite common in our Galaxy."[4] Several hundred such L dwarfs have subsequently been found by a variety of methods.

In 1989, Harvard astronomer David Latham and his colleagues reported a possible brown dwarf using a very different technique, the change in the line-of-sight, or "radial velocity," of a star with an unseen companion. The size of the tug indicated a companion to the star HD 114762, which they calculated could have a mass as small as .001 of the Sun, some 11 Jupiter masses. "Thus the unseen companion of HD 114762 is a good candidate to be a brown dwarf or even a giant planet," they concluded, allowing that there was less than a 1% chance that this companion could be massive enough to burn hydrogen stably. The uncertainty was due to the unknown orbital inclination of the object with respect to its star as viewed from Earth, the so-called "M sin (i) factor," where M is the mass and i is the inclination. Because of this factor, Latham and his colleagues cautioned that the object was most likely not an extrasolar planet (P 18), but a brown dwarf.[5] Even now, great uncertainty surrounds the nature of this object. It may be brown dwarf or an L dwarf star, and some astronomers still believe it may be the first extrasolar giant planet discovered. In any case, neither Becklin and Zuckerman's object nor Latham's were unambiguously the long-sought substellar brown dwarfs. Like extrasolar planets, brown dwarfs were still in danger of remaining hypothetical objects as the 1990s began.

On October 27, 1994, using an adaptive optics coronagraph on the 60-inch telescope at Mt. Palomar, Caltech astronomer Tadashi Nakajima and his colleagues imaged an even lower temperature object, Gliese 229B, orbiting the bright nearby M dwarf Gliese 229 (Fig. 10.1). "Here we report the discovery of a probable companion to the nearby star Gl 229, with no more than one tenth the luminosity of the least luminous hydrogen-burning star," they wrote. "We conclude that the companion, Gl 229B, is a brown dwarf with a temperature of less than 1,200 K, and a mass 20–50 times that of Jupiter." Astronomers Sam Durrance (also an astronaut!) and David Golimowski confirmed the Gliese 229B discovery with a now-famous Hubble Space Telescope image

---

[4] E. E. Becklin and B. Zuckerman, "A low-temperature companion to a white dwarf star," *Nature*, 336 (1988), 656–658.

[5] David W. Latham et al., "The unseen companion of HD114762 - A probable brown dwarf," *Nature*, 339 (1989), 38–40.

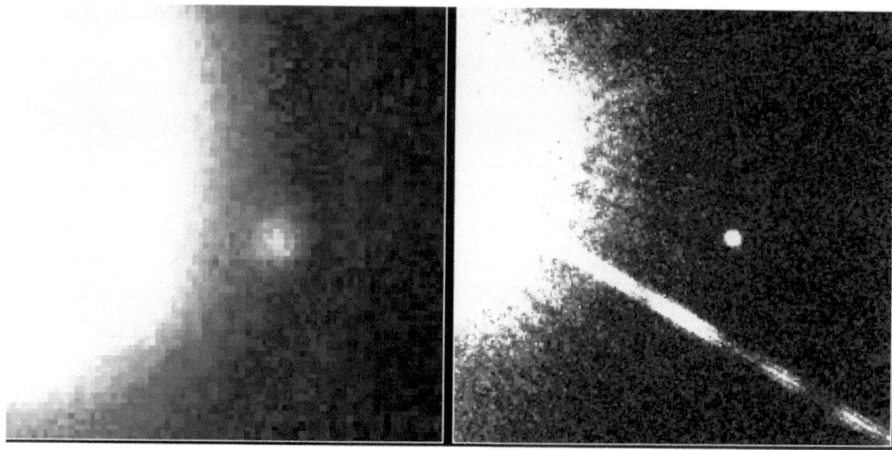

Fig. 10.1. The first unambiguous photographic discovery of a brown dwarf (Gliese 229B), observed (left) on October 27, 1994 with adaptive optics using the 60-inch reflector at Palomar, and confirmed (right) with the Hubble Space Telescope on November 17, 1995. Left, courtesy T. Nakajima (Caltech), S. Durrance (Johns Hopkins). Right, courtesy S. Kulkarni (Caltech), D. Golimowski (Johns Hopkins) and NASA

on November 17, 1995 (Fig. 10.1).[6] This turned out to be the first brown dwarf discovered and imaged, the prototype of the class of objects now known as "T dwarfs." The rapid acceptance of this object as the first unambiguous brown dwarf was due to the presence of methane it its atmosphere, found during follow-up observations with the Palomar 200-inch telescope.

In the meantime, astronomers in the Canary Islands led by Rafael Rebolo discovered an object known as Teide 1 in the Pleiades, a likely host for brown dwarfs because of its young age. Their claim was unambiguous: "Here we report the discovery of a brown dwarf near the centre of the Pleiades. The luminosity and temperature of this object are so low that its mass must be less than 0.08 solar masses, the accepted lower limit on the mass of a true star." Teide 1 is sometimes also referred to as the first verified brown dwarf.[7] Rebolo pioneered in the lithium test for brown

[6] T. Nakajima, B.R. Oppenheimer, S. R. Kulkarni et al, "Discovery of a Cool Brown Dwarf," *Nature* 378 (1995), 463–465; HST Release, November 29, 1995, "Astronomers Announce First Clear Evidence of a Brown Dwarf," http://hubblesite.org/newscenter/archive/releases/1995/48/text/.

[7] R. Rebolo, M. R. Zapatero Osorio and E. L. Martin, "Discovery of a brown dwarf in the Pleiades star cluster," *Nature* 377 (1995), 129–131.

dwarfs; just as the presence of methane is an indication of brown dwarf status because methane cannot survive in a star undergoing fusion, so stars also rapidly deplete lithium. The presence of lithium is therefore also a test for brown dwarfs, though older and more massive brown dwarfs may have burned their lithium.

In the year 2000, Hubble scientists announced the discovery of 50 brown dwarfs in the Orion Nebula's Trapezium cluster, followed by even more in Orion announced in 2018. Since then hundreds more have been verified, giving rise to the theory that brown dwarf formation is a universal process in star-forming regions.[8] They are often found by large area sky surveys in the infrared, including the 2MASS and Sloan Digital Sky Survey projects. On April 20, 2009, UK astronomers using the Gemini telescope on Mauna Kea announced a brown dwarf with a temperature of approximately 300° C, the coolest body ever detected outside the Solar System. Wolf 940b, with an estimated mass of 20 to 30 times that of Jupiter, circles an M dwarf star 40 light years from Earth in the constellation Aquarius. At least one planetary mass brown dwarf has been announced with eight Jupiter masses, using the Hubble and Spitzer Space Telescopes. Beginning in 1999, X-ray brown dwarfs were detected with the Chandra Telescope, suggesting changing magnetic fields. And at least four brown dwarfs are known to have planets orbiting around them.[9] In short, the field of brown dwarf research has blossomed to an extent undreamt of two decades ago.

---

[8] The Orion brown dwarfs are announced at Hubble Release, August 24, 2000, "Hubble Spies Brown Dwarfs in Nearby Stellar Nursery," http://hubblesite. org/newscenter/archive/releases/2000/19/ and more at HST Release, January 11, 2018 "Hubble Finds Substellar Objects in the Orion Nebula,"
  http://hubblesite.org/news_release/news/2018-03/42-brown-dwarfs.
[9] Wolf 940b is described at http://www.gemini.edu/node/11250. K. L. Luhman et al, "Discovery of a Planetary-Mass Brown Dwarf with a Circumstellar Disk," ApJ Lettters, 635 (2005), L93–L96; Robert Rutledge, Gibor Basri et al., "Chandra Detection of an X-Ray Flare from the Brown Dwarf LP 944-20," ApJ, 538 (2000), L141–L144; UC Santa Barbara Release, July 11, 2000, "First X-ray from brown dwarf observed," http://www.spaceref.com/news/viewpr.html?pid=2192. On a planetary companion to a brown dwarf see HST Release, April 6, 2000, "Small Companion to Brown Dwarf Challenges Simple Definition," http://hubblesite. org/news_release/news/2010-03/42-brown-dwarfs;

The abundance of brown dwarfs remains in question. Because of the difficulty of their detection, it has been conjectured they could account for a substantial fraction of baryonic dark matter, but others believe they could amount to no more than a billion solar masses, about 0.1% of the Galaxy's total mass. The results of a study of 233 nearby multiple-star systems by NASA's Hubble Space Telescope, announced in 2009, found only two brown dwarfs as companions to normal low-mass M stars. This lent some credence to the so-called "brown dwarf desert," the absence of brown dwarfs around solar-type stars, and extended that desert to the smallest stars in the universe. Nevertheless, other studies indicate that L and T dwarfs could rival the number of normal stars in the Galaxy, exceeding 100 billion. And in 2018 astronomers using the Hubble Telescope announced the discovery of a large population of brown dwarfs in the Orion Nebula, including 17 brown dwarf companions to red dwarf M stars, one brown dwarf pair, and one brown dwarf with a planetary companion.[10] The James Webb Space Telescope should discover many more, and will be able to study brown dwarf atmospheres.

The latest on brown dwarfs from the Hubble website is at http://hubblesite.org/news/42-brown-dwarfs. A compendium of L, T, and Y dwarfs as of 2012 is at the IPAC site http://spider.ipac.caltech.edu/staff/davy/ARCHIVE/index.shtml.

---

[10] R. E. Ryan et al., "Constraining the Distribution of L and T Dwarfs in the Galaxy," ApJ, 631 (2005), L159–L162. On the possible number of brown dwarfs in our Galaxy see https://arxiv.org/abs/1707.00277.

11. The Interstellar Medium Family

Class S 23: Cool Atomic Cloud (H I)

With this class of objects, we enter the Family of the interstellar medium. The interstellar medium consists of about 99% gas and 1% dust, and the gas component consists of three main classes of objects: cool atomic clouds composed mainly of neutral hydrogen (H I); hot ionized clouds also composed mainly of hydrogen, known as H II regions (S 24) and observed as emission nebulae; and cold molecular clouds (S 25) composed largely of hydrogen (H_2) with a sprinkling of other molecules.

About 25% of interstellar gas is composed of helium. Interstellar gas is extremely rarefied, with a density of about one atom per cubic centimeter. A "cold" H I region has a temperature of around 100 K, compared to 10,000 K for an H II region. It is detected not by optical emission lines as in H II regions, but by radio observations of the so-called 21-cm emission lines, or 21-cm absorption lines if a hotter object is in the background. About 95% of interstellar hydrogen is H I; the transition to the ionized H II requires a nearby star. H I and H II regions are therefore often found adjacent to each other near stars, and the extent of the neutral hydrogen converting to ionized hydrogen depends on the luminosity and temperature of the star. Among other reasons, H I regions are important because they trace the spiral arms of the Galaxy and provide a window on numerous other phenomena associated with the interstellar medium.

The first evidence of interstellar gas came in 1904, when the German spectroscopist Johannes Hartmann deduced "stationary absorption lines" of ionized calcium (Ca II) in the spectrum of the binary star delta Orionis, lines that did not shift back and forth through the binary orbit as did the spectral lines caused by the stars. After eliminating other possibilities, Hartmann wrote "We are thus led to the assumption that at some point in space in the line of sight between the Sun and delta Orionis there is

© Springer Nature Switzerland AG 2019
S. J. Dick, *Classifying the Cosmos*, Astronomers' Universe,
https://doi.org/10.1007/978-3-030-10380-4_11

a cloud which produces that absorption...[it is] very probable from the nature of the observed line, that the cloud consists of calcium vapor."[1] V.M. Slipher and Walter S. Adams supported this claim, but Hartmann's interpretation was not immediately accepted, and the nature of the calcium lines remained a puzzle for many years. Some astronomers thought the gas was near the star rather than in interstellar space. But in 1924 and 1925, J.S. Plaskett and Otto Struve argued that Hartmann's observations and others showed that vast clouds of calcium, perhaps thrown out by stellar prominences, exist throughout space. In 1926 Eddington showed conclusively from a theoretical standpoint that the lines came from interstellar gas.[2]

The discovery that cool gas clouds were primarily interstellar hydrogen hydrogen (Fig. 11.1), and that calcium was only a tracer, came slowly. Following the late-1920s discovery of Cecelia Payne, William McCrea, and others that stars were composed mainly of hydrogen, it was logical to consider hydrogen the most abundant

Fig. 11.1. Clouds of cold hydrogen rise above the molecular cloud at the edge of the Carina nebula where the starlight does not ionize the hydrogen. Credit for Hubble Image: NASA, ESA, N. Smith (University of California, Berkeley), and The Hubble Heritage Team (STScI/AURA); Credit for CTIO Image: N. Smith (University of California, Berkeley) and NOAO/AURA/NSF

---

[1] J. Hartmann, "Investigations of the Spectrum and Orbit of delta Orionis," ApJ, 19 (1904), 268-286: 274. This and the following paragraphs are adapted from Dick (2013), pp. 84–86.
[2] Otto Struve, "On the Calcium Clouds," PA, 34 (1926), 10.

element in the universe. By 1932, Bengt Strömgren showed that hydrogen probably dominated stellar cores as well. Once it was realized that the space between the stars was not empty, it took no great stretch of the imagination to imagine the existence of interstellar neutral hydrogen in those regions not too close to a star that would ionize the hydrogen. Already in 1934, Eddington made the suggestion that "in a normal region of interstellar space the hydrogen will be entirely un-ionized, and indeed in molecular form."

It was the work of Theodore Dunham and Otto Struve in the late-1930s that clearly identified the clouds as primarily neutral hydrogen. Dunham's work showed that the number of free electrons in space must be much larger than expected, and Struve deduced that "the most promising source is interstellar H, the existence of which—reasonable on general grounds—has recently been suggested by spectrographic observations at the McDonald Observatory," namely 22 regions of the Milky Way that show emission lines of hydrogen. Struve calculated that interstellar gas clouds would have a million times more hydrogen than calcium atoms. Because hydrogen provided free electrons to ionize the Ca II absorption lines observed by Hartmann and others, but was not hot enough to produce observable Balmer absorption lines characteristic of H II regions, for decades the gas was believed to be clouds of calcium rather than hydrogen.[3]

Only five years after Struve's publication indicating the abundance of neutral hydrogen in interstellar space, in 1944 the Dutch astronomer Hendrik van de Hulst predicted that neutral hydrogen should produce radiation at a frequency of 1420 MHz (21 centimeters) due to two closely spaced "hyperfine" energy levels in the hydrogen atom. This frequency was in the radio region of the spectrum, and radio astronomy was still very much in its experimental stages. Nevertheless, on March 25, 1951, the American astronomer Harold Ewen and the American physicist Edward Purcell, both at Harvard, detected this radiation from space emanating

---

[3] Otto Struve, "The Physical State of Interstellar Gas Clouds," PNAS, 25 (1939), 36–43; also Otto Struve and Velta Zebergs, *Astronomy of the Twentieth Century*, (New York and London: Macmillan, 1962), 375, and Struve "Note on Calcium Clouds," ApJ (1934). Eddington's work is "The density of interstellar calcium and sodium," MNRAS, 95 (1934), 2–11, and Theodore Dunham's is "Interstellar Neutral Potassium and Neutral Calcium," PASP, 49 (1937), 26-28, and "Forbidden Transition in the Spectrum of Interstellar Ionized Titanium," *Nature*, 139 (1937), 246–247. Saha commented that Dunham's work "forms a landmark in the story of interstellar investigations," *Nature* 139 (1937), 840.

from the interstellar medium. (Today their detector is displayed at the National Radio Astronomy Observatory (NRAO) in Green Bank, West Virginia.) Over the next three months, two other groups confirmed the 21-cm radiation with their own detections, the Dutch in May and the Australians in June. The American and Dutch results were published in the same issue of *Nature*, as well as a brief note on the Australian results.[4]

Because neutral hydrogen is found predominantly in the spiral arms of a galaxy, it has played an important part in the history of astronomy: it was observations of the 21-cm line that allowed the arms of the Milky Way to be mapped with radio telescopes in the 1950s. Early observations during the 1950s concentrated on emission from the galactic plane because of its brightness. Since that time, 21-cm emission studies have been a key probe of the structure and dynamics of the Galaxy, and the subject has undergone a renaissance since the large-scale surveys of the last decade. Such surveys now trace a dynamic interstellar medium with structures on all scales, demonstrating the recycling of matter between stars and the medium and revealing the Galaxy as "a violent, breathing disk surrounded by highly turbulent extra-planar gas."[5] In 1990 John M. Dickey and Felix "Jay" Lockman used radio observations to show that H I gas was also located in the galactic halo (G 12). In 2002, Lockman reported further the discovery of H I clouds in the

---

[4] H. I. Ewen and E.M. Purcell, "Observation of a Line in the Galactic Radio Spectrum: Radiation from Galactic Hydrogen at 1,420 Mc/sec," *Nature*, 168 (1951): 356, followed by C. A. Muller and J. H. Oort, "Observation of a Line in the Galactic Radio Spectrum: The Interstellar Hydrogen Line at 1,420 Mc./sec., and an Estimate of Galactic Rotation," 357. For details see Woodruff T. Sullivan, *Cosmic Noise: A History of Early Radio Astronomy*, (Cambridge: Cambridge University Press, 2009), chapter 16, "The 21 cm hydrogen line," 394-417. For recollections by radio pioneers see Gart Westerhout, "The Pioneers of H I," in *Seeing through the Dust: The Detection of H I and the Exploration of the ISM in Galaxies*," A. R. Taylor, T. L. Landecker and A. G. Willis, eds. (San Francisco: ASP, 2002), vol. 276, p. 3, and the adjacent series of historical articles. This volume also details how the field of 21 cm radio astronomy grew over the next 50 years.

[5] J. H. Oort, G. Westerhout, F. J. Kerr, "The Galactic System as a Spiral Nebula," MNRAS, 118 (1958), 379-389. The famous map of the galaxy is Figure 4. See also the previous 1954 H I map cited in note 5 of this paper. Peter M. W. Kalberla and Jurgen Kerp, "The H I Distribution of the Milky Way," ARAA, 47 (2009), pp. 27–61: 27.

Galactic halo. Lockman estimated that as much as half the mass of the neutral halo may be in the form of hydrogen clouds.[6] The 21-cm technique has since been applied to many other galaxies as well.

The neutral hydrogen 21-cm line is also famous because of its association with SETI, the Search for Extraterrestrial Intelligence. Because it is a prominent line emitted by the most abundant element in the universe, the physicists Giuseppe Cocconi and Philip Morrison put it forward in their famous paper in 1959 as a likely frequency (later dubbed "magic frequencies") on which extraterrestrials might communicate. And in 1960, Frank Drake made the first search for ETI, using the 21-cm wavelength on the 85-foot Tatel telescope at Green Bank, West Virginia.[7]

The most detailed map of neutral hydrogen in our Galaxy, released in 2016, is at https://www.mpifr-bonn.mpg.de/pressreleases/2016/13 and reported at https://www.skyandtelescope.com/astronomy-news/astronomers-map-milky-way-incredible-detail/. A map of neutral hydrogen in the spiral galaxy M101 is at http://galaxymap.org/drupal/node/202.

---

[6] A review of knowledge of H I in the Galaxy as of 1990 is at J. M. Dickey and F. J. Lockman, "H I in the Galaxy," ARAA, 28 (1990), 215–261, http://adsabs.harvard.edu/abs/1990ARA%26A..28..215D; Felix J. Lockman, "Discovery of a Population of H I Clouds in the Galactic Halo," ApJ, 580 (2002), L47–L50.

[7] G. Cocconi and P. Morrison, "Search for Interstellar Communications," *Nature*, 184 (September, 1959), 844. On the context see Steven Dick, *The Biological Universe: The Twentieth Century Extraterrestrial Life Debate and the Limits of Science* (Cambridge: Cambridge University Press, 1996), chapter 8.

Class S 24: Hot Ionized Cloud (H II)

Unlike H I regions consisting of cold neutral hydrogen (S 23), H II regions are clouds of hot ionized hydrogen where the electrons have become separated from their parent protons through the absorption of ultraviolet photons from a nearby star. This occurs especially around the hot O, B, and A stars; an O5 star would ionize hydrogen to distances of about 300 light years, while an A0 star might ionize a region one light year distant. H II regions are categorized by their size, ranging from ultracompact regions from a few tenths to a few tens of light years in diameter to giant regions up to several hundred light years in diameter. Smaller H II regions may contain a few solar masses of hydrogen, while the largest regions may contain the equivalent of tens to hundreds of millions of solar masses. H II regions are one of several types of emission nebulae, so-called because of the emission lines in their spectra, especially the hydrogen alpha line. Like the cooler H I regions in the radio spectrum, H II regions proved important for delineating the structure of the Milky Way Galaxy in the optical region.

Although called H II regions because ionized hydrogen is their most significant component, such regions also contain other elements. A typical H II region is composed of about 90% hydrogen, laced with helium, oxygen, nitrogen, and other trace elements. The often-spectacular colors of H II regions are caused by their chemical composition and degree of ionization. Red is due to the strong emission lines of the Balmer series of hydrogen, while green and blue are indicative of higher energy environments.

H II regions are strong sources of radio waves emitted by free electrons, but they are also spectacular objects in the optical region of the spectrum, providing some of the most beautiful images from the Hubble Space Telescope. The nearest and most famous H II region is the Orion Nebula (Fig. 11.2). Other well-known H II regions in our Galaxy are the Eagle Nebula with its eerie "pillars of creation" (Fig. 7.1) and the Eta Carina Nebula with its extremely massive stars (Fig. 8.11). Such regions also exist in other galaxies; the Large Magellanic Cloud, for example, harbors the gigantic Tarantula Nebula, also known as 30 Doradus (Fig. 11.8). At 600 light years in diameter, it is much larger than the Orion Nebula, perhaps the largest H II region in the Local Group of galaxies.

Knowledge of the composition and characteristics of emission nebulae did not come easily. Those objects we now know as H II regions were not observed until after the advent of the telescope.

Fig. 11.2. The Orion Nebula, 24 light years across, seen here in a composite of images taken by HST in 2004 and 2005. The Orion Nebula includes both H I and H II regions, depending on distance from a star, as well as reflection nebulae and molecular clouds. The H II region is centered on Theta Orionis C and has a temperature of about 10,000 K. Credit: NASA, ESA, M. Robberto (Space Telescope Science Institute/ESA), and the Hubble Space Telescope Orion Treasury Project Team

Although Galileo observed and resolved star clusters, his Italian compatriot Nicolas Claude Fabri de Peiresc, working in Paris, is credited with observing what turned out to be the first gaseous nebula, the Orion Nebula, in 1610. Christiaan Huygens studied it in more detail and produced the first recognizable drawing. Many such cloudy objects were eventually discovered, especially through the systematic searches of William Herschel in the late 18th century, but the question remained whether they were a fundamentally different class of object from star clusters, or whether they were so distant as to be star clusters that only appear cloudy (Fig. 11.3). As historian Michael Hoskin has shown, from 1784 to 1790 Herschel equated all nebulae with star clusters, but he changed his mind because of a crucial observation in 1790. That observation was of what he termed a "planetary nebulae" (S 17), a class of object he had thought were star clusters in the late stages of evolution. But in November 1790 Herschel observed a planetary nebula with a visible central star, and he classified both together

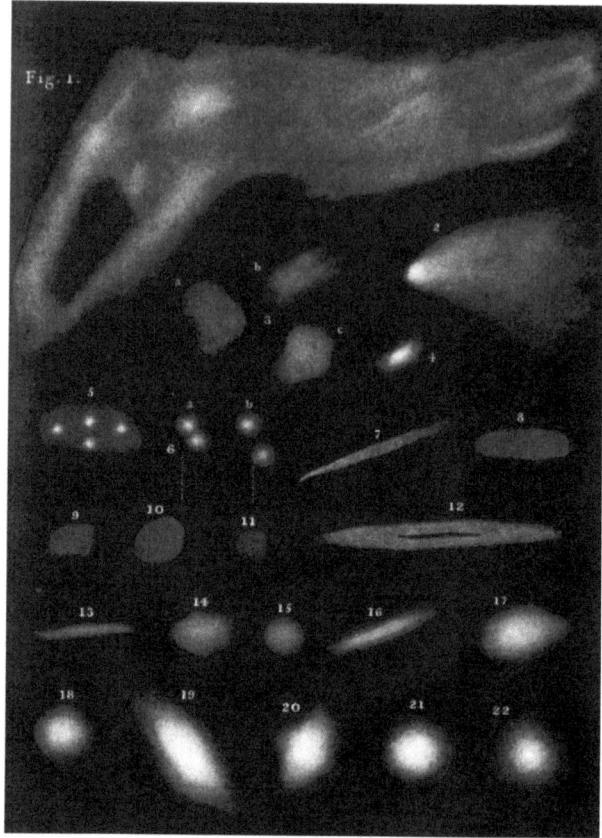

Fig. 11.3. Nebular morphologies observed by William Herschel from his 1811 paper on the subject. Herschel early on divided nebulae into eight classes based on appearance, and later 12 classes. They bear little resemblance to modern classes, but those labeled 5 and 6 are multiple and double galaxies, and 18, 21, and 22 are globular clusters of stars

as a "nebulous star." Seeing a star surrounded by nebulosity convinced him of the existence of true nebulosity, since he believed it demonstrated the star was condensing out of the nebula under the force of gravity.[8]

---

[8] Michael Hoskin, "William Herschel's Early Investigations of Nebulae: A Reassessment," JHA, vol. X (1979), 165–176. Herschel's article "On Nebulous Stars, properly so called," appears in Hoskin, *William Herschel and the Construction of the Heavens* (London: Oldbourne, 1963), pp. 118–129. The "nebulous star" is a planetary nebula known as NGC 1514.

Because of this observation, astronomers Simon Newcomb and Edward Holden in their astronomy textbook of 1881 credited William Herschel with "the first exact statement of the idea that, beside stars and star-clusters, we have in the universe a totally distinct series of objects, probably much more simple in their constitution."[9] The question remained open, however, until the British amateur astronomer William Huggins showed in 1864 that stars exhibit spectral lines, while some nebulae had a continuous spectrum. His excitement was palpable when he wrote "The riddle of the nebulae was solved. The answer, which had come to us in the light itself, read: Not an aggregation of stars, but a luminous gas. Stars after the order of our own sun, and of the brighter stars, would give a different spectrum; the light of this nebula had clearly been emitted by a luminous gas."[10] In the next four years, Huggins examined the spectra of about 70 nebulae and found that one-third of them were gaseous nebulae as opposed to star clusters. Thus were what we today call nebulae (S 23 through S 28) first separated from star clusters (S 34 and S 35) as a distinctive class of objects, which were further distinguished into many classes of objects as time went on. It took another 50 years, until 1925, to show definitively that not all fuzzy patches were nebulae or star clusters, and that "island universes" constituted yet another class (actually again many classes) that we know today as galaxies.

The same observations that proved the gaseous nature of the nebulae also hinted at their composition.[11] Huggins' first observations of nebulae were mainly of planetary nebulae rather than H II regions, but his subsequent observations of other nebulae showed emission lines of hydrogen, helium, carbon, nitrogen, and oxygen. But the predominance of hydrogen was not yet evident; the strongest emission line in planetary nebulae was later shown to be due to oxygen.

---

[9] Simon Newcomb and Edward S. Holden, *Astronomy for High Schools and Colleges* (New York: Henry Holdt and Company, 1881), third ed., revised, p. 459.

[10] Robert Smith, "Beyond the Galaxy: The Development of Extragalactic Astronomy 1885-1965, Part I," JHA, 39 (2008), 91–119; and Part II, JHA, 40 (2009), 71–107; and Smith, *The Expanding Universe: Astronomy's Great Debate, 1900-1931* (Cambridge: Cambridge University Press, 1982), pp. 3 ff.; Barbara Becker, *Unravelling Starlight* (Cambridge: Cambridge University Press, 2011).

[11] Huggins, "On the Spectra of Some of the Nebulae," PTRAS, 154 (1864), 437–444.

The next major step came in 1922, when astronomer Edwin Hubble announced that H II regions with their emission line spectra were found near hot O and B1 spectral type stars, while reflection nebulae (S 28) with continuous spectra were found near cooler stars.[12] He credited Henry Norris Russell with advancing the theory that nebulae with emission spectra, both diffuse and planetary, "are excited to luminosity by radiations from involved or neighboring stars." Russell had indeed advanced this theory, both in a 1921 article in the journal *Observatory* and in the *Proceedings of the National Academy of Sciences* the following year, stating that "the luminosity of gaseous nebulae is probably due to excitation of the individual atoms by radiations of some sort (ethereal or corpuscular) emanating from neighboring stars of very high temperature. In the Orion nebula the stars of the Trapezium (theta Orionis) appear to be the source of excitation."[13] Russell added that this did not mean that the luminous gas constituted most of the matter of the nebula, simply that it was the most excited by the stellar radiation. Hubble tested this theory in his landmark 1922 paper and found it to fit the observations. In 1926, Donald Menzel and Herman Zanstra independently explained the primary mechanism by which emission lines in a gaseous nebulae are produced as ionization by absorption of stellar ultraviolet radiation followed by recombination.[14]

As yet, however, the nature of the ionized gas was unknown, even though spectra showed hydrogen, helium, carbon, nitrogen, and oxygen lines. In 1934, Arthur S. Eddington had argued that in most regions of space, hydrogen exists in a non-ionized state (H I). In 1938 Otto Struve and Christian Elvey observed extended areas of the Galaxy exhibiting hydrogen Balmer emission lines and

---

[12] E. P. Hubble, "The source of luminosity in galactic nebulae." *ApJ*, 56 (1922) 400–438.

[13] Henry Norris Russell, *Observatory*, 44 (1921), 72, and Russell, "Dark Nebulae," PNAS, 8, (1922), 115–118. Dick (2013), pp. 82–83.

[14] D. H. Menzel, "The Planetary Nebulae," PASP, 38 (1926), 295–312; H. Zanstra, *Physical Review*, 27 (1926), 644. On Zanstra's work, which Otto Struve described as being based on "a truly revolutionary idea," see Struve and Zebergs, Astronomy of the 20th Century, pp. 386–390; R. O. Redman, "The award of The Gold Medal to Professor Herman Zanstra,"QJRAS, 2 (1961), 109–111, and Zanstra's summary of his own work at "The Gaseous Nebula as a Quantum Counter (George Darwin Lecture)," QJRAS, 2 (1961), 137–148.

argued that these clouds were mainly ionized hydrogen.[15] This was immediately followed up by Danish astronomer Bengt Strömgren, who studied H II regions in more detail and published his results in 1939. He originated the concept now called the "Strömgren sphere," the sphere of hydrogen that can be kept ionized around a star of given luminosity. In his classic 1948 paper, noting that Struve and Elvey's work published in 1938 had already shown that "From the observed strengths of interstellar emission lines, we know that hydrogen is by far the most abundant element in interstellar space," Strömgren further distinguished ionized H II regions from non-ionized H I regions.[16]

While today both the term "emission nebula" and "H II region" are used, the latter is more specific and was assured of continued usage after US Naval Observatory astronomer Stewart Sharpless, having published *A Catalogue of Emission Nebulae* in 1953, published his second catalogue of such nebulae in 1959 under the title *A Catalogue of H II Regions*. There he explained that "the term 'H II region' (Stromgren, 1948) is used here instead of the term 'emission nebula.' An H II region is an entity defined not only in terms of the ionized gas but also in terms of the hot stars which are responsible for the ionization."[17] These catalogues helped trace the spiral arms of the galaxy, several years before it was traced by radio observations of the cooler H I regions.[18]

[15] Struve and Elvey, "Emission Nebulosities in Cygnus and Cepheus," Ap J, 88 (1938), 364; Struve and Elvey, "Observations Made with the Nebular Spectrograph of the Mcdonald Observatory," series of three articles in ApJ, 89 (1939), 119ff and 517 ff., and vol. 90, 301 ff; Arthur S. Eddington, "The density of interstellar calcium and sodium," MNRAS, 95 (1934), 2–11.

[16] Bengt Strömgren, "The Physical State of Interstellar Hydrogen," ApJ, 89 (1939), 526; Strömgren, "On the Density Distribution and Chemical Composition of the Interstellar Gas," ApJ 108 (1948), p. 242.

[17] Stewart Sharpless, "A Catalogue of H II Regions," *Astrophysical Journal Supplement*, vol. 4 (1959), p. 257; A Catalogue of Emission Nebulae Near the Galactic Plane, ApJ, 118 (1953), p. 362.

[18] W. W. Morgan, S. Sharpless, and D. Osterbrock, "Some Features of Galactic Structure in the Neighbourhood of the Sun," AJ, 57, 3 (1952); and *Sky and Telescope*, 11, 134 (1952).

To add to the nebular nomenclature that students still find confusing, the term "diffuse nebula" is sometimes used to refer to any nebula with irregular outlines, including dusty reflection nebulae and gaseous emission nebulae. Moreover, diffuse nebulae are commonly mixtures of gas and dust clouds that become visible when they appear as emission, reflection or absorption (dark) nebulae, depending on their environment. This is certainly true of the Orion Nebula and the associated Orion Molecular Cloud out of which stars are forming. All these associated nebulae have in common the characteristic that they are stellar or pre-stellar nebulae, as opposed to post-stellar nebulae like supernovae remnants (S 19 and S 26) and planetary nebulae (S 17).

More on the Orion Nebula image at http://hubblesite.org/image/1826/news/3-nebulae. A spectacular 3D journey through the Orion Nebula, courtesy of the Hubble and Spitzer Telescopes, is at http://hubblesite.org/news_release/news/2018-04/3-nebulae.

Class S 25: Molecular Cloud

A molecular cloud is an interstellar gas cloud dense enough to permit formation of molecules, mostly hydrogen (H_2) but also laced with more complex molecules. An atomic cloud (S 23) becomes a molecular cloud when the density increases about a thousandfold. Hydrogen and more complex molecules are fragile, and this density helps shield them from ultraviolet radiation, which tends to tear them apart in the vicinity of a star, where the hot ionized clouds known as H II regions (S 24) are formed.

Molecular clouds form 25% of the interstellar medium, and at about 20 K they constitute the cold component. Because gravitational attraction within the cloud overcomes the small outward pressure due to low temperature, molecular clouds are the perfect place for the formation of protostars (S 1). They are also important to understanding molecular complexity in protoplanetary disks (P 1) surrounding some of those young stars. A molecular cloud also harbors interstellar dust (S 27), on which the gas molecules are believed to form.

The study of molecules in interstellar clouds, as well as in comets and other astrophysical environments, has given rise to the field of molecular astrophysics, sometimes called astrochemistry. Amazingly, it turns out interstellar molecular clouds are the sites not only of active chemistry, but also of active organic chemistry. Because molecular spectral lines are often in the microwave region of the spectrum, radio astronomy has been the chief technique for discovering molecules in space, particularly in the millimeter and submillimeter domains. Detections are sometimes also made in the infrared.

Molecular clouds have diameters ranging from less than one light year to about 300 light years and contain enough gas to form 10 to 10 million stars like our Sun. At least two types of molecular clouds are distinguished based on size: small molecular clouds less than a few hundred times the mass of the Sun, known as Bok globules, and giant molecular clouds (GMCs) that exceed the mass of millions of suns. The Dutch-American astronomer Bart Bok first drew attention to what are now called Bok globules in 1947; several dozen of them were already evident in photographic atlases such as E.E. Barnard's catalog of dark nebulae, but Bok singled them out for special attention as possible places of starbirth. In 1950, South African astronomer Andrew Thackeray found such globules in the open cluster IC 2944, also known as the Running

Chicken Nebula, and they have been an object of study as possible star formation sites ever since.[19]

Little is known about their nature, but they are only a few thousand astronomical units across and often associated with H II regions. Because star formation is taking place in these dark globules and because they are rich in molecules, they seem to be classified better as a type of molecular cloud rather than as a dark nebula. While the smaller Bok globules may give birth to double or multiple stars, giant molecular clouds over their long lifetimes may form stars by the hundreds, thousands, or millions. The Hubble Space Telescope famously revealed such a stellar nursery in the three gaseous pillars of the Eagle Nebula (Fig. 7.1).

Molecules were found in space long before molecular clouds were discovered. Despite the lack of expectations, the possible existence of molecules in space was deduced already in the 1930s when Mt. Wilson astronomers Paul W. Merrill, Theodore Dunham, Jr., and Walter S. Adams discovered spectral lines that could not be identified with atomic transitions.[20] In 1937, the Belgian spectroscopist Polydore Swings and Leon Rosenfeld (a younger associate of Neils Bohr) calculated that diatomic compounds such as OH, CH, NH, O_2, CO, and CN should occur in interstellar space, and in 1940 Andrew McKellar, an astronomer working at the Dominion Astrophysical Observatory in Canada, identified three astrophysical spectral lines with CH, CN, and NaH molecular spectra as produced in the laboratory. McKellar wrote:

> If these identifications are proved true, they are of considerable interest and importance in that they constitute the first definite evidence of the existence of molecules in interstellar space. Furthermore, they demonstrate the presence of carbon and

---

[19] Bart J. Bok and Edith F. Reilly, "Small Dark Nebulae," ApJ, 105 (1947), 255–257. A. D. Thackeray, "Some southern stars involved in nebulosity," MNRAS, 110 (1950), 524; Bo Reipurth, "Star formation in Bok globules and low-mass clouds. I - The cometary globules in the GUM Nebula," A & A, 117 (1983), 183–198. Dick (2013), pp. 88–89.

[20] P. W. Merrill, "Unidentified Interstellar Lines," PASP, 46 (1934), 206–207, and Merrill, "Stationary Lines in the Spectrum of the Binary Star Boss 6142," ApJ, 83 (1936), 126–128; T. Dunham Jr., and W. S. Adams, PASP, 9 (1937), 5; T. Dunham, Jr., "Interstellar Neutral Potassium and Neutral Calcium," PASP, 49 (1937), 26–28; Dunham, "Forbidden Transition in the Spectrum of Interstellar Ionized Titanium," *Nature*, 139 (1937), 246–247. See "The pioneering investigations in the field of the interstellar molecules, 1935–1942," *Astrophysics and Space Science*, 55 (1978), 263–265

nitrogen in interstellar space and provide direct observational basis for the view, held by astronomers for many years, that there must be an abundance of hydrogen in the vast spaces between the stars.[21]

After this discovery, however, almost a quarter century of inactivity passed. As astronomer James Kaler later wrote, "Most astronomers were fairly well convinced that interstellar molecules were not very important—difficult to make in the low densities and temperatures of space, and easily destroyed by high-energy stellar radiation."[22]

Nevertheless, theorists Charles Townes and Joseph Shklovskii predicted that some lines of interstellar molecules might appear in the radio spectrum, and in 1963 absorptions lines from the hydroxyl radical (OH) were identified in the supernova remnant Cassiopeia A.[23] In 1968, Townes and his colleagues found ammonia (NH_3) emissions toward the center of the Galaxy, followed by water one year later and the first organic molecule, formaldehyde (H_2CO.) The discovery of carbon monoxide (CO) emission in 1970 proved especially valuable, because it traces molecular hydrogen, otherwise unobservable at temperatures less than 100 K.[24] Using CO as a tracer, in 1975 Nick Scoville and P.M. Solomon, following the first survey of CO emission in the galactic plane, reported that a large fraction of interstellar hydrogen is in molecular form.

---

[21] Andrew McKellar, "Evidence for the Molecular Origin of Some Hitherto Unidentified Interstellar Lines," PASP, 52 (1940), 187–192; P. Swings and L. Rosenfeld, "Considerations Regarding Interstellar Molecules," ApJ, 86 (1937), 483–486. McKellar cites Otto Struve and C. T. Elvey as having supported the abundance of hydrogen by their observation of H alpha emission in a number of extended Milky Way regions, Struve and Elvey, "Emission Nebulosities in Cygnus and Cepheus," Ap J, 88 (1938), 364 (See entry S 24). See also Dick (2013), p. 87.

[22] Kaler, *Cosmic Clouds: Birth, Death and Recycling in the Galaxy* (Scientific American: New York, 1997), 104.

[23] S. Weinreb, A. H. Barrett, M. L. Meeks, and J. C. Henry, "Radio Observations of OH in the Interstellar Medium," *Nature*, 200 1963), 829–831. For the competitive environment see Alan H. Barrett, "The Beginnings of Molecular Radio Astronomy," in K. Kellermann and B. Sheets, *Serendipitous Discoveries in Radio Astronomy* (Green Bank: NRAO, 1983), 280–290.

[24] Alan Barrett, in Kellermann and Sheets, 286; George Carruthers, "Rocket Observation of Interstellar Molecular Hydrogen," ApJ, 161 (1970), pp. L81–L85; for a review see Carruthers, "Atomic and Molecular Hydrogen in Interstellar Space," SSR, 10 (1970), pp. 459–482, especially 476-480 for history of how molecular hydrogen was deduced in dark dust clouds.

Ever more complex molecules like ethyl alcohol were discovered, and by 1985, 68 interstellar molecules had been reported.[25]

Perhaps most astonishing of all was the reported detection in 2003 by Yi-Jehng Kuan and coworkers of interstellar glycine (NH_2CH_2COOH), the simplest amino acid, one of the building blocks of life. The detection was reported in three sources, including the hot molecular cloud in the galactic center known as Sagittarius B2 (Sgr B2), and was based on the observation of 27 lines in 19 different spectral bands. However, in an indication of the difficulty of identifying such complex molecules, in 2005 Lewis Snyder, one of the pioneers in the field of interstellar molecule detection, concluded that "key lines necessary for an interstellar glycine detection have not yet been found."[26] It is still considered unconfirmed as of 2018. At this level of complexity, the ensemble of lines is modeled, as opposed to single-line detection for less complex molecules, making identification more difficult. In 2018 astronomers reported the detection of benzonitrile, likely formed from the reaction of the cyanogen radical (CN) with benzene. This is important because large polycyclic aromatic hydrocarbons, organic compounds containing only carbon and hydrogen, are believed to form from benzene, and are possibly related to the origins of life.[27] PAHs are ubiquitous in the universe and were found in the famous Mars rock ALH 84001, one of the lines of evidence leading to the claim in 1996 that it contained nanofossils. PAHs have also been found in the atmosphere of Titan, the large moon of Saturn.

To date, about 200 molecular species have been identified in interstellar molecular clouds with up to 13 atoms, and the number and type of (generally similar) molecular species in comets is rapidly catching up. Comets are the potential "delivery system" or these molecules to Earth, because we know comets have impacted Earth in the past. In general, the more complex the molecule, the less its abundance.

---

[25] N. Z. Scoville and P. M. Solomon, "Molecular Clouds in the Galaxy," ApJ, 199 (1975), L105–L109; Lewis Snyder, "The Search for Biomolecules in Space," in K. I. Kellermann and G. A. Seielstad, eds., *The Search for Extraterrestrial Intelligence*," (NRAO, 1986), 39-50. This and the following paragraphs are from Dick (2013), p. 87.

[26] Lew Snyder et al, "A Rigorous Attempt to Verify Interstellar Glycine," ApJ, 619 (2005), 914–30. The original discovery paper was Yi-Jehng Kuan et al.,"Interstellar Glycine," ApJ, 593 (2003), 848–867.

[27] Christine Joblin and Jose Cernicharo, "Detecting the Building Blocks of Aromatics," *Science*, 359 (2018), 156–157.

Using CO as a tracer, in 1977 Solomon, David Sanders, and Scoville estimated about 3,000 Giant Molecular Clouds in the Galaxy, with dimensions 10 to 80 parsecs (30 to 250 light years); this may have been the first use of the term "Giant Molecular Cloud," extending the earlier use of "Molecular Cloud."[28] Molecular clouds are dominated by molecular hydrogen, followed by carbon monoxide, which is much easier to observe. The most abundant and famous molecular clouds in our Galaxy are the Orion Molecular Cloud and the Sagittarius B2 cloud near the center of our Galaxy (Fig. 11.4). Like H I and H II regions, molecular clouds are found primarily in spiral arms, but molecular hydrogen is concentrated much more to the center of the galaxy. Molecular clouds have also been found in other galaxies.[29] A typical spiral galaxy contains about 1,000 to 2,000 Giant Molecular Clouds and many more smaller ones.

Fig. 11.4. Combined imagery from the Hubble (yellow), Spitzer (red), and Chandra (blue and violet) space observatories of the central region of our Milky Way Galaxy. The Sagittarius B2 molecular cloud, several hundred light years from Sagittarius A, has a total mass about three million times that of the Sun. The entire image is about ½ a degree, the width of the full Moon. Credit: NASA, ESA, SSC, CXC, and STScI

---

[28] N. Z. Scoville and P. M. Solomon, "Molecular clouds in the Galaxy," ApJ, 199 (1975), L105–L109; P. M. Solomon, D. B. Sanders, and N. Z. Scoville, "Giant Molecular Clouds in the Galaxy: A Survey of the Distribution and Physical Properties of GMC's," BAAS, 9 (1977), 554; full article in Solomon, Sanders and Scoville, "Giant molecular clouds in the Galaxy - The distribution of CO-13 emission in the galactic plane," ApJ, 232 (1979), L89–L93.

[29] Yasuo Fukui and Akiko Kawamura, "Molecular Clouds in Nearby Galaxies," ARAA, 48 (2010), 547–580.

Astrochemistry remains a very active field of research.[30] Despite progress, one great mystery that remains are the so-called "diffuse interstellar bands," broad, shallow spectral lines in sharp contrast to the narrow emission or absorption features found in most atomic or molecular interstellar spectra. Paul W. Merrill discovered the first diffuse interstellar bands in 1938. About 200 have been identified since then. Their source is still largely unknown, but one of the leading theories is that they are molecular in origin, perhaps complex polycyclic aromatic hydrocarbons (PAHs).[31]

An updated list of interstellar molecules, divided into galactic (interstellar medium or circumstellar shells) and extragalactic and conveniently displayed by number of atoms, is at http://www.astro.uni-koeln.de/cdms/molecules. Other useful lists with discovery dates are at http://physics.nist.gov/PhysRefData/Micro/Html/tab1.html, and more information at http://www.astrochymist.org/astrochymist_mole.html and http://www.astrochymist.org/astrochymist_ism.html. The Cone Nebula is a good example of how different classes of the interstellar medium can coexist in the same location. Thus, the Hubble Space Telescope image at http://hubblesite.org/image/1189/news/3-nebulae shows the cold molecular hydrogen cloud in front of a faint H II region ionized by the star S. Monocerotis. The dark interior and the outer regions are populated with large hydrocarbons.

---

[30] See, for example, NASA's Cosmic Ice Laboratory at GSFC http://science.gsfc.nasa.gov/691/cosmicice/ Also the Astrophysics and Astrochemistry Lab at NASA Ames http://astrochemistry.org/

The International Astronomical Union has an Astrochemistry Working Group http://www.strw.leidenuniv.nl/iau34/

And see other labs at http://www.astrochymist.org/astrochymist_links.html and at http://www.strw.leidenuniv.nl/iau34/astrochemresearch.html

[31] George Herbig reviews the subject in "Diffuse Interstellar Bands," ARAA, 33 (1995), 19–75.

Class S 26: White Dwarf Supernova Remnant

Unlike supernovae that explode after the core collapse of a massive star and leave a remnant (S 19) around the resulting neutron star or black hole, and unlike a nova remnant (S 18) that surrounds its binary system, a white dwarf supernova remnant (Type Ia in astronomical lingo) results from accretion of material onto a white dwarf at the Chandrasekhar limit, and its explosion normally leaves no star behind, or a star in the process of being ejected from its system. It thus belongs to the Interstellar Medium Family rather than the Circumstellar Family (Chapter 9), justifying the placement of the two types of supernovae remnants into separate classes. However, possible stellar remnants have been reported for two white dwarf supernovae, so the question of Family placement remains open. In any case, white dwarf and core collapse supernovae result from very different mechanisms, and their physical constitution is different: an exploding core collapse object spews forth all the heavy elements that have been generated in the many fusion burning layers of its massive star, while a white dwarf is composed largely of carbon and oxygen.

Famous white dwarf supernova remnants include those from the supernova explosions of 1006, 1572, and 1604; the latter two were observed by Tycho Brahe and Johannes Kepler, respectively, and are known as Tycho's supernova and Kepler's supernova. The Crab Nebula, on the other hand, is an example of a core collapse supernova remnant (Fig. 9.6), dating from the explosion of 1054. White dwarf supernovae remnants show no hydrogen or helium lines; some show emission lines from the iron group in their later phases.

While the Crab Nebula remnants were discovered already in the 18th century (though not convincingly recognized as supernova remnants until 1942), the remnants for Type Ia white dwarf supernovae were more elusive. Three of the most famous were only recognized in 1943, 1952, and 1965, and their true nature was only gradually revealed. In 1943, Walter Baade identified what was then called "Nova Ophiuchi of 1604," observed by Kepler and others, as a Type I supernova based on its light curve. The same paper announced its remnant nebulosity, declaring that "the fact that it is a remnant of the former Supernova Ophiuchi seems to be established beyond doubt;" its radio detection was still uncertain in

Fig. 11.5. A composite image of the remnant of Tycho's supernova, the result of a Type Ia white dwarf explosion The circular blue line at the outer edge of the shock wave consists of ultra-energetic electrons seen in X-ray. Credit for X-ray: NASA/CXC/SAO; for infrared: NASA/JPL-Caltech; optical: MPIA, Calar Alto, O.Krause et al.

1968.[32] Kepler's supernova remnant (Fig. 8.14) has been observed in detail from the ground and space. Like other supernova, the ejecta from the supernova itself have swept up the surrounding interstellar gas into an expanding shell. A combined image from Hubble, Spitzer, and Chandra telescopes revealed this spherical shell of gas and dust 14 light years away, expanding at four million miles per hour (2,000 km per second). The material is rich in iron.

The remnant of Tycho's supernova (Fig. 11.5) was not discovered until R. Hanbury-Brown and Cyril Hazard observed it as a radio source in 1952, and its optical remnants were not discovered until the 1960s from Mt. Palomar. In 2004, the companion to Tycho's exploded white dwarf was found, rushing away from the scene of the explosion. Its velocity, three times that of stars its

---

[32] Walter Baade, "Nova Ophiuchi of 1604 as a Supernova," ApJ, 97 (1943), 119–128. See also R. Minkowski's review "Nonthermal Galactic Radio Sources," in *Nebulae and Interstellar Matter*, Barbara M. Middlehurst and Lawrence H. Aller, eds., Stars and Stellar Systems series, vol. 7 (Chicago: University of Chicago Press, 1968), pp. 623–666.

vicinity, was crucial to its association with the supernova event. "There was no previous evidence pointing to any specific kind of companion star out of the many that had been proposed. Here we have identified a clear path: the feeding star is similar to our Sun, slightly more aged...The high speed of the star called our attention to it," astronomer Pilar Ruiz-Lapuente stated.[33] Despite its high speed, because of its distance after 400 years the star appears only slightly off center from the expanding shell, but well on its way to ejection from the system. It is this phenomenon that might raise questions about a separate class of white dwarf supernovae remnants in the Interstellar Medium Class; the remnants would be in the Circumstellar Family (Chapter 9) until ejection. But at what point does it evolve from a circumstellar to an interstellar object?

The supernova of 1006 is believed to be the brightest supernova, and probably the brightest stellar event of any kind, ever recorded. Historian Bernard Goldstein first provided evidence for the supernovae in the southern constellation Lupus based on Egyptian and Far Eastern sources, and estimated its absolute magnitude as between -8 and -10. Based on new data in 2003, astronomer Frank Winkler estimated its brightness at -7.5, halfway between the brightness of Venus and the Full Moon. Historical records indicate it may have been visible for up to 2.5 years. The monks of the Benedictine Abbey of St. Gaul in Switzerland recorded a star "dazzling to the eyes. . .in the extreme limits of the south," meaning near the southern horizon, while the Egyptian physician and astrologer Ali ibn Ridwan described it as "2.5 or 3 times the magnitude of Venus,...and a little more than a quarter of the brightness of the Moon."[34]

After an unsuccessful search for its remnant in the optical spectrum by Baade, Frank Gardner and Doug Milne discovered the remnant as a radio source in 1965. Frank Winkler and F.N. Laird reported X-ray emission in 1976, and in the same year Sidney van

---

[33] R. Hanbury-Brown and C. Hazard, "Radio-Frequency Radiation from Tycho Brahe's Supernova (A.D. 1572)," *Nature* **170** (1952), 364–365; Pilar Ruiz-Lapuente *et al.*, "The binary progenitor of Tycho Brahe's 1572 supernova," *Nature* **431** (2004): 1069–1072. The quote is from HST Release, October 27, 2004, "Stellar Survivor from 1572 A.D. Explosion Supports Supernova Theory," http://hubblesite.org/news_release/news/2004-34

[34] Bernard R. Goldstein, "Evidence for a Supernova of A.D. 1006," AJ, 70 (1965), 105–114; Bernard R. Goldstein and Ho Peng Yoke, "The 1006 Supernova in Far Eastern Sources," AJ 70 (1965), 748–753; N. A. Porter, "The Nova of AD 1006 in European and Arab Records," JHA, 5 (1974), 99–104.

den Bergh reported faint optical filaments. In 2003, Winkler and his team found a shell of glowing hydrogen about the size of the full Moon where the supernova had been seen almost 1,000 years earlier. They calculated the filaments were moving at 2,900 km per second, over six million miles an hour— about 1% the speed of light. They also calculated its distance at 7,100 light years. The International Astronomical Union held a "Millennium Party" for the supernovae at its General Assembly in Prague in 2006, the same meeting that saw the demotion of Pluto to dwarf planet status.[35]

More information on the composite image of Tycho's supernova is at http://chandra.harvard.edu/photo/2009/tycho/, and the press release about its probable surviving companion to is at http://hubblesite.org/newscenter/archive/releases/2004/34/.

---

[35] F. F. Gardner and D. K. Milne, "The Supernova of A. D. 1006," AJ, 70 (1965), 754; R. Minkowski, "Supernova of +1006," AJ, 71 (1966), 371–373. P. Frank Winkler, Gaurav Gupta, and Knox Long, "The SN 1006 Remnant: Optical Proper Motions, Deep Imaging, Distance, and Brightness at Maximum," ApJ, 585 (2003), 324–335, including the history of its observation. David DeVorkin recounts the story of its identification as a supernova in his fascinating article "The A. D. 1006 Puzzle," *The Astronomy Quarterly*, 5 (1985), 71–86. In addition to a cake, the IAU's millennium event consisted more substantially of "Joint Discussion 9, Supernovae: one Millennium after SN 1006," with summary papers in *Highlights of Astronomy*, vol. 14, Karel A. van der Hucht, ed. (Cambridge, CUP: 2007), pp. 299–307.

Class S 27: Dark Nebula

A dark nebula, also called an absorption nebula because the gas component absorbs most or all of the light from the hotter star or emission nebula behind it, is composed largely of interstellar dust interspersed with gas. It becomes visible when it blocks light from background stars or emission nebulae such as H II regions (S 24). It differs from a reflection nebula (S 28) in that it either has no star nearby to scatter light off the dust particles, or the cloud is so thick that it absorbs rather than reflects starlight as it does in a reflection nebula. A thin, dark nebula may diminish light, giving rise to the phenomenon of the "extinction" of starlight. The same dust also reddens starlight. Such extinction and reddening also occur in dust thinly diffused throughout space, but the more detectable dark nebulae were discovered long before diffuse dust. Dark nebulae range in size from the famous Horsehead nebula in Orion to smaller dark nebulae known as Bok globules.

The mass of dust in our Galaxy amounts to only about 1% of the mass of interstellar gas and is less than .1% of the total mass of the Galaxy. Nevertheless, because of interstellar extinction and reddening, interstellar dust has an important effect on our understanding of galactic structure and plays a considerable role in processes in the interstellar medium. Dust grains provide the reaction site for the formation of complex molecules and constitute the building blocks for planet formation (P 1). While a certain amount of diffuse gas exists throughout the interstellar medium, a significant amount of it is tied up in dark nebulae and reflection nebulae, both of which include some gas but are largely composed of dust. Abundant dust has also been found in intergalactic space (G 16), giving rise to the phenomenon of intergalactic extinction and reddening.[36]

Because Bok globules appear to be much more complex than normal dusty dark nebulae, astronomers today often classify them as a type of molecular cloud (S 25). Astronomer James Kaler in his 1997 book *Cosmic Clouds* remarked that "the globules, which are rich in molecules like CO, claim the right to be called molecular clouds, albeit small ones," hardly a ringing endorsement, but one based on the most recent evidence and reflecting the principle that

---

[36] E. Xilouris et al., "Abundant Dust Found in Intergalactic Space," ApJ, 651 (2006), L107–L110.

classification should be based on physical constitution rather than mere appearance. Bok globules, originally termed "small dark nebulae" in Bok's original article, are thus an example of how an object can have an ambiguous classification, to the extent that some would place it in a different class of object based on new knowledge.[37]

The larger dark nebulae are most notable in the Milky Way that stretches across the dark night sky. Viewed by the naked eye, this river of light appears to be split by numerous dark clouds, later catalogued and given names like the Coalsack in the Southern Cross and the Horsehead Nebula in Orion (Fig. 11.6). The Coalsack, which appears many times larger than the full Moon, is so prominently silhouetted against the Milky Way that Vincente Yanez Pinzon spotted it in 1499 during an expedition to the South American coast. Other explorers also noticed it; it is sometimes

Fig. 11.6. The Horsehead Nebula in Orion, a dark nebula superimposed on an H II emission nebula. Credit: NASA, ESA, and the Hubble Heritage Team (STScI)

---

[37] James Kaler, *Cosmic Clouds: Birth, Death and Recylcing in the Galaxy* (New York: W. H. Freeman, 1997), p. 119; Bart J. Bok and Edith F. Reilly, "Small Dark Nebulae," ApJ, 105 (1947), 255–257.

called "Magellan's spot." The Horsehead Nebula, also known as Barnard 33 and (at a distance of 1,500 light years) more than twice the distance of the Coalsack, takes its name from its shape and is part of the Orion Molecular Complex. It is a good example of a dark nebula superimposed on an H II emission nebula (S 24). Much more distant still is the Cone Nebula at 2,600 light years. First discovered by William Herschel in 1785, it is also silhouetted against an H II region. Among other examples are the Dark Horse Nebula, the Pipe Nebula, and Rho Ophiuchi; at about 400 light years, the latter is one of the closest and most studied. Like other large dark clouds, Rho Ophiuchi is a large complex of clouds, including the main molecular cloud. Spitzer and Chandra observations in the infrared and X-ray have revealed 300 young stellar objects with a median age of only 300,000 years.[38]

Although observed already by naked eye observers and telescopic observers such as William Herschel, the American astronomer E.E. Barnard was the first systematically to gather photographic evidence of dark nebulae in the 1890s and early 1900s and was the first to catalogue such objects, producing his first catalogue in 1907.[39] By 1927, he had catalogued 349 dark objects. In 1962, Beverly T. Lynds used the Palomar sky Survey to catalogued more than 1,800 dark nebulae, including Bok globules. Today, thousands of dark nebulae have been catalogued. Already in 1922, Henry Norris Russell speculated that stars may be born in dark nebulae.[40]

Barnard also displayed the ambiguities of nomenclature that arise when a new class of object is discovered. Convinced that he had "conclusive evidence that masses of obscuring matter exist in space," he went on to write:

> What the nature of this matter may be is quite another thing. Slipher has shown spectroscopically that the great nebula about Rho Ophiuchi is probably not gaseous; that is, it does not have the regular spectrum of a gaseous nebula. The word

---

[38] Spitzer Release, February 11, 2008, "Young Stars in their Baby Blanket of Dust," http://www.nasa.gov/mission_pages/spitzer/multimedia/20080211. html

[39] E. E. Barnard, "On the Dark Markings of the Sky, with a Catalogue of 182 Such Objects," ApJ, 49 (1919), 1–24; E. E. Barnard, *A Photographic Atlas of Selected Regions of the Milky Way*, E. B. Frost and M. R. Calvert, eds., Carnegie Institution of Washington (1927)

[40] Henry Norris Russell, "Dark Nebulae," PNAS, 8, (1922), 115–118.

"nebula," nevertheless, remains unchanged by this fact, so that we are free to speak of these objects as nebulae. For our purposes it is immaterial whether they are gaseous or non-gaseous, as we are dealing only with the question of obscuration.[41]

Interstellar dust is of much more than passing importance. In 1930 Robert J. Trumpler demonstrated that light traversing dark, dusty clouds or thinner undetectable dust is subject to interstellar extinction and reddening. This dimming of starlight is most pronounced in the plane of the Milky Way, and Trumpler's meticulous observations of distant versus close star clusters showed that this dimming amounted to about one magnitude per kiloparsec, close to the modern value.[42] Eventually astronomers realized the implications of Trumpler's work: because extinction made objects appear dimmer (and therefore more distant) than they really were, taking this effect into account reduced the size of the Milky Way from 300,000 to 100,000 light years. Moreover, the same dust that causes interstellar extinction also causes interstellar reddening (not to be confused with spectral redshift). The dust scatters blue light preferentially, reddening the remaining light, an important effect when measuring the color indices of stars.

A Hubble gallery of dark nebulae is at http://hubblesite.org/news/13-dark-nebulae and http://hubblesite.org/newscenter/archive/releases/2002/01/image/a/. The Spitzer Space Telescope has imaged the rho Ophiuchus complex; see http://www.nasa.gov/mission_pages/spitzer/multimedia/20080211.html and a different view of the Eagle nebula at http://spitzer.caltech.edu/images/1708-ssc2007-01a-Cosmic-Epic-Unfolds-in-Infrared-The-Eagle-Nebula.More on the Horsehead nebula at http://hubblesite.org/image/3165/news_release/2013-12.

---

[41] Barnard, 1919, 2.

[42] R. J. Trumpler, "Preliminary results on the distances, dimensions and space distribution of open star clusters," *Lick Observatory Bulletin*, 14 (1930) 154–188.

Class S 28: Reflection Nebula

Reflection nebulae are clouds of dust that do not shine by their own light, but by light reflected from a nearby star. Such light is preferentially scattered in the blue rather than the red (the same phenomenon that gives us the blue skies of Earth), giving many reflection nebulae their bluish tinge. Red reflection nebulae, though rarer, also exist, especially around cooler stars. Reflection nebulae and gaseous H II regions (S 24) are often seen together where gas and dust are both present, and they are largely confined to within 20 degrees of the plane of the Galaxy. The nebula around the Pleiades open cluster (S 34) is a good example of a reflection nebula, in which a dust cloud is made visible by a chance proximity to the star cluster (Fig. 11.7). Reflection nebulae are quite common; one study of 500,000 stars in our Milky Way Galaxy revealed 500 associated reflection nebulae and indicated that at high galactic latitudes some were not associated with stars but were illuminated by the integrated light of the Milky Way.[43] Reflection nebulae are also observed in other galaxies.

Fig. 11.7. The Pleiades reflection nebula, about 440 light years distant. Credit: NASA, ESA, and AURA/Caltech

---

[43] S. Van den Bergh, "A study of reflection nebulae," AJ, 71 (1966), 990–998, including a catalog of reflection nebulae north of declination -33 degrees.

In 1859, using a four-inch refractor, Wilhelm Tempel discovered a nebula around Merope, one of the stars near the Pleiades. With the improvement of photography, in the 1880s Paul and Prosper Henry in Paris, and Isaac Roberts in England, first revealed the full extent of the nebula surrounding the Pleiades. Using analogy as an argument, many astronomers believed "all the clustered Pleiades constitute, and it were, a second Orion trapezium in the midst of a huge formation of which Tempel's nebula is but a fragment."[44] It was only in 1912, when V.M Slipher showed that the spectrum of the nebula around Merope was the same as that of the Pleiades star cluster and thus shined by reflected light, that "reflection nebulae" were first separated from other classes of nebulae. Slipher found absorption lines of hydrogen and helium, "a true copy of that of the brighter stars of the Pleiades... [containing] no traces of any of the bright lines found in the spectra of gaseous nebulae." Slipher speculated that it could not be gas, but "disintegrated matter similar to what we know in the Solar System, in the rings of Saturn, etc.," and which "shines by reflected star light."[45]

In 1922, Edwin Hubble further advanced this argument when he showed that emission nebulae occurred around hot O-type stars, while reflection nebulae were found around cooler stars, with the B1 spectral type harboring an equal number of emission and reflection nebulae.[46] This demonstrated that reflection nebulae are a mixture of gas and dust: the H II emission nebula around the hotter stars overwhelms the reflection nebulae that one would otherwise see from scattering in the dust, but in the cooler stars, the reflection nebula from the dust component dominates. If there is no nearby star, the dust blocks the distant starlight and the cloud is dark.

---

[44] Otto Struve, "Recent Progress in the Study of Reflection Nebulae," *PA*, 45 (1937), 9–22; Dick (2013), p. 78.

[45] V. M. Slipher, V. M., "On the spectrum of the nebula in the Pleiades," *Lowell Observatory Bulletin*, 1 (1912), 26–27.

[46] E. P. Hubble, "The source of luminosity in galactic nebulae." *ApJ*, 56 (1922) 400–438.

Henry Norris Russell identified reflection nebulae with dark nebulae and argued that "it appears probable that the aggregate mass contained in one of these great obscuring clouds must be very considerable—probably sufficient to form hundreds of stars—and that a sensible fraction of the whole mass must be in the form of dust less than .1 mm in diameter."[47] Today we know reflection nebulae, in addition to a gaseous component, are composed of interstellar dust grains, solid matter on the order of 1/1000 of a millimeter across composed mostly of carbon, silicon, and oxygen. Such dust grains are spread throughout interstellar space but are most evident by their absorption, which causes dark patches in our Milky Way Galaxy known as dark nebulae (S 27) and dark bands across other galaxies. The origin of interstellar dust grains is unknown, but they are clearly associated with young stars.

Dust grains, whether in clouds or more diffusely spread, are also responsible for the phenomena known as interstellar reddening and interstellar extinction. The dust grains absorb and scatter incoming light, removing it from the light that reaches us and making it appear redder. Dust grains also make light dimmer, causing interstellar extinction. The astronomer Robert Trumpler discovered these effects in 1930 when he noticed that distant star clusters were dimmer than expected.[48]

Yet another effect of interstellar dust is the polarization of starlight, discovered independently by William A. Hiltner and John S. Hall in 1949. As Otto Struve wrote, "The detection of interstellar polarization always will remain one of the most striking examples of a purely accidental discovery, such as Wilhelm Rontgen's discovery of X-rays in 1895. It required exceptional instrumental skill to detect the small variations of the light of a star as the analyzer was turned, but even more important as the quick realization that the effect was completely new and in no way foreshadowed by previous work."[49]

---

[47] Henry Norris Russell, "Dark Nebulae," PNAS, 8, (1922), 115–118: 117; James Kaler, *Cosmic Clouds: Birth, Death and Recycling in the Galaxy* (Scientific American: New York, 1997), pp. 65–69.

[48] R. J. Trumpler, "Preliminary results on the distances, dimensions and space distribution of open star clusters," *Lick Observatory Bulletin*, 14 (1930) 154–188.

[49] John S. Hall, "Observations of the polarized Light from Stars," *Science*, 109 (1949), 166–77, adjacent to W. A. Hiltner's paper "Polarization of Light from Distant Stars by Interstellar Medium," 165; Struve, *Astronomy of the Twentieth Century*, p. 372

One might well argue whether reflection nebulae should be a separate class from dark nebulae, since both are clouds consisting largely of dust with intermingled gas, and classification should be based on physical criteria rather than appearance. But as Kaler has pointed out, dark nebulae and reflection nebulae are distinguished by physical processes as well as by appearance. A dark cloud removed from the vicinity of a star merely blocks any starlight from background stars, whereas the dust in a dark cloud near a star is absorbing or scattering starlight as well as undergoing heating. A thick cloud is seen as a dark nebula, while a thinner cloud will appear as a reflection nebula, its brightness dependent on the amount of scattered and absorbed light.

NGC 1999 at http://hubblesite.org/newscenter/archive/releases/2000/10/image/a/ is another example of a reflection nebula. More on the Pleiades image is found at http://hubblesite.org/news_release/news/2004-20/107-illustrations. And more on reflection nebulae in general can be access by entering "reflection nebula" into the search field at http://hubblesite.org/news.

Class S 29: Stellar Winds

Just as our Sun emits a solar wind (P 16) consisting of charged particles from its upper atmosphere, so do a wide variety of other stars. In fact, all stars are believed to have stellar winds of one kind or another, including exotic specimens such as white dwarfs. While winds from solar type stars are similar to the Sun's in speed and composition, more massive O and B type stars have lower mass loss but with much higher velocities approaching 1,500 miles per second. Highly evolved post-main sequence stars such as red giants (S 7 and S 8) and supergiants (S 9), on the other hand, can eject large quantities of mass at much lower velocities, less than 10 miles per second. Mass loss from these evolved stars can be a billion times that of the Sun, resulting in circumstellar shells (S 16). Those extremely massive stars known as Wolf-Rayet stars, believed to be stars about to explode as supernovae, lose an enormous mass through their winds, equivalent to one Earth per year. Neutron stars also produce extremely high-speed winds, approaching the speed of light.

While stellar winds such as the Sun's are unlikely to affect the course of a star's evolution on the main sequence, high mass loss among giants and supergiants could determine whether a particular star explodes as a supernova or settles down to the more sedate life of a white dwarf (S 12). Interacting stellar winds may also result in the exotic shapes of the planetary nebulae (S 17). At the opposite end of stellar life, winds from young T Tauri stars (S 2) start to blow when hydrogen is first ignited, halting the infall of matter and sealing the initial mass of the star. The star-forming region 30 Doradus (Fig. 11.8) gives a good indication of the role of stellar winds in shaping the surrounding environment. Although most heavy elements are delivered to the interstellar medium by supernova explosions, most of the carbon, nitrogen, oxygen, and other elements derive from the dense stellar winds of red giants. Stellar winds are also believed to drive galactic winds (G 17), an important mechanism for recycling matter and energy in galaxies.

The mechanisms for stellar winds vary depending on the type of star. Winds from solar-type stars arise from the pressure of the very hot corona, whereas in higher mass stars radiation pressure is believed to drive the wind expansion. Multiple methods exist for observing stellar winds. Spectral lines, known as "P-Cygni profiles" after the prototype star, are sensitive indicators of mass loss

Fig. 11.8. Stellar winds in 30 Doradus, an intense star-forming region in the Large Magellanic Cloud some 170,000 light years from Earth. The stars in the cluster known as R 136 (the large blue blob left of center) are producing intense stellar winds sculpting the gas and dust in the surrounding neighborhood. This view is about 150 by 200 light years across. Credit: NASA, N. Walborn and J. Maíz-Apellániz (STScI), R. Barbá (La Plata Observatory, La Plata, Argentina)

from hot stars. Stars with high mass loss may show atomic emission lines in their spectra. Stars with ionized stellar wind emit an excess of continuum radiation from the infrared to the radio region of the spectrum, and winds from cool stars show molecular emission lines. Infrared radiation by dust may also be observed.[50]

Confirmation of the existence of the solar wind in the 1960s was a leading indicator that other stars might have their own stellar winds, but there was as yet no hint of the great variety of these winds. Armin Deutsch reported low-velocity mass loss from cool M supergiants as early as 1956. A full-blown concept of stellar wind was not long in coming after Eugene Parker's landmark paper on the solar wind in 1958; indeed, the following year astronomer A.G.W. Cameron, citing Parker's paper, wrote that stars must lose mass in order to produce white dwarfs and "one method of mass loss is by a 'stellar wind' analogous to the solar wind." In 1960, Parker himself expanded his concept of solar wind to stellar winds in another landmark paper. Here he concluded

---

[50] Henny J. G. L. M. Lamers, "Observations of Stellar Winds," *Astrophysics and Space Science*, 260 (October, 1998), 63–80.

that "we expect at least all main-sequence stars later than class F to possess *stellar winds*, in analogy with the solar wind, so that the phenomenon is of widespread importance in the mass balance of the Galaxy, as well as in the early evolution of most stars." The boldness of Parker's assertions can be gauged by William McCrea's statement in 1962 that "in recent years there has been much interest in the phenomenon of the so-called solar wind. Even in the case of the Sun, the existence of the phenomenon is still in doubt."[51] Nevertheless, theories of stellar wind developed rapidly in the 1960s, followed quickly by observations. In 1967, Donald C. Morton reported Aerobee rocket-ultraviolet observations of O and B supergiants, showing broad P-Cygni profiles indicating high velocity outflows.[52]

Observations of stellar winds have been greatly refined by spacecraft. The International Ultraviolet Explorer (IUE), active from 1978–1996, allowed important stellar wind spectral lines from evolved stars to be observed in the ultraviolet. Pioneering observations were also made with the Far Ultraviolet Spectroscopic Explorer (FUSE) and the Hubble Space Telescope STIS instrument. The latter allowed astronomers to measure how different parts of the wind move off a star and how this phenomenon differs in other galaxies. Theoretical and observational studies of stellar winds remain a very active field in astrophysics, in particular the mechanism driving the mass loss in different classes of stars.

More information on Fig. 11.8 is at http://hubblesite.org/image/1080/category/72-open-clusters.

---

[51] A. J. Deutsch, "The Circumstellar Envelope of Alpha Herculis," ApJ, 123 (1956), 210; A. G. W. Cameron, "Pycnonuclear Reactions and Nova Explosions," ApJ, 130 (1959), 916–940; E. N. Parker, "The Hydrodynamic Theory of Solar Corpuscular Radiation and Stellar Winds," ApJ, 132 (1960), 821–866. Parker followed this up with his article "The Stellar Wind Regions," ApJ, 134 (1961), 20–27. W. H. McCrea, "Evidences of Evolution in Astronomy," QJRAS, 3, 63 (1962), 63–79.

[52] Donald C. Morton, "Mass Loss from Three OB Supergiants in Orion," ApJ, 150 (1967), 535–542.

Class S 30: Galactic Cosmic Rays

Galactic cosmic rays (GCRs) are energetic particles with energies ranging from about 100 MeV to ten GeV (10^8 to 10^{10} electron volts) and more, corresponding respectively to 43% and 99.6% of the speed of light for protons. As energetic as they are, galactic cosmic rays are outdone by extragalactic cosmic rays (G18), with energies in the 10^{20} eV range, believed to be emanating from the cores of active galaxies (G 6 through G 9).

Galactic cosmic rays are believed to originate within our own Galaxy in the expanding shockwaves of supernovae explosions (S 11, S 19, S 26) that may leak into the intergalactic medium, while solar energetic particles (P 16) and anomalous cosmic rays (P 17) originate in or near the Solar System. Galactic cosmic rays, which are what are usually meant when the term "cosmic rays" is used, are composed of about 89% hydrogen nuclei (protons), 10% helium nuclei, and 1% nuclei of heavier elements. They have sometimes been dubbed "cosmic rain" or "cosmic bullets," both apt metaphors for particles approaching the speed of light ripping through the more sedate interstellar medium.[53]

Cosmic ray research is part of a broader field known variously as particle astrophysics, astroparticle physics, and high-energy physics. NASA's Goddard Space Flight Center, for example, hosts an Astroparticle Physics Laboratory, which undertakes research in cosmic ray and gamma-ray high-energy astrophysics. Most high-energy cosmic rays never reach the Earth because they get entangled in interstellar magnetic fields. Those that do reach Earth cannot point back to their sources because they have been redirected by the magnetic fields. But because high-energy cosmic rays may produce gamma rays, gamma-ray observatories such as Compton and Fermi have observed them indirectly (Fig. 11.9), and they do point back to the source where they were produced. While cosmic ray physics is an important part of space science and a concern for interplanetary spacecraft electronics, galactic cosmic rays are also a fundamental medical concern of interplanetary human spaceflight.

---

[53] Roger Clay and Bruce Dawson, *Cosmic Bullets: High Energy Particles in Astrophysics* (Reading, Mass.: Addison Wesley, 1997); Michael Friedlander, *A Thin Cosmic Rain: Particles from Outer Space* (Cambridge, Mass.: Harvard University Press, 2000).

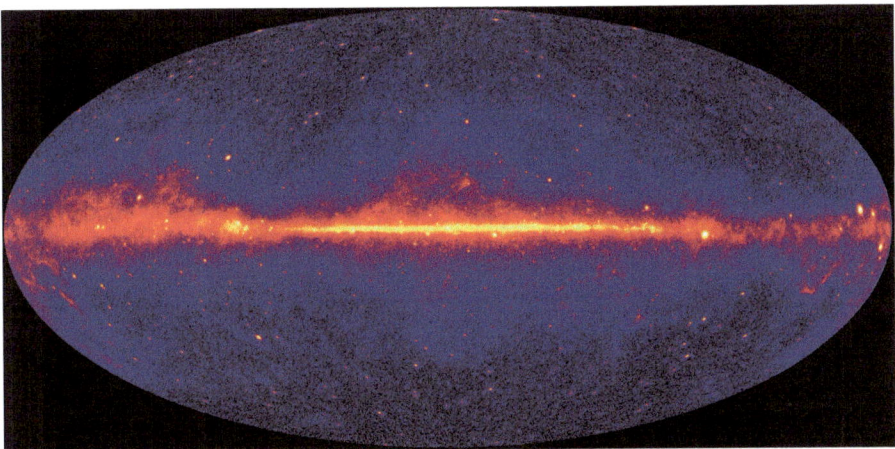

Fig. 11.9. The gamma-ray sky from five years of data with Fermi's Large Area Telescope. This is how the sky appears at energies greater than one billion electron volts. Some of the brightest sources are identified at https://svs.gsfc. nasa.gov/11342. Credit: NASA/DOE/Fermi LAT collaboration

The number of cosmic rays reaching Earth varies with their energy; counterintuitively, many more low-energy particles reach Earth than those with higher energy, in part because there are fewer high-energy particles. About 10,000 particles with 1 GeV energy reach Earth per square meter per second, while those with a thousand times more energy (1 TeV) arrive at a rate of only one per square meter per second. At the highest energies (extragalactic cosmic rays) this rate is reduced to one particle per square km per year. GCR activity is also modulated by solar activity, increasing during solar minimum and increasing during solar maximum. Historical data of beryllium 10 gathered from polar ice cores, and of carbon 14 from tree rings, shows that since 1950 GCR intensity has been at it lowest in 1,000 years, and that low solar activity over the last 10,000 years has often resulted in increasing intensity of GCRs. Studies using data from the IMP and Advanced Composition Explorer (ACE) spacecraft, and Pioneer, Voyager, and Ulysses at greater heliocentric distances, have provided a better understanding of cosmic ray modulation.[54]

---

[54] F. B. McDonald, W.R. Webber, D.V. Reames, "Unusual time histories of galactic and anomalous cosmic rays at 1 AU over the deep solar minimum of cycle 23/24," *Geophysical Research Letters*, 37 (2010), L18101.

As described in entry P 17, Victor Hess first detected galactic cosmic rays in 1912, and their nature as particles or rays was much debated into the 1930s, especially by the two Nobel Prize winners Robert Millikan and Arthur H. Compton, with the latter championing the particle side that we now know to be true. Progress in the new field of particle physics depended on new techniques for observing cosmic rays, and two of them proved fundamental for further progress in the 20th century: Hans Geiger's "Geiger counter" and Charles Wilson's cloud chamber.[55] Together with the ionization chamber and later the fluorescence technique and observation of the Cerenkov effect, cosmic ray research became a robust field of study in the 20th century.

An early discovery that proved key to cosmic ray research occurred when in 1927–1929 the French physicist Pierre Auger and Dimitry Skobeltzyn found the first evidence for "showers" of secondary particles produced in the Earth's atmosphere by cosmic rays.[56] After early work using Geiger counters to show the effect of the Earth's magnetic field on cosmic rays, Auger also found that if detectors were spaced apart, even at separations of 150 meters, the showers could still be observed. Auger recalled his discovery as follows:

> It was later on that I wanted to evaluate the extent of these showers by studying cosmic ray showers created in a lead screen by the observation of coincidences in two or three counters as a function of their separation. It was a surprise to observe coincidences when the counters were separated by more than one meter. Suspecting a new phenomenon, I decided to go whole hog, if I may so express myself, and thanks to the technical help of Roland Maze, we placed one of the counters in another building, more than one hundred and fifty meters away on rue Pierre Curie where my laboratory was. And there were still coincidences! It was the discovery of "cosmic ray showers." By pursuing this work at high altitudes, in order to increase the cosmic ray intensity, I showed that the showers covered more than a hectar in extent, and hence the number of

---

[55] Peter Galison, *Image and Logic: A Material Culture of Microphysics* (Chicago: University of Chicago Press, 1997)

[56] P. Auger *et al.*, "Extensive Cosmic-Ray Showers," Rev. Mod. Phys. 11 (1939): 288–291.

particles making up the showers was such that the energy of the primary particle which originated the "giant shower" must have been more than a million billion electron volts.[57]

This discovery of what is now known as extended air showers (EAS) was the key to the construction of large arrays now used to detect cosmic rays (see G 18). It meant that larger and larger arrays of detectors could be built to detect the increasingly lower numbers of cosmic rays found at higher and higher energies. One such array used to detect extended air showers of particles caused cosmic rays is the Akeno Giant Air Shower Array, located about 80 miles west of Tokyo. It is spread over an area of about 40 square miles and uses 111 scintillation detectors, each two square meters in area. One of the highest energy cosmic rays yet reported came from another type of detector, the "Fly's Eye detector" at the University of Utah. The problems of detection at such facilities can hardly be overemphasized. The Utah group compared their problem to that of detecting "a blue 5-watt light bulb streaking through the sky at the speed of light against a backdrop of starlight, atmospheric airglow, and man-made light pollution. In addition, sporadic sources of lights such as lightning, auroras, airplane, and smokestack strobe lights...create a certain visual havoc."[58]

In one way Millikan has been vindicated, since gamma rays are now known to be produced when cosmic rays interact with gas in our galaxy, either when accelerating protons collide with gas atoms or when fast-moving electrons fly past the nuclei of gas atoms. Astronomers can therefore infer the presence of cosmic rays from the glow of this gamma-ray emission and pinpoint their sources. The Fermi Gamma Ray Observatory (formerly known as GLAST) is producing landmark results in this area. In 2009 and 2010, Fermi reported that galactic cosmic rays were producing gamma rays due to acceleration in supernova remnants, including Cassiopeia A, W51C, W44, and IC 443. Younger supernova remnants seem to have stronger magnetic fields and the highest energy cosmic rays. In 2011, Fermi scientists reported that massive stars in OB associations near the star-forming region Cygnus X (not to be confused with the black hole X-ray source Cygnus X-1) are "likely the source of a substantial fraction of galactic cosmic rays." These stars produce cosmic rays when they result in core-collapse supernovae or black holes. The supernovae may

---

[57] Lars Persson, "Pierre Auger: A Life in the Service of Science," *Acta Oncologica*, 35:7 (1996), 785–787

[58] Friedlander, *Cosmic Rain*, 120.

form "superbubbles," in which the supernovae and shocks formed by high-velocity winds may accelerate cosmic rays.[59]

Since 2004, the European-funded, ground-based High Energy Stereoscopic System (HESS in honor of Victor Hess) has also produced groundbreaking work with its four telescopes located in Namibia. It can work at even higher energies—1,000 times that of Fermi and up to tens of TeV—using a technique pioneered in 1968 by the American astronomer Fred Whipple, realized today as VERITAS, the Very Energetic Radiation Imaging Telescope Array System, located at the Whipple Observatory in Arizona. When a cosmic ray hits the Earth's atmosphere, it produces a cascade of secondary particles that form an air shower. These particles emit a faint blue light known as Cerenkov radiation that illuminates an area of about 250 meters on the ground. Using its five Cerenkov telescopes in tandem, HESS can locate the source of the gamma ray. In its Census of Galactic Sources, HESS has found that most gamma rays emanate from pulsar wind nebulae, followed by unresolved winds, supernovae, and binary stars. These all have a connection to the evolution of heavy stars. Pulsars are more common sources than supernovae because they last much longer and because the electrons present in pulsar winds are a more efficient mechanism of gamma-ray production.

The Goddard Astrophysics program is described at http://science.gsfc.nasa.gov/sed/index.cfm?fuseAction=home.main&&navOrgCode=661. More gamma-ray maps are at https://svs.gsfc.nasa.gov/11342 and the latest news from the Fermi Observatory, including cosmic rays, is at https://fermi.gsfc.nasa.gov. Fermi celebrated its tenth anniversary of operation in 2018. On cosmic rays and the Cass A supernova remnant see https://www.nasa.gov/mission_pages/GLAST/science/cosmic_rays.html.

---

[59] A. A. Abdo et al., "Fermi-LAT Discovery of Extended Gamma-ray Emission in the Direction of Supernova Remnant W51C," ApJ, 706 (2009), L1; A. A. Abdo et al., "Fermi-Lat Discovery of GeV Gamma-Ray Emission from the Young Supernova Remnant Cassiopeia A," ApJ, 710 (2010), 92; A. A. Abdo et al, "Gamma-Ray Emission from the Shell of Supernova Remnant W44 Revealed by the Fermi LAT," *Science*, 327 (2010), 1103–1106; NASA Release, February 16, 2010, "NASA's Fermi Closes on Source of Cosmic Rays,"
http://www.nasa.gov/mission_pages/GLAST/news/cosmic-rays-source.html# . In total 41 SNR were found to be "potential" gamma-ray sources in the Fermi LAT catalogue, "Fermi Large Area Telescope First Source Catalog," http://arxiv.org/pdf/1002.2280v1, p. 46. The 2011 observations are reported in M. Ackermann et al., "A Cocoon of Freshly Accelerated Cosmic Rays Detected by Fermi in the Cygnus Superbubble," *Science*, 334 (2011), 1103–1107.

12. The Stellar Systems Family

Class S 31: Binary Star

The Family of stellar systems ranges from binary stars to globular clusters with millions of stars. Binary stars are two stars that are gravitationally bound in a mutual orbit. They are sometimes also termed double stars, but that designation may also include stars that only appear close together in the sky. It is estimated that at least 80% of all stars are members of binary or multiple star systems. Strictly speaking, binaries are multiple stars (S 32), but in common astronomical usage multiple stars are systems with three or more stars.

Binaries are in a separate category both for historical purposes and because of their great numbers and importance to astronomy. This separation is made clear by the designation of the International Astronomical Union's commission on "Binary and Multiple Star Systems," of which many of the world's experts on binary stars are members. The principle database of astrometric double and multiple star information is the Washington Double Star Catalog, maintained at the US Naval Observatory. As of 2018, it contains data for more than 146,000 systems. Of these about 27,000 are physical systems, about 5,000 are optical and the remainder are unknown. Published orbits exist for more than 2,000 pairs. The Hipparcos catalogue provides charts and data for more than 12,000 binary or multiple systems, and the Gaia spacecraft expects to chart tens of millions. A knowledge of binary stars is important for many reasons, not least as a method for measuring stellar masses using Newton's version of Kepler's laws. In other words, if the orbital period and separation of the stars can be determined, so can their masses.

The earliest binary star phenomena detected were the novae (S 18) seen already by the ancients, but no one at the time knew binary stars were involved in that spectacular phenomenon. Already in 1767, the British astronomer John Michell argued that stars appearing very close together in the sky might

© Springer Nature Switzerland AG 2019
S. J. Dick, *Classifying the Cosmos*, Astronomers' Universe,
https://doi.org/10.1007/978-3-030-10380-4_12

be gravitationally bound.[1] The term binary was first used in this context by William Herschel in 1802:

> "If, on the contrary, two stars should really be situated very near each other, and at the same time so far insulated as not to be materially affected by the attractions of neighbouring stars, they will then compose a separate system, and remain united by the bond of their own mutual gravitation towards each other. This should be called a real double star; and any two stars that are thus mutually connected, form the binary sidereal system which we are now to consider."[2]

Herschel observed some 700 "double stars" and was able to observe their motion together, arguing that they were physical doubles. Castor was the first such system he recognized. The first binary orbit was computed in 1827 by the French astronomer Felix Savary, the beginning of thousands now available.

Binary stars are usually classified according to the method of observation as visual (distinctly seen with the naked eye through the telescope), spectroscopic (distinguished by spectral line shifts due to the stars' mutual motions), photometric (by light variations), and astrometric (by positional changes). They are also classified as wide or close binaries and as eclipsing or non-eclipsing, among other characteristics. But following the principle of classification by physical characteristics, binary stars come in an astonishing array of types depending on which types of stars are paired and their proximity to each other. If a star is not close enough to exchange material with its partner, it may circle without much ado; the bright star Sirius, an A type main sequence with a white dwarf, is such a binary. But close binaries can interact, and because of the many different kinds of stars that can pair and interact, such pairings give rise to a large number of different phenomena.

Among the most spectacular interacting binaries are novae and Type Ia supernovae (S 11), both of which result from pairing a white dwarf with a stellar companion. During the 19th and early 20th centuries, novae were believed to be caused by collisions between stars or other objects; it was not realized until the 1950s and 1960s

---

[1] John Michell, "An Inquiry into the Probable Parallax, and Magnitude of the Fixed Stars, from the Quantity of Light Which They Afford Us, and the Particular Circumstances of Their Situation," PTRAS, 57 (1767), 234, reprinted in Bartusiak, 107–110.

[2] William Herschel, quoted in Robert Grant Aitken, The Binary Stars (New York: Dover, 1964), p. ix. For more context see Dick (2013), pp. 116–120.

that novae (short for "stella nova," meaning "new star") are the result of such a pairing. Merle Walker's remarkable discovery in 1954 of the binary nature of Nova DQ Herculis spurred speculation that all novae might be binaries, and a classic study of 10 novae by Robert P. Kraft in 1964 argued just that. "Since seven of the ten objects studied are certainly binaries," he wrote, "the hypothesis is advanced that membership in a certain type of close-binary system is a necessary condition for a star to become a nova."[3] In the case of the nova, the white dwarf undergoes increased nuclear fusion without exploding, resulting in much less brightness than a supernova and an accordingly smaller nova remnant (S 18). Supernovae can release a million times more energy than novae.

White dwarf pairings can also result in other exotic phenomena. A normal main sequence star paired with a white dwarf can produce a type of binary known as a cataclysmic variable (Fig. 9.9).[4] A post-main sequence Mira-type pulsating red giant paired with a white dwarf can result in an eruptive variable that may brighten by 10 magnitudes, which then dims after a few decades. The V407 Cygni system is a particularly interesting example because the Fermi Space Telescope has recently detected gamma-ray radiation emanating from this nova.[5] The Homunculus Nebula in Eta Carina (Fig. 8.11) is now believed to be caused by a binary that consists of a luminous blue variable supergiant and a white dwarf (S 10).

A normal star paired with an exotic object like a white dwarf, neutron star, or black hole, in which one star accretes matter from the other, may also result in an X-ray binary. The enigmatic SS 433 is an eclipsing X-ray binary in which the primary star is a stellar black hole or neutron star, and the secondary is a late A-type star. Because it has jets (S 20) and an accretion disk, it is also termed a microquasar, the first to be discovered. X-ray binaries are classified as low, intermediate, and high mass, depending on the mass of the normal star. A low-mass binary could be a main sequence or giant star paired with a neutron star or black hole. A high-mass

[3] Hilmar Duerbeck, "Novae: An Historical Perspective," in Michael F. Bode and Adeurin Evans, eds., *Classical Novae* (Cambridge: Cambridge University Press, 2nd edition, 2008), p. 3; Robert P. Kraft, "Binary Stars among Cataclysmic Variables. III. Ten Old Novae," ApJ, 139 (1964), 457.

[4] HST Release, May 22, 1995, "Cataclysmic Variable Star," http://hubblesite.org/image/292

[5] The Fermi-LAT Collaboration, "Gamma-Ray Emission Concurrent with the Nova in the Symbiotic Binary V407 Cygni," *Science*, 329 (2010), 817–821.

binary could be a supergiant paired with a white dwarf, neutron star or black hole. Cygnus X-1, a black hole orbiting a blue supergiant, is the most famous.

Even more exotic X-ray binaries exist, including binary black holes, which were first reported in 2009. In 2017, supermassive binary black holes were reported orbiting each other separated by only one light year, with an orbital period of about 100,000 years. It is therefore not surprising that they will eventually merge. As discussed in entries S 13 and S 14 on neutron stars and black holes, the first detection of gravitational waves in 2015 came from the merger of two black holes in a binary system system, and the second detection in 2017 from the merger of binary neutron stars.

At a more sedate level, a rare type of binary pairs a failed star known as a brown dwarf (S 22) with a normal star. A study of 233 nearby multiple-star systems by NASA's Hubble Space Telescope found only two brown dwarfs as companions to normal stars, but undoubtedly many exist throughout the Galaxy.

The Washington Double Star Catalog can be accessed at http://ad.usno.navy.mil/wds/. The Fourth Catalogue of Interferometric Measurements of Binary Stars, also maintained until 2018 at the Naval Observatory, contains measurements made by high-resolution techniques such as speckle interferometry for more than 93,000 stars. It can be accessed at http://ad.usno.navy.mil/wds/int4.html. The Hipparcos catalogue is at https://www.cosmos.esa.int/web/hipparcos/double-and-multiple-stars, and Gaia data can be accessed at https://gea.esac.esa.int/archive/.

Class S 32: Multiple Star

Multiple stars are gravitationally bound systems with three or more stars. Although stars can sometimes appear to be multiple by virtue of their proximity in the sky, these are not true multiples, in the same way the apparent double stars are not true binary stars (S 31). Such non-physical systems are sometimes referred to as "optical doubles" or "optical multiples." Most multiple star systems contain three stars, with higher numbers successively less likely; one typical early survey of 135 bright F and G solar-type stars found 42 singles, 46 doubles, nine triples, and two quadruples.[6]

The dividing line between multiple stars and associations of stars (S 33) is arbitrary, but the latter is often considered to comprise ten or more. Like associations, most multiple stars likely formed out of the same molecular cloud (S 25), though it is possible some formed as a result of close encounters. Because of dynamical considerations, multiple stars generally exhibit hierarchical nesting of their orbits. A triple star system, for example, will normally consist of a binary pair with a more distant companion. A sextuplet might consist of three binary pairs, of which one pair is in a more distant orbit.

Triple star systems became well known in the late 18th century when William Herschel began compiling his catalogues of double and multiple stars. Begun in 1779, his survey resulted in catalogues in 1782 (269 systems), 1784 (484 systems), and 1821 (145 systems), totaling more than 800 systems. In 1825, his son John Herschel published a catalogue of 380 known double and triple stars.

Alpha Centauri is the nearest triple star system, located at 4.37 light years. It consists of Alpha Centauri A, a G2 main sequence star similar to the Sun; Alpha Centauri B, a slightly cooler K1 main sequence star; and Alpha Centauri C, a much fainter M6 main sequence dwarf star. The latter is also known as Proxima Centauri, since at 4.2 light years it is closest to Earth. It has a mass of only 12% of the Sun, about 129 Jupiter masses, and orbits Alpha Centauri A and B at a distance of 13,000 astronomical units, about a quarter of a light year. The A and B components, on the other hand, orbit each other in about 80 years. The Jesuit priest Jean

---

[6]H. A. Abt and S. G. Levy, "Multiplicity among solar-type stars," ApJ Supplement Series, 30 (1976), 273–306.

Richaud first discovered the binary nature of Alpha Centauri in 1689; the fainter and more distant Proxima Centauri was not discovered until 1915 by Robert Innes.[7] Some uncertainty still exists as to whether Proxima Centauri is gravitationally bound to Alpha Centauri A and B. Because of the proximity of this star system, it has been a high priority target for the search for planets. A planet was already suspected in 2013 and was confirmed by a European Southern Observatory team in 2016. Proxima Centauri b orbits very close to its parent star (.05 astronomical units) with a period of 11.2 days and a mass about 1.3 times that of Earth. Because Proxima Centauri is a dwarf star, it appears to lie in the habitable zone where liquid water could exist.

Algol, also known as Beta Persei, is another famous triple star system at 92.8 light years, 20 times more distant from Earth than Alpha Centauri. It consists of the binary pair Algol A and B, a massive main sequence star and a less massive subgiant, respectively. Separated by only .06 astronomical units, both the A and B components are orbited by Algol C at distance of about 2.7 astronomical units. This means that the binary pair is more than 10 times closer to each other than the Earth is to the Sun, with the C companion 2.7 times more distant than the Earth from the Sun. Algol is the prototype of an eclipsing binary (B eclipsing A and vice versa), and one of the first variable stars discovered. The binary pair is believed to have exchanged matter over the lifetime of the stars; in particular, the subgiant was once more massive, but as it began to expand toward its red giant stage, material spilled over onto its companion—a scenario that may still be occurring.

More distant still is the well-known pole star Polaris at 430 light years from Earth, the nearest and brightest classic Cepheid variable star. The Polaris system (Fig. 12.1) consists of a massive bright giant or supergiant designated Polaris A, orbited by a close dwarf star at 18.5 AU designated Polaris Ab, and a 1.5 solar mass main sequence star at 2,400 AU, about two billion miles. William Herschel discovered the B component already in 1780, and in 1929 the smaller closer component was discovered indirectly using spectral analysis. In 2006, the Hubble Space Telescope produced the first direct image of the close component.[8]

---

[7] N. Kameswara Rao et al., "Father J. Richaud and early telescope observations in India," *Bulletin of the Astronomical Society of India*, 12 (1984), 81–85.

[8] N. R. Evans, G. Schaefer, H. Bond et al. "Direct detection of the close companion of Polaris with the Hubble Space Telescope," AJ, 136 (2008), 1137–1146; the abstract was first published in BAAS, 37 (2005).

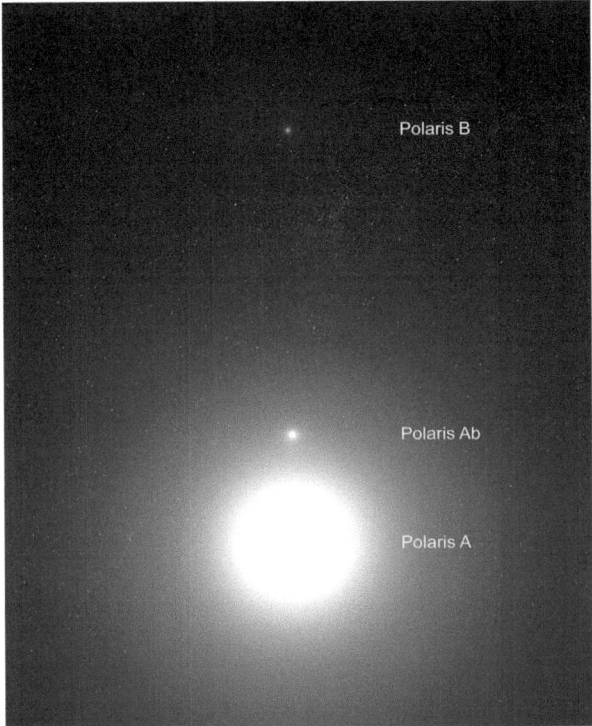

Fig. 12.1. Artist's conception of the Polaris system. Polaris B is about 240 billion miles from Polaris A. Credit: NASA, ESA, G. Bacon (STScI)

Moving beyond triple stars we come to Castor, one of the two "twins" in the constellation Gemini, the closest known sextuplet system, 52 light years distant. It was discovered to be binary already in 1678 by G. D. Cassini; the A and B components are now known to have an orbital period of 445 years and an average separation of about 100 AU. Each of those components was discovered to be a spectroscopic binary. Castor C is an eclipsing double and therefore a variable star, also known as YY Geminorum. The pair consists of nearly identical M dwarfs, separated from the AB components by 1,000 AU, orbiting the quadruple in 14,000 years.

The only other known sextuplet within the local volume of 150 light years is the Mizar system, about 80 light years distant from Earth. Mizar, the second star from the end of the Big Dipper's handle, has a storied history that illustrates the piecemeal and lengthy nature of astronomical discovery. It was known since antiquity that Mizar appeared very close to another star, Alcor, so close that separating the two was and still is a test of keen eyesight. Already in 1616, Benedetto Castelli, a protégé of Galileo, detected with an early telescope that Mizar itself was a double star; at his request, Galileo himself made the same observation in January, 1617.[9] Mizar A and B were the first known binary pair and the first to be photographed (along with Alcor) in 1857. In 1890, spectroscopy revealed the Mizar A was itself a binary, and in 1908 Mizar B was also revealed as a binary. This pair of binaries constituted a quadruplet system, with Alcor appearing relatively far away.

That was not the end of the story for the Mizar system. In 2009, astronomers reported that Alcor, a relatively young A5 main sequence star twice as massive as the Sun, is itself a binary, separated by 28 astronomical units from Alcor B, an M class red dwarf a quarter the mass of our Sun. Moreover, the Alcor system appears to be gravitationally bound to the Mizar system, making the entire system a sextuplet. Recent observations indicate that the Mizar binaries and the Alcor binary are separated by about 75,000 astronomical units, more than one light year.[10] The ROSAT spacecraft has detected X-ray emission from Alcor, probably due to activity on Alcor B. "Finding that Alcor had a stellar companion was a bit of serendipity," said astronomer Eric Mamajek, leader of the team that found the star simultaneously with another group. "We were trying a new method of planet hunting and instead of finding a planet orbiting Alcor, we found a star."[11] The use of adaptive optics on the Multiple Mirror Telescope in Arizona was crucial to the discovery.

These are only representative examples of the great variety of multiple star systems known exist throughout the Galaxy and in other galaxies, destined like families to live their lifetimes together due to the circumstances of their birth. Because of its proximity to our Solar System, the alpha Centauri system has been the location of many science fiction novels, most recently the Chinese writer Cixin Liu's *Three Body Problem*. The trilogy describes the results of a METI (Messaging Extraterrestrial Intelligence) program gone

---

[9] Leos Ondra, "A New View of Mizar," http://www.leosondra.cz/en/mizar/

[10] E. E. Mamajek et al., "Discovery of a Faint Companion to Alcor Using MMT/AO 5 μm Imaging," AJ, 139 (2010), 919–925; Neil Zimmerman, Ben Oppenheimer et al., "Parallactic Motion for Companion Discovery: An M-Dwarf Orbiting Alcor," ApJ, 709 (2010), 733 ff.

[11] "Faint Star Orbiting the Big Dipper's Alcor Discovered," *Science Daily*, December 10, 2009, at http://www.sciencedaily.com/releases/2009/12/091210000851.htm; "First Known Binary Star is Discovered to be a Triplet, Quadruplet, Quintuplet, Sextuplet System," *Science Daily*, December 13, 2009, at http://www.sciencedaily.com/releases/2009/12/091210092005.htm

wrong, as the Trisolarians (so called after the alpha Centauri triple star system) send an invasion fleet to Earth. The novel follows the progress of the 1,000-ship fleet—and resulting events on Earth—as the aliens journey at one hundredth the speed of light over 4.5 light years, taking 450 years to arrive on Earth. The Trisolarian invasion is motivated by the unstable situation of their home planet with a chaotic orbit around three stars.

The Multiple Star Catalogue by A. Tokovinin contains more than 1,000 multiple systems with three or more components. The discovery of new systems, some with exotic components, is not only a favorite pastime of astronomers but also a route to learning more about stars and stellar evolution.[12] A catalog of 2,000 hierarchical multiple star systems with three to seven components is at http://www.ctio.noao.edu/~atokovin/stars/. That site also shows a diagram of the famous Castor system. Recent X-ray images and videos of the Alpha Centauri system from Chandra are at http://chandra.harvard.edu/photo/2018/alphacen/, and more images of the Polaris system are at http://hubblesite.org/images/news/release/2006-02.

---

[12] A. Tokovinin, "Multiple Star Catalogue: A Catalogue of Physical Multiple Stars," A&A Supplement, 124 (1997), 75–84.

Class S 33: Association (OB)

An association of stars is a group of tens to hundreds of young stars that are close together because they were formed from the same giant molecular cloud (S 25). Like molecular clouds, they are therefore concentrated along the Galaxy's spiral arms. Because the star density is small, they are only loosely bound by gravity and are dispersed over a few tens of millions of years by differential galactic rotation. Due to their common origin, associations tend to consist of young stars that have not yet had time to disperse from their natal cloud. A considerable fraction of star formation in the Galaxy may occur in OB associations, which contain many protostars (S 1). Associations are distinguished from open star clusters (S 34), which are much older, larger, and more gravitationally bound.

The Dutch astronomer Adriaan Blaauw and the Armenian astronomer Viktor Ambartsumian first recognized stellar associations in 1946 and 1947. The latter gave them their name and distinguished two types: OB and T associations. The Canadian astronomer Sidney van den Bergh later distinguished an R association for those that illuminate reflection nebulae.[13] The international standard for stellar association nomenclature is the constellation, followed by the association type and a number, as in Scorpius OB2. OB associations are historically important because they have been used to trace the spiral structure of the Andromeda and Milky Way galaxies, and they continue to be important for addressing fundamental problems of star formation.[14]

Not surprisingly, OB associations contain hot stars of spectral type O and B. These associations have overall ages in the range of 30 to 50 million years, and many of their low-mass members are still in the pre-main sequence stage (S 2 and S3), only a few million or tens of millions of years old. In 1961 Blaauw explained why some O and B stars originating in OB associations are travel-

---

[13] Adriaan Blaauw, "A study of the Scorpio–Centaurus cluster," *Publications of the Kapteyn Astronomical Laboratory at Groningen*, 52 (1946 Ph.D. Thesis), and V. A. Ambartsumian, in *Stellar Evolution and Astrophysics* (Yerevan: Armenian Academy of Sciences, 1947); Sydney Van den Bergh, "A study of reflection nebulae," AJ, 71 (1966), 990–998; W. Herbst, "R associations. I - UBV photometry and MK spectroscopy of stars in southern reflection nebulae," AJ, 80 (1976), 212–226.

[14] Dick (2013), p. 119.

ing very rapidly, the so-called "runaway stars." Blaauw suggested they had originally been members of binary systems, and when one member became a supernova, the other member "ran away," either at its orbital velocity or with added velocity from the explosion.[15] This theory is supported by the frequency of binary O and B stars while still in associations, and the lack of binary runaway stars.

Examples of OB associations are Perseus OB1, Orion OB1, Scorpius OB2, and Carina OB2. Scorpius OB2, also known as Scorpius-Centaurus or Sco-Cen, is the nearest OB association, located about 400 light years from the Sun. It is much studied because it is the nearest region of massive star formation. These studies have revealed over 500 probable members of the association, including numerous high and intermediate-mass stars, as well as hundreds of low-mass stars and dozens of brown dwarfs, clustered in three subgroups with ages of about five million and 16 million years. Spitzer Telescope results indicate that at least 35% of Sco-Cen members have dusty debris disks (S 15).[16]

Adriaan Blaauw first identified the Orion OB1 association in 1964 and subdivided it into four regions, lettered a through d, located near Orion's Belt and Sword. In addition to several dozen hot giant O and B stars, it contains numerous lower mass stars, protostars, and debris disks, all less than 10 million years old. In the 1970s, Wayne Warren and James Hesser carried out an extensive survey of the entire association, which is part of the Orion Molecular Cloud Complex.[17] The Perseus OB1 association is notable not only for its remarkable "h and Chi Persei" double open cluster but also because it atypically has no giant molecular cloud

[15] Adriaan Blaauw, "On the origin of the O- and B-type stars with high velocities (the "run-away" stars), and some related problems," *Bulletin of the Astronomical Institutes of the Netherlands*, 15 (1961), 265–290.

[16] Thomas Preibisch and Eric Mamajek, "The Nearest OB Association: Scorpius-Centaurus (Sco OB2)," in Bo Reipurth, ed., *Handbook of Star Forming Regions*, vol. II (San Francisco: Astronomical Society of the Pacific, 2008), online at http://arxiv.org/pdf/0809.0407v1; Carpenter et al, "Circumstellar Disks in the Upper Scorpius OB Association," ApJ, 705 (2009), 1646 ff.

[17] A. Blaauw, ARAA, 2 (1964), 213 ff.; W. H. Warren and J. E. Hesser, "A photometric study of the Orion OB 1 association. I - Observational data.," ApJ Supplement Series, 34 (1977), 115–206; J. Hernandez et al., "Spitzer Observations of the Orion OB1 Association: Second-Generation Dust Disks at 5–10 Myr," ApJ. 652 (2006), 472–481.

associated with it. Studies show four remnant molecular clouds in data from the IRAS spacecraft and the Sloan Digital Sky Survey II. The same studies also found pre-main sequence stars in the form of intermediate mass Herbig AeBe stars (S 3) and classical T Tauri stars, implying that there has been recent star formation at considerable distances from the association's most luminous stars.[18]

About a dozen OB associations are known to be located within 2,000 light years of the Sun, many of them recognized based on measurements with the Hipparcos satellite. Many lower mass associations have also been found in the Large Magellanic Cloud (Fig. 12.2) and the Andromeda Galaxy. Associations are undoubtedly very numerous but are notoriously difficult to locate in our own Galaxy because they can extend over large areas of

Fig. 12.2. One of hundreds of low-mass associations found in the Large Magellanic Cloud, this one designated LH 95. Credit: NASA, ESA, and the Hubble Heritage Team (STScI/AURA)-ESA/Hubble Collaboration; Acknowledgment: D. Gouliermis (Max Planck Institute for Astronomy, Heidelberg)

---

[18] Hsu-Tai Lee and Jeremy Lim, "On the Formation of Perseus OB1 at High Galactic Latitudes," ApJ, 679 (2008), 1352–1363.

the sky, up to several hundred square degrees for the closest ones. They therefore mingle with foreground and background stars, and special techniques are needed to pick them out, like needles in a haystack.

T associations contain young pre-main sequence T Tauri stars (S 2); about 20 are known. The nearest and most studied is the Taurus-Auriga T association, about 400 light years from the Sun. One study has found that 94 of its 217 members are pre-main sequence stars.[19] Other examples include the associations in TW Hydrae and Vela. T Tauri stars are also found in OB associations such as Sco-Cen.

R associations, defined by van den Bergh in 1966 as those stars embedded in reflection nebulae, have proven useful as spiral arm tracers like their cousins, the OB associations. R associations are sometimes considered intermediate mass analogs of the T and OB associations.

[19] C. Bertout and F. Genova, "A kinematic study of the Taurus-Auriga T association," 460 (2006), 499–518; C. Bertout and Siess, "The Evolution of Stars in the Taurus-Auriga Association," A&A, 473(2007), L21–L24.

Class S 34: Open Cluster

An open cluster (also known as a galactic cluster) is a group of hundreds to thousands of stars formed in the same molecular cloud (S 25) and still loosely bound gravitationally. Open clusters are distinguished from globular clusters (S 35), which are much more massive and very tightly gravitationally bound; open star clusters are the metaphorical new towns of a galaxy, and globular clusters its aging cities. Our Sun was likely born in an open cluster that dispersed across the Galaxy over a billion years, leaving us to our solitary existence. About 1,100 open clusters are known in our Galaxy, and they are always found in its disk, in stark contrast to globulars. Because these clusters tend to disperse in less than a billion years, their stars are relatively young, but not as young as star associations (S 33) still in their birth throes.

The Hyades, the Coma, the Pleiades, and the Beehive/Praesepe clusters are among the closest open clusters visible to the naked eye. H and Chi Persei is a prominent double cluster in the constellation Perseus, a famously beautiful sight through even a small telescope. Open clusters are important to astronomy for determining cosmic distances, for stellar evolution data and for a variety of other research. Cluster ages are easily determined, because the cluster's main sequence will burn away from top to bottom. In other words, its massive stars will burn out first. Thus, an open cluster like the Pleiades containing massive hot B stars is about 100 million years old; the Hyades lacks B stars and is likely closer to a billion years old. This gives each cluster a characteristic Hertzsprung-Russell diagram (S 4), usually expressed as a color-magnitude diagram for these purposes. In this way, not only was the relationship of subgiant to main sequence stars first deciphered in the early 1950s, but the existence of subgiants (S 7) also demonstrated that stars evolved off the main sequence and became giants or supergiants through the connecting subgiant stage, clearly visible on the diagram. The same effect is seen in the older globular clusters. Both open clusters and globulars thus played an important role in our unraveling of stellar evolution.

The detection of the brighter open clusters, in the sense of first being seen if not understood, occurred in prehistoric times. In classical literature, Homer mentions the Pleiades in the *Odyssey*, where it is said of Odysseus "The master mariner steered his craft, sleep never closing his eyes, forever scanning the stars, the Pleiades and the Plowman [Bootes] late to set and the Great Bear that mankind also calls the Wagon." In the Bible, the Book of Job reads "Canst thou bind the sweet influences of Pleiades, or loose the bands of Orion?" The Greek poet Hesiod and the Roman poet Ovid also mention the Pleiades.[20] Ancient mariners considered the heliacal rising of the Pleiades a sign that the sailing season had begun.

Only in the mid-18th century, however, were the stars comprising the bright clusters widely believed to be physically associated. Open clusters were already distinguished from globular clusters by their shape in Charles Messier's 1781 *Catalogue des Nébuleuses & des amas d'Étoiles* [catalogue of nebulae and star clusters]. 33 of the 103 objects in the Messier catalogue are open clusters.[21] William Herschel is generally credited with resolving what he called "round nebulae" (globulars) into stars, thus distinguishing them as a class of star cluster different from open clusters. The classic early 20th century study of open clusters was undertaken by Robert Trumpler and published in 1930. Using data from 100 open clusters both close and distant, Trumpler famously showed the absorption of starlight by interstellar dust. The effect drastically lowered the estimated size of our Galaxy from Harlow Shapley's estimate of 300,000 to 100,000 light years.[22]

The Hyades, consisting of 300 to 400 stars at a distance of 151 light years in the constellation Taurus, is the nearest known open cluster. Believed to be about 625 million years old, it is visible with the naked eye and known since antiquity. The Coma cluster contains about 40 stars at a distance of 288 light years and is believed to be about 450 million years old. Even more distant, the famous Pleiades, M45 in the Messier catalogue, is also known as the Seven Sisters from its seven brightest stars, which precede Orion's rising in the northern hemisphere winter night sky. The entire cluster (Fig. 11.7) contains about 1,000 stars at a distance of about 440 light years. In telescopic photographs it is seen surrounded by a bluish reflection nebulae (S 28), comprised of dust through which

---

[20] *The Odyssey*, Book V, 297–300, Robert Fagles, trans., 160; Book of Job, verse 38:31, Ovid, XIII, 293.

[21] Charles Messier, 1781. Catalogue des Nébuleuses & des amas d'Étoiles [Catalogue of Nebulae and Clusters of Stars]; *Connaissance des Temps* for 1784 (published 1781), pp. 227–267.

[22] R. J. Trumpler, "Preliminary results on the distances, dimensions and space distribution of open star clusters," *Lick Observatory Bulletin*, 14 (1930) 154–188.

Fig. 12.3. Open cluster NGC 290 in the Small Magellanic Cloud. Image Credit: European Space Agency & NASA; Acknowledgment: E. Olszewski (University of Arizona)

the cluster is now passing. Its hot blue stars are less than 100 million years old. Slightly more distant Praesepe, the Beehive cluster, is one of the first objects Galileo observed with his telescope in 1609. The distance of its thousand stars is not well known, but probably exceeds 500 light years. The Orion Trapezium cluster, discovered by Galileo with one of his early telescopes in 1617, is a young cluster spread over about 1.5 light years and illuminating the surrounding famous nebula. Open clusters are also found in other galaxies, including the Magellanic Clouds roughly 200,000 light years away (Fig. 12.3).

One of the most unusual open clusters was first reported in 2006 (Fig. 8.9), discovered as one of many newfound clusters during the Two Micron All-Sky Survey (2MASS). 20,000 times the mass of the Sun and harboring an estimated 20,000 stars, it is 20 times more massive than typical open clusters in our Galaxy. Located about 18,900 light years from Earth, it is a source of X-rays and gamma-rays, a sure sign of intense activity. The cluster includes 14 red supergiants, highly unusual objects near the end of their lives that will eventually explode as supernovae. They are believed to have between eight and 25 times the mass of the Sun and are likely six and 15 million years old.[23] The previous record holder for red supergiants was held by NGC 7419, shown in 2003 to have five such supergiants.

Several classification systems have been devised for open clusters, but the Trumpler classification, based on the number, magnitude, and concentration of the cluster's stars, is the most widely used. A Roman number from I to IV indicates the degree of concentration, with I the tightest. A number from 1 to 3 indicates the range of brightness, with 1 indicating a small range. The richness of the cluster is indicated by p (poor), m (moderate), or r (rich), and an "n" is added if nebulosity is present. The Pleiades in the Trumpler scheme is classified as II 3 r n.[24]

The latest Hubble news and images on star clusters can be seen at http://hubblesite.org/news/23-star-clusters. The Legacy ExtraGalactic UV Survey (LEGUS), the most comprehensive ultraviolet-light look at nearby star-forming galaxies, includes a catalog of 8,000 massive star clusters ranging from one to 500 million years old. The 2MASS gallery of open clusters is at http://www.ipac.caltech.edu/2mass/gallery/images_open.html. Comparison with other Messier clusters is at http://www.ipac.caltech.edu/2mass/gallery/messiercat.html. Two open clusters in the Small Magellanic Cloud are imaged at http://hubblesite.org/newscenter/archive/releases/2006/17/. See also http://hubblesite.org/image/1900/news_release/2006-17 and http://hubblesite.org/news_release/news/2016-03/23-star-clusters.

---

[23] Donald F. Figer et al, "Discovery of an Extraordinarily Massive Cluster of Red Supergiants," ApJ, 643 (2006), 1166–1179; HST Release, January 9, 2006, "A Hidden, Massive Star Cluster Awash with Red Supergiants," http://hubblesite.org/image/1848

[24] R. J. Trumpler, "Preliminary results on the distances, dimensions and space distribution of open star clusters," *Lick Observatory Bulletin*, 14 (1930) 154–188.

Class S 35: Globular Cluster

If open stars clusters are analogous to young towns, globular clusters are its large cities. Named for their spherical appearance,
"globulars" are concentrations of tens of thousands to millions of
stars. Unlike more sparsely populated open clusters (S 34), globulars are very old and are not concentrated in the plane of the
galaxy. In the Milky Way Galaxy, most of the younger globulars
are concentrated near the galactic nucleus, while the older ones
reside in the halo (G 12). Globulars are useful for numerous studies in astronomy; in 1917, Harlow Shapley used the distribution
of globulars to determine the location of our Solar System at the
periphery of our Galaxy. They are also used for studies of galaxy
formation and chemical and stellar evolution, since all the stars
in a given cluster are at about the same distance and have about
the same age but a range of stellar masses. Because more massive
stars have short lifetimes, all globular clusters lack stars above the
cooler G types. By contrast, open clusters still harbor hotter stars,
just how hot depending on the cluster age.

The question of when globular clusters were "discovered"
as a separate class of objects is tied up with the general problem
of the nature of the nebulae (see S 24), which came to a head in
the late 18th century. Bright globulars such as M22 in Sagittarius
and Omega Centauri in the Southern Hemisphere could be seen
with the naked eye and were well-known in the late 17th century,
but they were fuzzy objects not distinguished from other nebulae. Nicholas-Louis de La Caille published 42 nebulae and clusters of the Southern Hemisphere in 1755, while in the Northern
Hemisphere Charles Messier's *Catalogue des Nébuleuses & des
amas d'Étoiles* [catalogue of nebulae and star clusters (1774–
1781)] implies by its very title that nebulae and star clusters
are two different classes of objects. The catalog of 1781, the last
edition published in Messier's life, in fact contains a total of 28
"round nebulae" as he referred to them, and one (M4) that he
resolved into stars.

The question was whether all "round nebulae," or indeed all
nebulae in general, are resolvable into stars given a large enough
telescope, or whether they were indeed two separate classes of
object. As we have seen in our discussion of the hot ionized hydrogen clouds known as emission nebulae (S 24), that question was
answered by William Herschel around 1790. Herschel discovered
37 new globular clusters during his sweeps of the sky and was the
first to resolve virtually all of them into stars.

Herschel coined the term "globular cluster" in the discussion adjacent to his second catalog of 1,000 deep sky objects (1789). There he described them as consisting of "a number of lucid spots, of equal luster, scattered over a circular space, in such a manner as to appear gradually more compressed towards the middle; and which compression, in the clusters to which I allude, is generally carried so far, as, by imperceptible degrees, to end in a luminous center, a resolvable blaze of light." After arguing that these were groups of stars and not chance arrangements, Herschel wrote, "The next step in my argument will be to shew that these clusters are of a globular form...from the above-mentioned appearances, we come to know that there are globular clusters of stars nearly equal in size, which are scattered evenly at equal distances from the middle, but with an increasing accumulation towards the center." Herschel went on to assert "I think myself plainly authorized to conclude that they are thus formed by the action of central powers."[25]

In the 1869 tenth edition of John Herschel's *Outlines of Astronomy*, he uses the term "globular cluster" to convey the idea of "a globular space filled full of stars, insulated in the heavens, and constituting in itself a family or society apart from the rest, and subject only to its own internal laws. It would be a vain task to attempt to count the stars in one of these *globular clusters*." Herschel contrasted such globulars with those star clusters of irregular figure and with nebulae in general, but he remained uncertain whether they were fundamentally different classes of object. The issue of real nebulae without stars was finally resolved in 1864, when the British amateur astronomer William Huggins showed that stars exhibited spectral lines, while nebulae had a continuous spectrum.[26]

We now know that our Milky Way Galaxy contains more than 200 globular clusters. In 1975, Lodewijk Woltjer investigated their

[25] Michael Hoskin, "William Herschel's Early Investigations of Nebulae: A Reassessment," JHA, vol. X (1979), 165–176; the discussion in his second catalogue is William Herschel, "Catalogue of a Second Thousand of new Nebulae and Clusters of Stars; with a few introductory Remarks of the Constructions of the Heavens," PTRAS, 79 (1789), 212–226, reprinted in Hoskin, *William Herschel and the Construction of the Heavens* (London: Oldbourne, 1963), 106–115, quotation on pages 107–110. For context see Dick (2013), pp. 65–71.

[26] John Herschel, *Outlines of Astronomy* (New York: P. F. Collier & Son, 1902; 10th edition, 1869), vol. II, p. 780; Robert Smith, *The Expanding Universe: Astronomy's Great Debate, 1900–1931* (Cambridge: Cambridge University Press, 1982), pp. 3 ff.

distribution and classified them as halo globulars or disk globulars. Accurate proper motions have confirmed that they move in highly elliptical orbits over long periods of time, taking them out into the galactic halo and in toward the galactic center.[27] They are composed of low metal old population II stars (S 36).

Among the most famous globulars of the Milky Way Galaxy is the bright northern Hemisphere object M13, located 25,000 light years away in the constellation Hercules. In 1974 Frank Drake and colleagues beamed one of the first messages for extraterrestrial intelligence toward M13; traveling at the speed of light, no reply is possible for 50,000 years. Omega Centauri (Fig. 12.4) is the most massive globular in the Galaxy, containing 10 million stars, located at a distance of 17,000 light years but nevertheless visible to the naked eye in the Southern Hemisphere. It contains more than a million stars and measures 600 light years in diameter.

Fig. 12.4. Omega Centauri globular cluster of some two million stars. Credit: NASA, ESA, and the Hubble Heritage Team (STScI/AURA); Acknowledgment: A. Cool (San Francisco State University) and J. Anderson (STScI)

---

[27] L. Woltjer, "The Galactic Halo: Globular Clusters," A&A, 42 (1975), pp. 109–118; M. Geffert et al. "Absolute Proper and Space Motions of Globular Clusters," in *Proceedings of the ESA Symposium, Hipparcos* (ESA SP-402, 1997), pp. 579–582; P. Brosche et al., "Space Motions and Orbits of Globular Clusters," ibid., pp. 531–536.

Omega Centauri may contain an intermediate mass black hole (S 14) equivalent to some 40,000 solar masses at its core. Both Omega Centauri and M13 were discovered by Edmond Halley, the first in 1677 during his expedition to St. Helena, and the latter in 1713.

The closest known globular cluster is M4, 7,000 light years distant and discovered in 1746 by Philippe de Chesaux. It is one of the smallest and sparsest globulars known. In 2003, the Hubble Space Telescope found a planet, dubbed the oldest known planet in the universe, in the M4 globular. 2.5 times the mass of Jupiter, it orbits a neutron star (S 13) and a companion white dwarf (S 12) and is thus classed as a pulsar planet (P5).

Globular clusters have also been found in many galaxies, including more than 500 in the nearby Andromeda Galaxy (M31), 16 in the Large Magellanic Cloud, and even four accompanying the Fornax Dwarf Spheroidal Galaxy about 470,000 light years away. And the closest galaxy to ours, the Sagittarius Dwarf Spheroidal Galaxy, has the globular cluster M54 embedded at its center. Much further afield, in 2008 the Hubble Space Telescope detected over 11,000 globular clusters in 100 galaxies in the Virgo supercluster (G 18) of over 2,000 galaxies, about 54 million light years distant. Most of the globulars are older than five billion years. The study indicated that globulars tend to form in dense areas of rapid star formation, rather than uniformly across all galaxies. In other words, star cluster formation depends on the environment, so that dwarf galaxies near the Virgo supercluster center contain more globulars than those farther away. Moreover, the study found evidence of galactic cannibalism in the form of an abnormal number of globulars around the giant elliptical galaxy M87, located near the Virgo supercluster's center. It is now known that M87 itself harbors 12,000 globular clusters. Dwarf galaxies within 130,000 light years of M87 have an abnormally low number of globulars, suggesting that M87 has snatched away some of their globulars. Evidence indicates that globular clusters sort out stars by mass, so that heavier stars sink to the cluster's core, leaving lighter stars at the periphery. [28]

Hubble news and images of globular clusters are at http://hubblesite.org/news/55-globular-clusters; On Omega Centauri see HST Release April 2, 2008, "Starry Splendor in Core of Omega Centauri," http://hubblesite.org/image/2278/news_release/2008-14.

---

[28] HST Release, August 5, 2008, "Globular Clusters Tell Tale of Star Formation in Nearby Galaxy Metropolis," http://hubblesite.org/news_release/news/2008-30. For an overview of globulars beyond our Galaxy see Steve Gottlieb, "In Search of Extragalactic Globulars," S&T, 136 (November, 2018), 34–40.

Class S 36: Population

Like the citizens of countries, stars may be divided into broad "populations" based on location. Unlike citizens, however, stellar populations are often highly correlated with age and chemical composition. Although astronomers sometimes colloquially refer to various populations of stars based on a variety of other properties, the term is more properly reserved, in concept and in use, to three well-defined stellar populations.

To a first approximation, Population I stars are the youngest, Population II are intermediate in age, and Population III are the oldest. More precisely, population I stars include young, blue, massive, and short-lived O and B type stars found in stellar associations (S 33) and open clusters (S 34), as well as cooler Sun-like stars. They are predominantly located in the spiral arms, where most of the action of star formation takes place. Population II stars are older stars that originated with, or were incorporated into, the Galaxy more than 10 billion years ago. They are found within the halo (G 12) and central bulge of the Galaxy, and in globular clusters (S 35). The first generation of stars to form is now called Population III, composed only of hydrogen and helium produced by the Big Bang. Astronomer Allan Sandage called the idea of stellar populations "one of those grand unifying themes in science, appearing every century or so, that ties together a number of intricate details and diverse parts of a particular field."[29]

The concept of stellar populations originated with the German astronomer Walter Baade while observing the Andromeda Galaxy and two of its companion galaxies. Baade, director of the Hamburg Observatory in Germany, came to the United States in 1931 to work at Mt. Wilson Observatory, located in the San Gabriel Mountains overlooking Pasadena and Los Angeles. In 1943, during the wartime blackouts at the Observatory, Baade was for the first time able to observe details of the Andromeda Galaxy (M31) and its companions, M32 and NGC 205. In particular, pushing the limits of his photographic plates, Baade was able to resolve individual stars in these galaxies and distinguish two kinds of populations: the brightest O and B stars found in the disk part of the galaxies, and the red giants found in the spheroidal component. Because of

---

[29] Allan Sandage, "The Population Concept, Globular Clusters, Subdwarfs, Ages, and the Collapse of the Galaxy," ARAA, 24 (1986), 421–458.

the distance of the Andromeda galaxy, the observations were very difficult, and Baade pushed his photographic plates to their limits. Nevertheless, he was confident in his conclusions: "Although the evidence presented in the preceding discussion is still very fragmentary," he wrote in a landmark 1944 paper, "there can be no doubt that, in dealing with galaxies, we have to distinguish two types of stellar populations, one which is represented by the ordinary HR diagram (type I), the other by the HR diagram of the globular clusters (type II). Characteristic of the first type are highly luminous O-and B-type stars and open clusters; of the second globular clusters and the short-period Cepheids."[30] Baade's type I and type II populations quickly became known as Population I and Population II stars.

Baade also pointed out that as early as 1926 the Dutch astronomer Jan Oort had found evidence of two populations in our own Galaxy by separating high velocity from low velocity stars. Oort had indeed concluded in that year, "there are two distinct classes of stars, the members of one of which all move systematically towards a limited part of the sky with respect to the centre of gravity of the other class. The former class consists mainly, if not entirely, of stars with velocities higher than about 65 km/sec."[31] We now know that Population II stars do indeed have eccentric orbits compared to the more circular orbits of stars in the disk. They dip in and out of the galactic disk, while the Sun, remaining in the disk, takes about 240 million years to move around the Galaxy, the so-called "Cosmic Year." The concept of stellar populations, however, only took root with the work of Baade.

Baade's work had profound implications, not only for Galaxy evolution but also for the scale of the universe. This is because it turns out there are also two populations of Cepheid variables, those of Population I found in the Galaxy's disk being much more luminous than those of Population II, found in the

---

[30] Walter Baade, "The Resolution of Messier 32, NGC 205, and the Central Region of the Andromeda Nebula," ApJ, 100 (1944), 137–146; reprinted in part in Lang and Gingerich, 744–749, and in full in Abt, ApJ *Centennial Issue*, 349–358, with commentary by Dmitri Mihalas, pp. 359–361. On Baade see Donald E. Osterbrock, *Walter Baade: A Life in Astrophysics* (Princeton: Princeton University Press, 2001); Dick (2013), pp. 119–120.

[31] Jan Oort, "The Stars of High Velocity," *Publications of the Kapteyn Astronomical Laboratory at Groningen*, 40 (1926), 1–75. The quotation here is from Oort's summary, "Asymmetry in the distribution of stellar velocities," *The Observatory*, 49 (1926), 302–304.

globular clusters. Hubble had used the period-luminosity relation of Population I Cepheids for his distance determinations to the Andromeda Galaxy, whereas he was actually observing Population II stars (now known as W Virginis stars), which have a different period-luminosity relationship. Because they are intrinsically fainter, Hubble had underestimated Andromeda's distance. By 1952, Baade realized the estimated distance to Andromeda (now known to be 2.5 million light years), and the size of the universe, should be doubled.

Following a groundbreaking paper by Lawrence Aller and Joseph W. Chamberlain in 1951 on subdwarfs (S 6), it was realized that not all stars had the same composition as the Sun. Gradually, it was realized that stellar populations could be separated in this way also based on what astronomers call the "metallicity" of the stars, where "metals" means anything other than hydrogen and helium. (As one wit remarked, the astronomer's periodic table is much simpler than the chemist's!). Under this definition, older generations of stars in the galactic halo have lower metallicity than their younger Population I counterparts in the disk. The very first "Population III" stars are believed to have been born about 100 million years after the Big Bang, the first generation of stars ever produced. Extremely massive (perhaps 100 solar masses), they were composed only of hydrogen and helium and in theory should no longer exist because high-mass stars evolve rapidly. The second generation of stars, the first and most extreme Population II stars, was likely born about a billion years after the Big Bang and contain more metals because they incorporate heavier elements created in the older stars. Because many of them have less than one solar mass, they evolve very slowly and may still exist, as do many subsequent generations of even more metal-rich Population II stars. Population I stars have the highest metallicity of all, having gone through numerous generations.

Metallicity is most often defined as the ratio of the mass of elements heavier than helium to the mass of all elements in the object and usually measured in practice by observing the ratio of iron to hydrogen [Fe/H] in a particular star, as compared to the same ratio for the Sun. The Sun has a metallicity of about 1.8% by mass. Expressed as a logarithmic scale, on which the Sun is near 0, a star with a metallicity -1 has 10% of the Sun's metals, -2 has 1%, -3 has 0.1% (one thousandth of the Sun's metals), and so on. By this definition, extreme Population II stars have recently been found with as little metallicity as -5.4, between 1/100,000[th] and one millionth of the Sun, but up to 2% for later generations.

Population I stars, having even more metals to incorporate, range in metallicity from about 1 to 4%. It follows that the more metals in a star, the more complex its spectrum. Conversely, metal-poor stars are discovered by their lack of numerous spectral lines. Extremely metal-poor stars are very rare, like looking for a needle in the haystack of stars.

While stellar populations generally correlate with age, "age" is a relative term. Our Sun is a Population I star about five billion years old—halfway through its lifetime. Moreover, since the beginnings of the universe, many generations of stars have been born, as massive stars with short lifetimes explode as supernovae and spew their metals into the interstellar medium, where they are incorporated into molecular clouds, which then form protostars. The metallicities of galaxies can also be compared, the Large Magellanic Cloud having a metallicity 40% that of the Milky Way (therefore younger), and the Small Magellanic Cloud only 10% of the Milky Way (much younger). Populations are therefore important because they help us understand the evolution of galaxies and the universe. Moreover, one of the search strategies for planetary systems (P 18) is to look for stars of high metallicity, since planets such as the Earth form by the accretion of heavy elements.

Because of the difficulty of the relevant observations, only recently have astronomers used the term "Population III stars" to refer to the very first stars that formed in the universe.[32] In November, 2005 the Spitzer Space Telescope claimed to have detected Population III stars, though the evidence was circumstantial and, while very metal poor, likely not the first generation. The idea that Population III stars were massive and short-lived, perhaps 50 to 300 times the mass of the Sun, is still very much open to current research. Meanwhile, the search for the earliest and oldest stars continues. Following the final servicing mission for the Hubble Space Telescope, in 2009 astronomers announced the discovery of galaxies formed only 600 million years after the Big Bang, 13.1 billion years ago. These galaxies contained stars that had already been burning for 300 million years. The James Webb Space Telescope, scheduled to launch in 2021, should be able to see back even further to the beginning of time. Spectra, albeit ones with very few lines for such metal-poor objects, would be needed to prove the existence of such stars.[33]

---

[32] Volker Bromm and Richard B. Larson, "The First Stars," ARAA, 42 (2004), pp. 79–118.
[33] Robert Irion, "The Quest for Population III," *Science*, 295 (2002), 66–67; *Science* (2010), p. 258.

So far only so-called "extreme Population II stars" have been detected with certainty. A concerted search for such stars in the halo of our Galaxy using high resolution spectroscopy has turned up 150 with metallicities less than -3, $1/1000^{th}$ that of the Sun.[34] They may have been deposited in the halo by dwarf spheroidal galaxies that formed long ago and interacted with the Milky Way Galaxy. Some galactic stellar streams (G 10) are believed to be the shredded remains of such galaxies. The study of extreme population II stars, sometimes termed "stellar archaeology," thus yields a wealth of information about the history of our Galaxy.

Two illustrations showing the evolution of two populations of stars in globular clusters are at http://hubblesite.org/image/3200/category/107-illustrations and http://hubblesite.org/image/1970/news/107-illustrations. On the stellar populations in our galaxy see http://hubblesite.org/news_release/news/2012-25/107-illustrations.

---

[34] Anna Frebel, "Bright Metal-poor Stars from the Hamburg/ESO Survey. I. Selection and Follow-up Observations from 329 Fields," ApJ, 652 (2006), 1585–1603; A. Frebel, "Stellar archaeology: Exploring the Universe with metal-poor stars," *Astronomische Nachrichten*, 331 (2010), 474–488 for a review of literature on extremely metal-poor stars.

Part III
The Kingdom of the Galaxies

Galaxies NGC 4302 (seen edge-on at left) and NGC 4298 (almost face on at right), two spiral galaxies 55 million light years from Earth. They are located in the Virgo cluster and are possibly interacting binary galaxies. Hubble Space Telescope image released April 20, 2017 for the 27th anniversary of its launch. Credit: NASA, ESA, and M. Mutchler (STScI). The Hubble Ultra Deep Field shown in the frontispiece of this book gives a good idea of the amazing extent of the Kingdom of the Galaxies

With the Kingdom of the Galaxies we enter a realm not recognized until a century ago, when certain "nebulae" thought to be within our own Galaxy were found to be extremely distant "extragalactic" systems of stars—galaxies in their own right. This third relatively new Kingdom encompasses both previous Kingdoms, harboring stars and planets, clustering into systems and hierarchies of systems more readily than even stellar systems. The profusion of what we now know are hundreds of billions of galaxies beyond our own brings us into the realm of cosmology, bearing on the structure of the universe and the questions of its origins and meaning.

Although there had been plenty of earlier debate about the nature of the nebulae, it was only American astronomer Edwin Hubble who in 1924 demonstrated that most nebulae outside the plane of the Milky Way Galaxy are actually extragalactic systems of stars. It was also Hubble who in the 1920s and 1930s distinguished four classes of galaxies based on their morphology: ellipticals, lenticulars, spirals, and irregulars—a system still used today in its basic outlines. In the century that followed, astronomers have studied the origin, nature, and evolutionary relationships of these classes of galaxies, beginning with our own impressive spiral Milky Way Galaxy and extending billions of light years almost to the Big Bang. We can now observe protogalaxies forming within 500 million years of the origin of the universe, and we recognize that even mature galaxies such as our own are still accreting matter, cannibalizing, shredding, and accreting anything that comes near. In short, the Kingdom of the Galaxies represents evolution on the largest scales of the universe.

Beginning in the 1940s, the nascent Kingdom of the Galaxies became stranger yet. All four previous classes of galaxies were considered "normal" in the sense that their radiation output could be understood based on known principles. But in 1943 the American astronomer Carl Seyfert defined a class, now known as Seyfert galaxies, that had anomalously bright nuclei in the optical region of the spectrum. This was the first indication of something unusual going on at the cores of certain galaxies, today known to be due to supermassive black holes.

Capitalizing on a new window into the universe, only three years later the British physicist J. Stanley Hey discovered the first discrete radio source, later known as Cygnus A and shown to be a galaxy pouring out radiation in the radio region of the spectrum at a rate more than 1,000,000 times greater than our own Galaxy. This was a power so astonishing that astronomers were described as having to practice hard before breakfast at believing such an observation could be real. But it was, and 16 years later even more powerful radio objects were discovered, labeled quasars. Quasars are now known to be the cores of extremely distant galaxies, at billions of light years. Finally in 1972, astronomers declared yet another class of anomalously bright galaxies, now known as blazars, characterized by rapid variations in intensity at radio, infrared, and visual wavelengths. They too found it hard to believe what they were seeing, and hard to understand how any reasonable physical model could account for the observations. What all of these discoveries over three decades have in common, from Seyfert galaxies in 1943 to blazars in 1972, is that these galaxies are believed to harbor supermassive black holes that are actively accreting matter. Taken together this Subfamily of galaxies is known as "active galaxies," as opposed to "normal galaxies" like our own Milky Way, which also harbors a supermassive black hole, but is not accreting massive amounts of matter. It turns out the four classes of active galaxies may in fact be one class, differentiated only by the viewing angle they present to Earth. But over the years differentiating four classes has proven useful, and utility is one of the hallmarks of classification.

Just as circumstellar objects exist, so do circumgalactic objects in the form of satellites, jets, and galactic halos. It would come as a surprise to most that at least 59 satellite galaxies orbit our own Milky Way Galaxy, dominated by irregular, elliptical, and spheroidal dwarf galaxies. What is even more surprising is the recent discovery that many satellite galaxies are in the process of being shredded by the gravitational field of our Milky Way, creating stellar streams stretching across the sky in arcs ranging up to 100 degrees and more. These streams are sometimes given colorful names like "river of stars," "gravity's rainbow," and, when they intersect, a crisscrossing "field of streams." The discovery of these streams began in earnest only 20 years ago, and now comprises a major field of study made possible by giant star-charting projects such as the Sloan Digital Sky Survey, the Dark Energy Survey, the Gaia satellite, and the upcoming Large Synoptic Survey Telescope. Such studies are important as a kind of "galactic archaeology," revealing a great deal about the history of the galaxies they now encircle and how these galaxies were assembled. They are also shedding light on clumps of dark matter, invisible except for their effect on the evanescent streams.

By analogy with the Kingdom of the Stars, the intergalactic medium includes three Subfamilies of objects consisting of gas, dust, and energetic particles. Among the classes of objects in these galactic subfamilies are the Warm Hot Intergalactic Medium (WHIM), Lyman alpha Blobs, the galactic wind, and extragalactic cosmic rays. Each of these classes has its own story. The WHIM, which consists largely of hydrogen gas and constitutes the bulk of the intergalactic medium, is now believed to account for most of the "missing mass" of normal matter (not to be confused with the missing non-baryonic dark matter). Lyman alpha blobs are large concentrations of hydrogen gas found in the early universe, spanning an area up to 400,000 light years. Galactic winds, analogous to solar and stellar winds, consist of streams of charged particles, dust and gas that blow off of galaxies. They are important for recycling matter and energy between galaxies and the intergalactic medium. And whizzing through it all are the highest energy particles of all, extragalactic cosmic rays, traveling at nearly the speed of light from active galaxies and sources unknown.

Finally, just as gravity forms systems in the planetary and stellar Kingdoms, so it does in the Kingdom of Galaxies, ranging from binaries to clusters and superclusters. We will relate the discovery stories of each of these classes, some of the largest structures in the universe, culminating in filaments and voids on the scale of hundreds of millions to billions of light years. This is the realm of cosmology, of redshift surveys, theories of cold dark matter, and of space telescopes penetrating into the deepest regions of space and the farthest reaches of time. We know a great deal more than a century ago, but what we don't know reminds us that plenty of mysteries remain to be solved—with even more yet to be revealed.

13. The Protogalactic Family

Class G 1: Protogalaxy

Like protoplanets (P 1) and protostars (S 1), galaxies must be born. Understanding their birth process is likewise fraught with difficulties and, even more than in astronomy's other two Kingdoms, hampered by extreme distance. The details of galaxy formation remain one of the great unsolved problems of astrophysics, and therefore a dynamic field of study exists, informed by observation, theory, and simulations. The observations, both ground- and space-based, range from nearby mature galaxies and extremely distant young galaxies present within a few hundred million years of the Big Bang. Theoretical modeling and simulations have grown increasingly important over the last three decades, to such an extent that in some cases even some astronomers cannot tell the difference between simulation and reality.[1] Nonetheless, the problem is complicated because the different structural components of galaxies may have been formed at different times. For example, bulges of spiral galaxies likely formed first, followed by disks gradually assembling around them. And different classes of galaxies likely did not form the same way as spirals. In a dynamic interplay, observers and modelers now feed on each other's results, and both await the results of the James Webb Space Telescope which will look back to the beginning of the universe.

The earliest galaxies existed at least 500 million years after the Big Bang, so in a technical sense these were the first protogalaxies. Unlike protostars, in which the dividing line between protostar and star is considered to be the end of accretion and the onset of stellar winds, there is no consensus on the dividing

---

[1] For the latest on galaxy formation simulations see Adrian Cho, "The Galaxy Builders," *Science*, 360 (2018) 954–957. For an overview of galaxy formation studies see William C. Keel, *The Road to Galaxy Formation* (Chichester, UK: Springer, 2007).

© Springer Nature Switzerland AG 2019
S. J. Dick, *Classifying the Cosmos*, Astronomers' Universe,
https://doi.org/10.1007/978-3-030-10380-4_13

line between a protogalaxy and a galaxy. Indeed, it is now known that mature galaxies such as our own are still accreting matter, even after 12 billion years. Given this situation, the definition of a protogalaxy is somewhat nebulous, so to speak. By analogy with stars, one might wish to associate the end of the protogalaxy stage with the onset of galactic winds (G 17). Galactic winds, however, appear to depend on active galactic nuclei caused by accretion onto supermassive black holes (S 14), as well as starburst activity and supernovae explosions. Given the uncertainties, an operational definition might be that a protogalaxy ends and a galaxy begins when stars light up the galaxy, and the wind from star formation and supernovae, as well as black hole accretion, exceeds the gravity of the host galaxy. This begs an entire range of questions about how different galaxy types form and their relationship to each other, problems that are the subject of active research.

For many years galaxies were believed to have formed from the collapse of a "protogalactic cloud" composed solely of hydrogen and helium, similar to star formation but on a much larger scale. But from the beginning the protogalactic cloud collapse model had its problems, including lack of an explanation for the persistence of spiral arms, as pointed out by James Jeans already in 1926.[2] In 1958, the Armenian astronomer Viktor Ambartsumian also argued that galaxy formation was more complex than protogalactic collapse allowed, in part due to the many features of galaxies such as jets (G 11) arising from the ejection of materials.[3] Nevertheless, based primarily on the observed motions of the stars in our Galaxy, in a classic paper in 1962 astronomers Olin Eggen, Donald Lynden-Bell, and Allan Sandage continued to argue that spiral galaxies formed through the collapse of a large gas cloud. From the theoretical side the Russian physicists Rashid Sunyaev and Yakov Zeldovich offered their "pancake theory" in the 1970s, and a method to test it with 21-cm observations of neutral hydrogen. This "top down" theory accounted for the then-known properties of our Galaxy's halo (G 12) and disk stars. On this scenario,

---

[2] J. H. Jeans, *Astronomy and Cosmogony* (Cambridge: Cambridge University Press, 1926); Lang and Gingerich, 763.

[3] Viktor Amazaspovich Ambartsumian, "On the Evolution of Galaxies," in *La structure et l'evolution de l'univers*, R. Stoops, ed (Brussels: Coudenberg, 1958), reprinted in part in Lang and Gingerich, 763–773.

also called the "monolithic collapse" theory, star formation occurs at the same time that the galaxy gains most of its mass.[4]

But there was an alternative to the top-down theory, logically named the "bottom-up" theory. In 1978 Leonard Searle and Robert Zinn proposed such a theory, whereby small galaxies merge to form larger ones. Based on observations of red giants and globular clusters in our own Galaxy, they concluded that the facts "are consistent with the hypothesis that the loosely bound [globular] clusters of the outer halo have a broader range of age than the more tightly bound clusters and originated in transient protogalactic fragments that continued to fall into dynamic equilibrium with the Galaxy for some time after the collapse of its central regions had been completed."[5] These fragments may also be referred to as subgalactic objects (G 13).

The bottom-up theory that individual galaxies form from protogalactic fragments was foreshadowed in the early 1970s by Princeton astronomer James Peebles, who argued that on a larger scale galaxies form first and then cluster. Perhaps the same was true of the formation of each galaxy, in which subgalactic objects would cluster into galaxies.[6] Peebles' large-scale analog was given more credence in 1978 when astronomers Simon White and Martin Rees published an influential article arguing that most of the material in the universe condensed early in its history into "dark objects" that subsequently underwent what they called "hierarchical clustering," giving rise to the large scale distribution of galaxies we presently observe. In their view, each stage of the hierarchy—including the formation of individual galaxies in the first instance—forms and collapses, and the luminous content of galaxies forms from the cooling and fragmentation of residual

---

[4] Ya. B. Zeldovich, "Gravitational instability: An approximate theory for large density perturbations," A&A, 5 (1970), 84–89; Ia. B. Zedlovich et al, "Giant voids in the universe," *Nature*, 300 (1982), 407–413; R. A. Sunyaev and Ia. Zeldovich, "On the possibility of radioastronomical investigation of the birth of galaxies," O. J. Eggen, Donald Lynden-Bell, A. R. Sandage, A. R., "Evidence from the Motions of Old Stars that the Galaxy Collapsed," ApJ, 136 (1962), 748.

[5] L. Searle, and R. Zinn, "Companions of Halo Clusters and the Formation of the Galactic Halo," ApJ, 225 (1978), 357–359.

[6] James Peebles, "Structure of the Coma Cluster of Galaxies," AJ, 75 (1970), 13–20; Peebles, "The Gravitational-Instability Picture and the Nature of the Distribution of Galaxies," ApJ, 189 (1974), L51–L53.

gas within the dark matter. Thus, while the overall scheme of both galaxy and cluster formation was a bottom-up hierarchical clustering, there was a collapse component within this clustering. "Every galaxy," they wrote, "thus forms as a concentrated luminous core embedded in an extensive dark halo."[7]

The theory was worked out in more detail in 1984 by George Blumenthal, Sandy Faber, Joel Primack and Martin Rees, became known as the cold dark matter (CDM) theory. Since then, one of its fundamental tenets has been that dark matter halos (G 12) form hierarchically by a series of mergers with smaller halos. This also implies that the luminous stellar halo is formed from disrupted, accreted dwarf galaxies, and that both the galactic halos and the accretion might be observed. More recent evidence, however, indicates that a significant part of the growth of spiral galaxies of Milky Way mass comes from snacking rather than gorging— in other words, by merging of smaller masses of gas rather than by merging galaxies comparable in size. Rather, the bulges and disks of these galaxies may increase their masses in part by cool gas funneled by cosmic filaments (G 24). The Atacama Large Millimeter/submillimeter array (ALMA) has begun to search for these streams. However, more massive galaxies such as ellipticals may still gorge.[8]

Observations of nearby galaxies like Andromeda and the ability of telescopes and spacecraft to probe ever more distant regions of the universe and look further back in time have provided evidence for the bottom-up CDM picture of galaxy formation (Fig. 13.1). Observations of the Cosmic Microwave Background by the COBE and WMAP spacecraft showed primordial temperature fluctuations 300,000 years after the Big Bang, which may have given rise

---

[7] S. D. M. White and M. J. Rees, "Core Condensation in Heavy Halos: a two-stage theory for galaxy formation and clustering," MNRAS (1978), 183, 341–358. A popular account of events leading to this paper is given in Marcia Bartusiak, *Through a Universe Darkly* (New York: Harper Collins, 1993), pp. 299–311.

[8] George R. Blumethal, S. M. Faber, Joel R. Primack and Martin J. Rees, "Formation of Galaxies and Large-Scale Structure with Cold Dark Matter," *Nature*, 311 (1984), 517–525; Pieter G. van Dokkum et al., "The Assembly of Milky Way-like Galaxies Since z~2.5," ApJ Letters, vol. 771: L35 (July, 2013). https://arxiv.org/abs/1304.2391, 2013.

Fig. 13.1. Galaxy building blocks from the Hubble Ultra Deep Field, evidence of the "bottom-up" theory of galaxy formation. The numbers refer to distance expressed in redshift, equivalent to about a billion years after the Big Bang. Images taken in 2003 and 2004, released in 2007. Credit: NASA, ESA, and N. Pirzkal (STScI/ESA)

to denser regions that became the seeds of galaxies. Studies of the microwave background indicate that following initial cooling, the universe was reheated and reionized by newborn stars during a "reionization epoch" at a redshift range between six and 14, corresponding to less than a billion years after the Big Bang. Using results from the Hubble Ultra-Deep Field and the Great Observatories Origins Deep Survey, in 2006 astronomers Rychard Bouwens and Garth Illingworth found evidence for hierarchical galaxy formation occurring between 700 and 900 million years after the Big Bang, at a redshift of about seven. They found about five times fewer galaxies at 700 million years than at 900 million years. The following year astronomers led by Daniel Stark of Caltech, using the Keck II telescope on Mauna Kea and the technique of gravitational lensing, observed galaxies at a redshift of nine, only 500 million years after the Big Bang.[9]

In 2010, in the wake of Hubble's last servicing mission, astronomers reported finding five galaxies 13 billion light years distant, dating to 600 million years after the Big Bang. Thus, between 300,000 years and 500 to 600 million years after the Big Bang, galaxies began to form. These galaxies observed by the Hubble Space Telescope in 2010 are small compared to the Milky Way Galaxy, about 5% of its size and less than 1% of its mass. Their stars appear to be 300 million years old, so its constituent stars must have formed within a few hundred million years of the Big Bang. Astronomers again interpreted the new images as evidence of the progressive buildup of galaxy assembly, whereby small objects accrete mass to form bigger galaxies, so that these primordial galaxies are the "building blocks" of today's larger galaxies—compatible with the "hierarchical formation" theory.[10] The observation of even more distant objects, 250 to 400 million years after the Big Bang, awaits the launch of the James Webb Space Telescope in 2021.

Closer to home observations of galactic halos, including the halo of our own Milky Way Galaxy, also provide evidence of the bottom up CDM mechanism for galaxy formation.[11] In light of the observations from the Hubble Space Telescope among others, the theory is now widely accepted, though this is only part of the story because the mechanism for the formation of the small primordial galaxies, either through protogalactic cloud collapse, gas accretion, star capture, or some combination thereof, is still open to question. It is also likely that different classes of galaxies, and on a different scale even individual galaxies, may have had different

[9]Rychard J. Bouwens and Garth D. Illingworth, "Rapid evolution of the most luminous galaxies during the first 900 million years," *Nature*, 443 (2006), 189–192, and Mansori Iye et al, "A galaxy at a redshift z = 6.96," *Nature*, 443 (2006), 186–188; Daniel Stark et al., "A Keck Survey for Gravitationally Lensed Lyα Emitters in the Redshift Range 8.5<z<10.4: New Constraints on the Contribution of Low-Luminosity Sources to Cosmic Reionization," ApJ, 663 (2007), 10–28.

[10]Hubble Space Telescope Press Release, "Hubble Reaches 'Undiscovered Country' of Primeval Galaxies," January 5, 2010, at http://hubblesite.org/newscenter/archive/releases/2010/02/full/

[11]Andromeda results and history may be found in S. C. Chapman, R. Ibata et al, "Kinematically Selected, Metal-Poor Stellar Halo in the Outskirts of M31," ApJ, 653 (2006), 255–266. http://iopscience.iop.org/0004-637X/653/1/255/pdf/64747.web.pdf

origins, depending on the balance of gas accretion and the merger history of a particular galaxy. Some mechanisms may have dominated different eras of the universe; observations indicate the early universe was dominated by spirals and irregulars, and that ellipticals may have resulted from mergers of these objects.

In 2008, after evidence was presented of the hierarchical formation of galaxies occurring even today, one astrophysicist remarked that "I don't think monolithic collapse is yet dead in everyone's mind, but I think the majority of astronomers have come around in the last few years, particularly with direct evidence...that these galaxies, which obviously have very old stars, show signs of still forming."[12] As if to confirm this statement that the idea of monolithic collapse was not yet dead, in 2010, Hubble astronomers reported "tentative evidence" that the huge globular cluster 47 Tucunae in the Milky Way's bulge was the same age as its halo, implying that both were formed together rather than by accretion. Calling the evidence still "very preliminary," the discovery again demonstrates that the protogalactic process is a subject of active research.[13]

The Hubble Space Telescope has been an essential tool in uncovering the mechanisms of galaxy formation. Already in 1996 astronomers using the telescope were reporting images of early building blocks of today's universe: http://hubblesite.org/newscenter/archive/releases/1996/29/text/. Ten years later, they presented evidence of a "Galaxy in the Making" http://hubblesite.org/newscenter/archive/releases/2006/2006/45/ and reported hundreds of young galaxies in the early universe http://hubblesite.org/newscenter/archive/releases/2006/2006/12/. They have also reported the arrested development of a galaxy: http://hubblesite.org/news_release/news/2018-17/4-galaxies.

---

[12] Kim Vy-Tran et al, "The Late Stellar Assembly of Massive Cluster Galaxies via Major Merging," ApJ Letters, 683 (2008), L17–L20. The astrophysicist quoted was Romeel Dave at the University of Arizona at Clara Moskowitz, "Galaxy Formation: A Clumpy Affair," http://www.space.com/science astronomy/080915-mm-galaxy-formation.html

[13] R. Goldsbury, H. B. Richer et al., "The ACS Survey of Galactic Globular Clusters. X. New Determinations of Centers for 65 Clusters," AJ, 140 (2010), 1810–1837, http://iopscience.iop.org/article/10.1088/0004-6256/140/6/1830/pdf. Thorsten Naab and Jeremiah P. Ostriker, "Theoretical Challenges in Galaxy Formation," ARAA, 55, pp. 55–109 (2017), preprint at https://arxiv.org/pdf/1612.06891.pdf.

14. The Galaxy Family

Class G 2: Elliptical Galaxy

In stark contrast to spiral galaxies like our own Milky Way, elliptical galaxies are almost featureless aggregations of stars distinguished by an ellipsoidal or spherical morphology. The ellipsoidal morphology was identified early on, even before these objects were realized to be external galaxies. In his classification of galaxies published in 1926, Edwin Hubble delineated three classes of galaxies: ellipticals, spirals (G 4), and irregulars (G 5), and ten years later in his famous book *The Realm of the Nebulae* he referred to extreme ellipticals as "lenticulars," now often recognized as a fourth class (G 3).[1] In his famous "tuning fork" diagram of galaxy types (Fig. 14.1), ellipticals were placed on the handle because Hubble thought they were early objects before acquiring their more mature morphologies as spirals. However, today many astronomers believe that at least some ellipticals form when smaller disk galaxies collide and merge. Thus, while Hubble distinguished this class of objects, he did not understand its place in cosmic evolution. Even today, and despite a long history of observation, perhaps the only consensus is that "we have not yet determined with certainty the physical mechanisms that differentiate galaxies into classes."[2]

Despite elaborations, Hubble's fundamental system remains largely in use today. Though based on morphology rather than physical characteristics (because morphology was largely the only observable characteristic in Hubble's time aside from luminosity), galaxy morphology does correlate with many physical properties,

---

[1] Edwin P. Hubble, "Extra-Galactic Nebulae," ApJ, 64 (1926), 321–369, reprinted in part in Lang and Gingerich, pp. 716–724; Hubble, *The Realm of the Nebulae*, (New Haven: Yale University Press, 1936), pp. 39–48. To be more precise, in his 1926 paper Hubble separated galactic from extragalactic nebulae, and divided the latter into regular and irregular galaxies, classifying ellipticals and spirals as regulars.

[2] Michael R. Blanton and John Moustakas, "Physical Properties and Environments of Nearby Galaxies," ARAA 47 (2009), 159–210.

© Springer Nature Switzerland AG 2019
S. J. Dick, *Classifying the Cosmos*, Astronomers' Universe,
https://doi.org/10.1007/978-3-030-10380-4_14

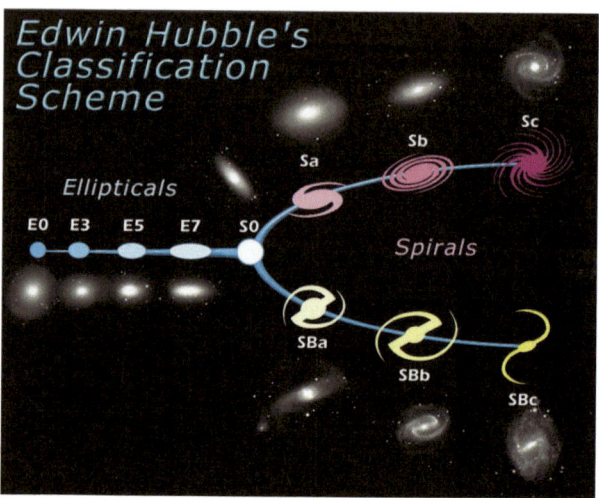

Fig. 14.1. Hubble's tuning fork diagram of galaxy classification. Spiral are at top, barred spirals at bottom; letters from a to c characterize the compactness of the spiral arms. Credit: NASA & ESA

including luminosity, color, mass, and star formation rates. Just as the Harvard system of stellar spectral types was elaborated and supplemented in the Yerkes MKK system years later, so was the so-called Hubble sequence elaborated by astronomers such as Gerard de Vaucouleurs, Sidney van den Bergh, and Hubble's successor Alan Sandage.[3] Although our understanding is as yet incomplete, Hubble's classification and its extensions have remained remarkably intact tools as modern galaxy surveys have been undertaken (Fig. 14.2). These surveys include the Galaxy Evolution Explorer (GALEX) in the ultraviolet, the Sloan Digital Sky Survey (SDSS) in the optical, the Two Micron All-Sky Survey (2MASS) and the Spitzer Infrared Nearby Galaxy Survey (SINGS) in the infrared, and the 21-cm radio surveys such as the H I Parkes All-Sky Survey (HIPASS). Redshift surveys that first revealed the large-scale structure of filaments and voids (G 24) in the universe provide a third dimension.

---

[3] The literature on galaxy classification is very large, but for an overview see Ronald J. Buta, Harold G. Corwin, Jr., and Stephen C. Odewahn, *The de Vaucouleurs Atlas of Galaxies* (Cambridge: Cambridge University Press, 2007), and Sidney van den Bergh, *Galaxy Morphology and Classification* (Cambridge, 1998).

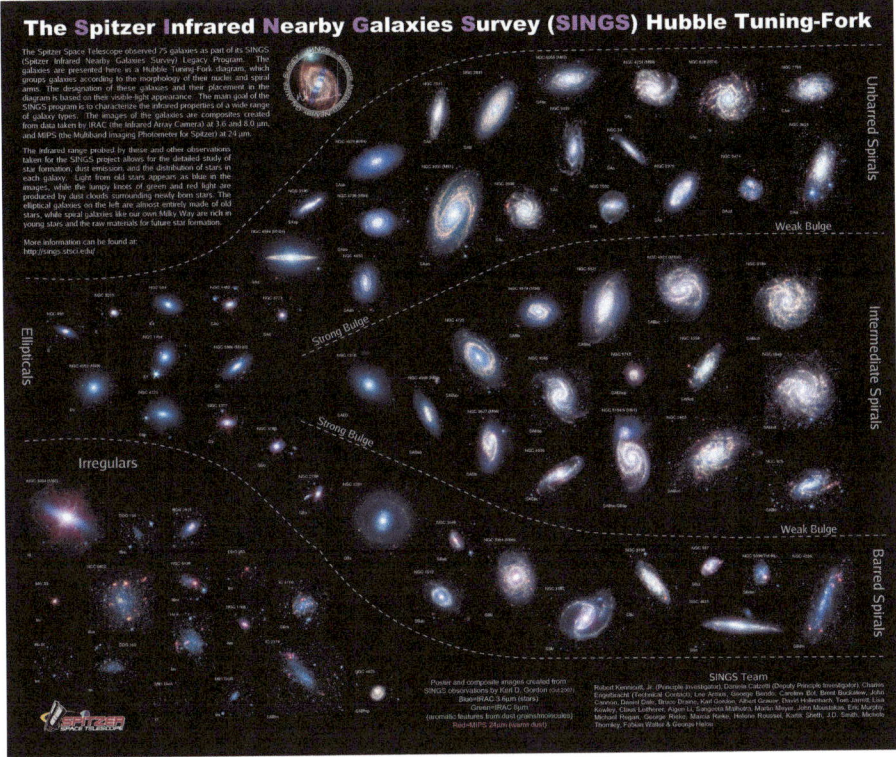

Fig. 14.2. 75 nearby galaxies from NASA's Spitzer Space Telescope Infrared Galaxy Survey, arranged on Hubble's tuning fork classification. Credit: NASA/JPL-Caltech/K. Gordon (STScI) and SINGS Team

During the telescopic era, elliptical galaxies had long been observed as fuzzy objects inseparable from many other types of nebulae. We now know that Charles Messier's 18th century catalogue of nebulae and star clusters, for example, contained six of them, and four more were later added to the extended catalog of 110 objects. It was Hubble himself who showed in 1924 (officially announced New Year's Day, 1925) that some of the "nebulae" were extragalactic, even though he himself still liked to call them nebulae, as evidenced in the title of his classic work *The Realm of the Nebulae*.

The question of the nature of ellipticals and their relation to other classes of galaxies has been explored continuously since Hubble's tuning fork classification. Hubble himself wrote in 1936, "Elliptical nebulae are highly concentrated and show no indications of resolution into stars. The luminosity falls rapidly away from bright, semistellar nuclei to undefined boundaries … Small patches of obscuring material are occasionally silhouetted against the luminous background, but otherwise these nebulae present no structural details."[4] We now know that ellipticals are dominated by older, low mass, and metal poor population II stars (S 36) moving in all directions in elongated randomly oriented orbits, and that ongoing star formation is rare, ellipticals having mostly depleted their interstellar matter. Although they are typically surrounded by numerous globular clusters (S 35) and harbor supermassive black holes (S 14) at their centers, because they are structureless they seem simpler than other classes of galaxies. Because of their older stars, ellipticals are still sometimes referred to as "early-type" bulge galaxies, as opposed to "late-type" disk spirals, harking back to Hubble's original designation but for different reasons.

As seen in Figs. 14.1 and 14.2, by morphology Hubble and his successors classified elliptical galaxies from E0 (spherical like M87) to E7 (elongated), shading into the lenticular galaxies. Round elliptical galaxies are believed to be the oldest ellipticals, forming from galaxy mergers two to three billion years ago and exhibiting no star formation. While small ellipticals are the most common galaxies in the universe, particularly in clusters of galaxies, ellipticals in general represent only ten to 15% of galaxies in the *local* universe. In contrast to spirals, they were less common in the early universe than now. Astronomer Alan Dressler showed in 1980 that ellipticals and lenticulars are the most prevalent galaxy classes in the centers of rich galaxy clusters.[5]

Elliptical galaxies may also be classified by luminosity, namely high luminosity and low luminosity. High luminosity ellipticals are further divided into two subtypes: those with and without nuclear cores. Those with cores are the most luminous. Among the lower luminosity elliptical types, there are three subtypes: low surface brightness galaxies such as the dwarf and spheroidal ellipticals, including NGC 205; high surface brightness, concentrated galaxies such as M32 (seen in Fig. 14.6 to the lower left of M31); and ultracompact dwarfs with radii as small as 30 light years. Even these general subtypes are still the subject of much interpretation.[6]

---

[4] Hubble, *The Realm of the Nebulae*, p. 39; Dick (2013), pp. 134–135.
[5] A. Dressler, "Galaxy morphology in rich clusters - Implications for the formation and evolution of galaxies," ApJ, 236 (1980), 351–365.
[6] Blanton and Moustakas, ARAA, 189; J. Kormendy, D. B. Fisher et al,

The sizes and luminosities of ellipticals can vary enormously, from so-called cD galaxies that can be 100 times more luminous than the Milky Way, to normal or giant ellipticals with luminosities a few times that of the Milky Way, to compact and dwarf ellipticals (dE) 1/10th the Milky Way's luminosity, and dwarf spheroidal ellipticals (dSph) so diffuse as to be barely visible. The envelopes of the relatively rare cD galaxies such as IC 1101 (Fig. 14.3) can stretch hundreds of thousands of light years, whereas giant ellipticals measure tens of thousands of light years across, and dwarf ellipticals are commonly 10,000 light years or less in extent. Their masses range from a few million to a trillion solar masses. Elliptical galaxies therefore range from the most to the least massive galaxies in the universe.

The nearest known elliptical galaxy, discovered in 1994, is the Sagittarius dwarf elliptical galaxy, at distance of only about 50,000 light years from the center of the Milky Way and some 70,000 light years from Earth. Another nearby well-known dwarf elliptical at 2.65 million light years is M32, a satellite of M31, the Andromeda

Fig. 14.3. The massive elliptical galaxy IC 1101, one of the largest known with up to 100 trillion stars. It is ambiguously classified as E/S0, elliptical, or lenticular. Credit: NASA/ESA/Hubble Space Telescope

---

"Structure and Formation of Elliptical and Spheroidal Galaxies," *ApJ Supplement*, 182 (2009), 216–309.

Galaxy seen in Fig. 14.6. In 2018, researchers used simulations to show that some two billion years ago, M31 cannibalized a large progenitor galaxy of most of its 25 billion solar mass, leaving the remnant M32 with only a few billion solar masses.[7]

The nearest giant elliptical galaxy is Maffei 1, about ten million light years away and a member of the nearest galaxy group, IC 342 Maffei (G 21). Because it lies in the plane of the Galaxy and is obscured by stars and dust, it was difficult to detect and was only discovered in 1968 by Paolo Maffei during a search for T-Tauri stars (S 2) and diffuse nebulae using the Schmidt telescope at Asiago Observatory. Classified as an E3 elliptical, it is best viewed in the infrared and has been observed by infrared telescopes such as WISE. In 2000, it was also observed by the Hubble Space Telescope, which detected its first globular clusters.[8]

Perhaps the best-known giant elliptical is M87, about 55 million light years away (Fig. 14.4). It has a diameter of 120,000 light

Fig. 14.4. Central core of the giant elliptical galaxy M87, the most massive object in the Messier catalog. The protruding jet indicates matter emanating from a massive black hole. Credit: Tod R. Lauer, Sandra M. Faber/NASA

---

[7] University of Michigan Press Release, "The Milky Way's long-lost sibling finally found," https://news.umich.edu/the-milky-ways-long-lost-sibling-finally-found/; Richard D'Souza and Eric F. Bell, "The Andromeda galaxy's most important merger about 2 billion years ago as M32's likely progenitor, *Nature Astronomy*, 2, 737–743 (2018).

[8] R. Buta and Marshall L. McCall, "Maffei 1 with the Hubble Space Telescope," AJ, 125 (2003), 1150–1163. http://iopscience.iop.org/1538-3881/125/3/1150/pdf/202281.web.pdf

years, contains 2.4 trillion solar masses, and harbors about 12,000 globular clusters (compared to about 200 for the Milky Way). Like most giant ellipticals, M87 harbors a supermassive black hole and is ejecting a jet (G 11) of matter extending 5,000 light years from its center. It is a radio galaxy (G 7), one of the first detected, also known as Virgo A.

Despite, or perhaps because of, their seeming simplicity, elliptical galaxies have played an important role in cosmological investigations such as the expansion rate of the universe. In his book *Voyage to the Great Attractor*, Alan Dressler describes how in the 1980s he and "the seven samurai" astronomers surveyed elliptical galaxies and ended up discovering the Great Attractor, a supercluster of galaxies.[9] Despite their deceivingly simple appearance, elliptical galaxies are now known to be complex objects both in terms of their dynamics and their assembly history. Many mysteries remain, and they are the subject of robust research among cosmologists.

A representation of the de Vaucouleurs system is at http://www.astronomy.com/~/media/import/files/pdf/9/0/1/september_2010_we_zoo.ash%20x. The Chandra X-ray Observatory's gallery of normal and starburst galaxies is at http://chandra.harvard.edu/photo/category/galaxies.html. WISE images of Maffei 1 and 2 are at http://wise.ssl.berkeley.edu/gallery_Maffei_1_2.html. A plethora of galaxy catalogs catering to various needs exists. The NASA/IPAC Extragalactic Database (NED) at https://ned.ipac.caltech.edu is a good entrée—an online database of more than 200 million extragalactic objects. There are also many galaxy Atlases compiled for various purposes. A good overview is given in Jean-Rene Roy, *Unveiling Galaxies: The Role of Images in Astronomical Discovery* (Cambridge University Press, 2018), pp. 205–246. Sandage and Bedke's *Carnegie Atlas of Galaxies* (1994) illustrates the modified Hubble classification scheme and is available at http://publicationsonline.carnegiescience.edu/publications_online/galaxy_atlas_1/default.html.

---

[9] Alan Dressler, *Voyage to the Great Attractor* (New York: Vintage Books, 1994).

Class G 3: Lenticular Galaxy

Lenticular galaxies are gravitationally bound conglomerations of stars with a rotating disk and a central bulge, but no spiral arms or dust lanes. In many ways they are intermediate in nature between elliptical and spiral galaxies, devoid of gas and dust like ellipticals but with some structure like spirals. They appear to be aging galaxies that have used up most of their interstellar medium, and they consist mostly of population II stars (S 36). In his 1936 book *Realm of the Nebulae*, Edwin Hubble identified lenticulars with the most flattened form of ellipticals, known as E7, and hypothesized what he called an S0 class between ellipticals and spirals, as seen in Fig. 14.1. It is this S0 class that we today we refer to as lenticulars—an interesting case of adjusting a classification system as more data became available. Hubble's prototype of an E7 elliptical, NGC 3115, is now recognized as an early S0 system because new observations clearly reveal a thin disk. In fact, three of 12 of Hubble's prototypes have been reclassified since 1936 based on better evidence.[10]

When Hubble proposed an intermediate S0 class of galaxies, none had been observed. But eventually they were observed, and in Gerard de Vaucouleurs' 1959 extension of the Hubble sequence of galaxies, lenticulars are further typed as barred (SB0) or unbarred (SA0). De Vaucouleurs later considered all S0 galaxies to be "lenticulars." Nevertheless, even today it is uncertain whether galaxies such as M84 and M86 should be classified as highly elongated E7 ellipticals or S0 lenticulars. Perhaps because of these classification difficulties, some sources list only three types of galaxies: ellipticals, spirals, and irregulars. But as one review concluded, "The lenticular population has received far too little attention, either being tossed in along with the elliptical galaxies or ignored altogether. The E/S0 separation is not an arbitrary one—the edges, rings, and bars are clear indications of type. We need well-defined and consistent methods of identifying S0s in the new massive surveys." Lenticulars deserve study due to their interesting intermediate nature. In light of this, and because the term is widely used

---

[10] Hubble, 1936, p. 40; Buta et al. (2007), pp. 8–9; Dick (2013), pp. 135–137.

in astrophysical literature and incorporated into de Vaucouleurs' system, here we consider it a distinct class.[11]

Like ellipticals, lenticular galaxies are often found in rich galaxy clusters, and may represent spirals transformed through some unknown process as they fall into the cluster. As early as 1951, Lyman Spitzer and Walter Baade suggested that dynamical processes within clusters might be responsible for this transformation. They showed that dense clusters of galaxies, such as the Coma and Corona clusters, contain large numbers of S0 galaxies, and theorized that collisions between galaxies in such dense clusters would sweep interstellar matter out of the galaxies, thereby accounting for the lack of spiral disks with star formation. Alan Dressler, Gus Oemler, and their "Morphs" collaboration have also shed light on the general mechanisms for the observed relationship between galaxy type and the density of clusters of galaxies, the so-called morphology-density relationship discovered by Dressler in 1980. The case for an evolutionary relationship between spiral galaxies and lenticulars has received additional support from recent observations with the Spitzer Space Telescope.[12]

As mentioned in the previous entry, the largest known galaxy, IC 1101, is ambiguously classified as E/S0, elliptical or lenticular. It is about one billion light years distant that may contain 100 trillion stars, compared to 400 billion in our own galaxy. The closest known lenticular galaxy NGC 404, located just beyond our Local Group at a distance of about ten million light years, has provided some insight into the nature of lenticulars. The galaxy has been referred to as "the ghost of Mirach" because it lies hidden in the glare of the red giant star Mirach. In 2008 it was observed in the ultraviolet during an all-sky survey by the Galaxy Evolution Explorer spacecraft, which found a mysterious ring surrounding it consisting of newer stars. Earlier radio observations with the Very Large Array in New Mexico showed a ring of hydrogen that

[11] Michael R. Blanton and John Moustakas, "Physical Properties and Environments of Nearby Galaxies," ARAA 47 (2009), 159–210: 201; Buta et al. p. 8; Dick (2013), 137.

[12] Lyman Spitzer, Jr. and Walter Baade, "Stellar Populations and Collisions of Galaxies," ApJ, 113 (1951), 413–418; Alan Dressler et al., "Evolution since z = 0.5 of the morphology-density relation for clusters of galaxies," ApJ, 490 (1997), 577–591; J. E. Geach et al., "The Nature of Dusty Starburst Galaxies in a Rich Cluster at z = 0.4: The Progenitors of Lenticulars?" ApJ, 691 (2009), 783.

may be attributed to a collision with a neighboring galaxy 900 million years ago. The ultraviolet observations show star formation is occurring in this ring, so that the galaxy collision may have rejuvenated NGC 404, giving rise to a hybrid galaxy. Subsequent Hubble Space Telescope observations also concluded that a "gas accretion event" reignited the galaxy about 500 million years ago.[13] About 5% of lenticulars are polar ring galaxies, in which a ring of gas, dust, and stars exists at right angles to the disk of the galaxy.

Another lenticular galaxy studied in considerable detail with the Hubble Space Telescope is NGC 1533, a barred lenticular (SB0) located at a distance of about 62 million light years in the Dorado galaxy group. Astronomers reported that it consists overall of evolved red stars, but that it appears to be in transition from late to early type, in their words "it is red, but not quite dead." It has a large system of globular clusters and shows faint spiral structures with luminous blue stars in isolated H II regions (S 24). The authors conclude that these characteristics suggest the galaxy is in the late stages of transition in morphology from a barred spiral type SBa to a barred lenticular type SB0, and thus perhaps a key to understanding morphology-density relation in galaxy clusters. While the large number of ellipticals and lenticulars in rich clusters has often been explained as infalling spirals being transformed into S0s by the cluster environment, the authors point out, "Recent evidence at intermediate redshift indicates that the transitions begin outside the clusters in small group environments through galaxy-galaxy interactions. Following infall the intracluster medium then serves to expedite the process by removing the remaining cool gas. With NC 1533, we have a close-up view of this transition in the Dorado group."[14]

The radio galaxy Centaurus A, the 5th brightest radio source in the sky, is a famous lenticular about 13 million light years distant. Most often classed as an S0 peculiar galaxy, but sometimes as

---

[13] David Thilker et al, "NGC 404: A Rejuventated Lenticular Galaxy on a Merger-Induced Blueward Excursion into the Green Valley," ApJ, 714, L 171; "Ghost of Mirach" materializes in Space Telescope Image," *Science Daily*, Nov 3, 2008, http://www.sciencedaily.com/releases/2008/11/081102211541. htm; B. F. Williams et al, "The Advanced Camera for Surveys Nearby Galaxy Survey Treasury. VI. The Ancient Star-forming Disk of NGC 404," ApJ, 716 (2010), 71.

[14] Regina Barber DeGraaff et al., "A Galaxy in Transition: Structure, Globular Clusters, and Distance of the Star-Forming S0 Galaxy NGC 1533 in Dorado," ApJ, 671 (2007), 1624–1639.

Fig. 14.5. The Spindle Galaxy, an S0 lenticular, has a flat stellar disk and a large ellipsoidal bulge, viewed here edge-on. Its diameter is about 60,000 light years, and it is 50 million light years distant. Credit: NASA, ESA, and the Hubble Team (STScI/AURA). Acknowledgment: W. Keel (University of Alabama, Tuscaloosa)

an elliptical, it is also an active radio galaxy (G 7) and a starburst galaxy with active star formation because it is being devoured by a spiral galaxy. Other famous lenticular galaxies include the Spindle Galaxy, also known as NGC 5866 and M102 (Fig. 14.5), the Cartwheel Galaxy, as well as Messier objects M84, M85, and M86. The Spindle Galaxy (not to be confused with Hubble's prototype E7 galaxy sometimes also called the Spindle galaxy) is located some 50 million light years distant. The Cartwheel Galaxy in Sculptor, discovered by Fritz Zwicky in 1941, is some 500 million light years away. It has a mass of three to five billion solar masses, and its morphology appears to be the result of a spiral galaxy colliding with another galaxy, giving rise to its unusual and spectacular features observed by the Hubble Space Telescope, Chandra, Spitzer, the Galaxy Evolution Explorer, and ground-based telescopes. It is a dramatic example of interacting galaxies (G 20).

For the Cartwheel galaxy see http://www.nasa.gov/mission_pages/galex/galex-20060111.html and http://hubblesite.org/image/256/category/91-astronomical.

Class G 4: Spiral Galaxy

Spiral galaxies are gravitationally bound conglomerations of stars characterized by complex structure, including flat disks, central bulges, and spiral arms wound with various degrees of tightness. The spiral arms are active regions of star formation. We are partial to this class of galaxy since we live in one, the Milky Way Galaxy. Aside from a variety of stars, spiral galaxy disks contains cool atomic hydrogen gas clouds (S 23), molecular clouds (S 25), hot ionized gas clouds known as H II regions (S 24), and dust in the form of dark nebulae (S 27) and reflection nebulae (S 28). The central bulge contains the densest region of stars, the nucleus, which in many spirals harbors a supermassive black hole. The entire structure is surrounded by a visible and dark matter halo (G 12), which contains globular clusters (S 35) and mysterious dark matter. The bulge and halo taken together are sometimes referred to as the spheroidal component, in contrast to the disk component. This basic structure of spiral galaxies was not fully known until about 50 years ago. In his 1958 lectures, *Evolution of Stars and Galaxies*, Walter Baade divided the structure of the Milky Way Galaxy into "the disk, the surrounding halo, and the central nucleus"—a description that was just beginning to become commonplace.[15]

Spiral galaxies range in mass from one billion to one trillion solar masses and in size from 15,000 to 300,000 light years in diameter, with the thickness only 1/50th of the diameter. More than half of all observed galaxies in the local universe are spirals, and half of those are "barred spirals," showing a bar extending from the bulge where the spiral structure begins. While it is tempting to make barred spirals a separate class, because lenticulars also may have bars, among other reasons, it is prudent to rank barred spirals one taxon level down, as a type of spiral, as did Edwin Hubble in his 1926 landmark paper.

In Gerard de Vaucouleurs' extension of the Hubble sequence in 1959, still used today, spiral galaxies are divided into three types: normal (SA), barred (SB), and intermediate (SAB), the latter containing weak bars. A graphical representation of this system in terms of the number of galaxy types would be lemon-shaped, with spirals and lenticulars forming the broad central part of the

---

[15] Walter Baade, *Evolution of Stars and Galaxies*, (Cambridge, Mass.: Harvard University Press, 1963; paperback edition MIT Press, 1975) p. 267.

lemon, and ellipticals, and irregulars at each end.[16] In 2005, the Spitzer Space Telescope confirmed that the Milky Way is actually a barred spiral, an SAB(rs)bc in de Vaucouleurs' system. Even more quantitative morphological measures are now being used, though they are still the subject of considerable controversy and debate.

Spiral galaxies present themselves at many viewing angles to Earth, the full spiral structure of the disk appearing best when viewed perpendicular to the plane of the disk. Both the disk and bulge of spiral galaxies are rotating, in the case of stars in the disk of our Galaxy taking roughly 250 million years to complete one nearly circular orbit at the Sun's location in the disk. Bulge stars have larger random velocities, and stars and globular clusters of the halo follow random eccentric orbits. Among large galaxies of the universe, about 80% are either spirals or lenticulars (G 3), while among small galaxies the smaller ellipticals (G 2) predominate. In contrast to ellipticals, spirals tend not to be found in the centers of rich clusters of galaxies but rather in the general field of galaxies.

Although we live in a spiral galaxy, this was not always obvious: it is not easy to visualize the structure of an object from the inside. We now know that the disk stars of our own Galaxy appear in the sky as the Milky Way, and that the galactic center is in the Southern sky toward Sagittarius. This Milky Way was one of the first phenomena Galileo observed with his telescope in 1610, resolving it into stars. But the first to speculate that this was a large flat disk of stars were the 18th century natural philosophers Thomas Wright of Durham (1750) and Immanuel Kant (1755). A few decades later, William Herschel became the first to base this conclusion on actual star counts. 27 out of 110 objects in Messier's extended catalogue are spiral galaxies.[17]

The discovery of spiral structure awaited larger telescopes. It was William Parsons, the Third Earl of Rosse in Ireland, who first detected this form of object in 1850 with his six-foot-diameter speculum metal reflector, twice the size of William Herschel's. His

---

[16] Michael R. Blanton and John Moustakas, "Physical Properties and Environments of Nearby Galaxies," ARAA 47 (2009), 159–210: 175 ff; R. A. Benjamin, E. Churchwell et al., "First GLIMPSE Results on the Stellar Structure of the Galaxy," ApJ, 630 (2005), L149–L152.

[17] Richard Berenzden, Richard Hard and Daniel Seeley, *Man Discovers the Galaxies* (New York: Science History Publications, 1976), pp. 7–14; Charles A. Whitney, *The Discovery of Our Galaxy* (New York: Alfred A. Knopf, 1971), pp. 77–132. Relevant parts of Wright and Kant are reprinted in Bartusiak, 168–188. On the Messier objects see Ronald Stoyan, *Atlas of Messier Objects: Highlights of the Deep Sky* (Cambridge University Press, 2008).

observations and sketches of M51, now known as the Whirlpool Galaxy, were presented to the Royal Society of London in 1850, with the following unequivocal words: "It will be at once remarked, that the spiral arrangement so strongly developed in 51 Messier, is traceable, more or less distinctly, in several of the sketches." Rosse had first detected the spiral structure of M51 in the spring of 1845.

In the following spring, Rosse detected another spiral, M99, and suspected several more from John Herschel's southern catalogue. By the time of the publication of his results in 1850, Rosse had detected 14 spirals, though he cautioned they were "comparatively difficult to be seen, and the full power of the instrument is required to bring out the details ... 51 Messier is the most conspicuous object of that class." Moreover, he cautioned, "we are in the habit of calling all objects spirals in which we have detected a curvilinear arrangement not consisting of regular re-entering curves; it is convenient to class them under a common name, though we have not the means of proving that they are similar systems." Despite these laudable cautions, Rosse was nevertheless the first to discover and declare a new class of object—spiral nebulae. [18]

Rosse himself did not know the true nature of his spiral structures as conglomerations of stars. Even in 1887, when Isaac Roberts made the first long-exposure photographs of the Andromeda Nebula and revealed its spiral structure, its distance was so uncertain that Roberts and other astronomers believed it to be a solar system in formation. And when in 1899 the German spectroscopist Julius Scheiner identified familiar stellar features in the spectrum of the same object, declaring that "the previous suspicion that the spiral nebulae are star clusters is now raised to a certainty," the idea of it being a system of stars was too radical for most. In 1917, V. M. Slipher showed that spectral lines in most spiral nebulae are redshifted, an observation he interpreted as them moving away from us. Finally in 1925, Hubble announced that spiral nebulae, in particular the Andromeda Nebula M31 and its companion M33, were extragalactic in nature—in other words, spiral galaxies consisting of stars, not spiral nebulae.[19] The origins

---

[18] Earl of Rosse, "Observations on the Nebulae," PTRAS, 140 (1850), 499–514, reprinted in part in Bartusiak, 191–195. See Dick (2013), pp. 129–130, from which parts of this paragraph are taken.

[19] Robert Smith, "Beyond the Galaxy: The Development of Extragalactic Astronomy 1885–1965, Part I," JHA, 39 (2008), 91–119; and Part II, JHA, 40 (2009), 71–107; and Smith, *The Expanding Universe: Astronomy's Great Debate, 1900–1931* (Cambridge: Cambridge University Press, 1982), pp. 3 ff; Bartusiak, *Archives*, pp. 407–414. On Scheiner's observations see J. Scheiner,

and persistence of spiral structure is still a subject of considerable debate. In 1925, Bertil Lindblad suggested the density wave
theory, which holds that the arms persist as a density wave moves
through the galactic disk. C.C. Lin and Frank Shu further developed the theory in 1964.[20]

Among the most famous spiral galaxies are those originally
catalogued by Messier in the 18th century. M31, the Andromeda
Galaxy, has a storied place in astronomical history because of
its size and proximity to our own Galaxy as the most prominent
member of the "Local Group" (Fig. 14.6). It was in M31 and nearby

Fig. 14.6. The Andromeda Galaxy, with a close-up of its nucleus harboring
a 100-million solar mass black hole. The circular object at lower center is
the dwarf galaxy M32, and M110 (NGC 205) is a dwarf elliptical galaxy at
upper right. Credit: Hubble Image of the nucleus: NASA, ESA, and T. Lauer
(National Optical Astronomy Observatory); the Andromeda Galaxy: WIYN/
KPNO Image: T. Rector and B. Wolpa (NOAO/AURA/NSF)

"Über das Spectrum des Andromedanebels," AN, 148 (1899), 325, English
translation in Scheiner, "On the spectrum of the great nebula in Andromeda"
ApJ, 9, (1899), 149–150.

[20] C. C. Lin and F. H. Shu, "On the spiral structure of disk galaxies," ApJ, 140
(1964), 646–655.

Fig. 14.7. M51, the Whirlpool Galaxy, the prototype of the class of spiral galaxies. It is about 30 million light years distant and is interacting with the dwarf galaxy NGC 5195 at upper right. Credit: NASA and European Space Agency

galaxies that Walter Baade first distinguished the two stellar populations (S 36), now a fundamental astrophysical concept. Aside from the prototype M51 Whirlpool Galaxy (Fig. 14.7), among other striking nearby spirals are M33, the Triangulum Galaxy; M74, like M51 also viewed nearly face-on; and M81, Bode's Galaxy. M101, the Pinwheel Galaxy, is a magnificent galaxy viewed head-on from our vantage point. It is 170,000 light years in diameter, with perhaps a trillion stars. M104, the Sombrero Galaxy, is known for its large bulge, bright nucleus, and the dark dust lane that gives it its name. The lemon-shaped graphical representation of the de Vaucouleurs system is at http://www.astronomy.com/~/media/import/files/pdf/9/0/1/september_2010_we_zoo.ashx.

Class G 5: Irregular Galaxy

Irregular galaxies originally comprised all those galaxies that did not fit into other classification categories as first defined by Edwin Hubble in 1926 and 1936. In *Realm of the Nebulae* (1936), Hubble, still using his favored term "nebulae" rather than "galaxies," commented that whereas the "regular nebulae"—spirals and ellipticals—"are characterized by rotational symmetry around dominating central nuclei...about one nebula in forty is irregular in the sense that one or both characteristics are absent."[21] Already, he recognized the Magellanic Clouds as conspicuous examples. Today, irregulars are defined as galaxies that either have little structure (Irr-I) or lack any structure whatsoever (Irr-II), and about one-third of all galaxies are classified as irregular. They are more common at higher redshift, as one moves back to earlier epochs in the universe. Irregular galaxies vary in mass from 100,000 to 10 million solar masses and in size from 1,000 to tens of thousands of light years. Irregular galaxies are often rich in gas and dust and may represent ellipticals or spirals that have been distorted over time by interactions with other galaxies. Some irregular galaxies are also starburst galaxies or interacting galaxies, and a few are the "peculiar galaxies" that appear Halton Arp's *Atlas of Peculiar Galaxies* (1966).

Irregular galaxies are well represented among the nearest galaxies in our Local Group. The Magellanic Clouds were for a long time believed to be the nearest galaxies to our own. Based on Harlow Shapley's work on Cepheids, Hubble considered the Large Magellanic Cloud (LMC) to be at about 85,000 light years, and the Small Magellanic Cloud (SMC) at about 95,000 light years from the Milky Way. We now know them to be at about 160,000 and 200,000 light years, respectively. While Hubble originally classed the Magellanic Clouds as "typical irregular nebulae—highly resolved, with no nuclei and no conspicuous evidence of rotational symmetry," de Vaucouleurs recognized rotational symmetry and some spiral structure. He suggested they were in fact part of the spiral sequence, and his classification system added the Sm and Im stages to account for these characteristics in the Magellanic Clouds as well as in other similar galaxies. In his system, they are

---

[21] Edwin Hubble, *Realm of the Nebulae* (1936), p. 40; Dick (2013), pp. 135–137.

thus typed as Irr/SB (s)m, the "B" indicating characteristics of a barred spiral, and the "s" indicating an s-shaped spiral emerging directly from a central bulge or the ends of a bar.[22]

The irregular characteristics of the Magellanic Clouds may arise from tidal interactions with each other and with the Milky Way. Recent Hubble observations show the LMC pulling out a large amount of gas from the SMC. Even in Edwin Hubble's time some 3,000 variable stars had been observed in the two Clouds, as well as numerous globular clusters and open clusters. In addition to its normal clusters, the Large Magellanic Cloud is now known to contain a young massive compact star cluster at the center of the 30 Doradus Nebula, the largest H II region (S 24) in the Local Group.[23]

The nearest irregular galaxy, believed to be the nearest of any galaxy to the Milky Way, is the Canis Major dwarf galaxy, only about 25,000 light years from the Sun. Containing about a billion solar masses, it was discovered only in 2003 by the Two-Micron All-Sky Survey (2MASS). It is being cannibalized by the Milky Way galaxy, giving rise to a string of material known as the Monoceros ring (G 10). Another dwarf irregular that is much studied is NGC 1569 (Fig. 14.8), also known as Arp 210, about 11 million light years distant. This galaxy, which is moving towards Earth rather than away from it like most galaxies, shows central star formation and weak spiral features. It contains two well-known super star clusters comparable in size to globular clusters but younger, and 45 other star clusters have recently been resolved using the Hubble Space Telescope.[24] As an irregular of the Magellanic kind, it is classified as IB(s)m, the "s" referring to the presence spiral features.

NGC 1569 is also an example of a starburst galaxy, where star formation is occurring at an unusually high rate, often because of interactions between galaxies (G 20). M82 (Fig. 14.9) is an example of an irregular galaxy (the only one in the Messier catalogue) interacting with a nearby spiral, M81. M82 is considered the prototype starburst galaxy, birthing stars at a rate ten times that of our

---

[22] Hubble, 1936, 131–137; Buta et al. (2007), pp. 15–16.

[23] On the LMC and SMC interactions see HST PR, "Hubble Solves Cosmic 'Whodunit' with Interstellar Forensics," March 22, 2018, http://hubblesite.org/news_release/news/2018-15/4-galaxies

[24] Deidre Hunter et al., "The Star Clusters in the Starburst Irregular Galaxy NGC 1569," AJ, 120 (2000), 2383–2401.

Fig. 14.8. The irregular galaxy NGC 1569, showing massive starbursts, was found in 2008 to be 11 light years distant, four million light years farther than previously thought. Image Credit: NASA, ESA, the Hubble Heritage Team (STScI/AURA), and A. Aloisi (STScI/ESA)

Fig. 14.9. The irregular galaxy M82, an active starburst galaxy some 12 million light years distant. Red hydrogen gas clouds dominate in this image. Credit: NASA, ESA, and the Hubble Heritage Team (STScI/AURA)

Milky Way. IC 10, another nearby Local Group irregular, is the nearest starburst galaxy and is perhaps an example of a "blue compact galaxy," classified as an Im/BCD or IBm. One and the same galaxy, such as NGC 1569, may be peculiar, starburst, interacting and irregular, but not all irregulars fall in all those categories. Thus, while blue compact galaxies and luminous infrared galaxies are two types of starburst galaxies, they need not be classified as irregulars. All irregulars, though, are irregular for a reason, gravitational interactions being a the primary one.

Arp's *Atlas of Peculiar Galaxies* can be accessed at http://ned. ipac.caltech.edu/level5/Arp/Arp_contents.html.

Class G 6: Seyfert Galaxy

With the Seyfert galaxies we enter the first class in that realm of objects that emit anomalously large amounts of radiation compared to normal galaxies—a Subfamily known as active galaxies. In addition to Seyferts, active galaxies include radio galaxies (G 7), quasars (G 8), and blazars (G 9), and they are associated with all classes of normal galaxies, though certain classes of active galaxies predominate in certain classes of normal galaxies. Classification of active galaxies is difficult, historically contingent, and subject to change, and recent work indicates that at least some classes of active galaxies may be one and the same kind of object viewed at different angles from Earth. On the principle that physical nature rather than accidental viewing angle from Earth should be one of the chief criteria for class status, if such a "unified theory" of active galaxies is demonstrated, some or all of the classes of this Subfamily may collapse into one.

Astronomer Carl Keenan Seyfert (1911–1960) defined as a class what came to be called "Seyferts" in 1943, based on a galaxy's bright optical nuclei, or more specifically on "high excitation nuclear emission lines superposed on a normal G-type spectrum," that is to say, superimposed on a solar-type spectrum.[25] The spiral galaxy M77 (NGC 1068) is the prototype for Seyfert galaxies, being the nearest and brightest Seyfert located at a distance of some 60 million light years (Fig. 14.10). Its peculiar properties were first observed 1908, but it was not declared the prototype of a class until 1943, when Seyfert named it as one of 12 galaxies with similar properties. Seyfert also named 3C 84, known as Perseus A and NGC 1275, as a member of the class; it was the first radio source to be identified with a Seyfert galaxy.

Also on Seyfert's original list was NGC 4151, located at about 50 million light years away, one of the brightest and best known Seyferts, having been observed by both the Hubble and Chandra space observatories. In 2000, astronomers using Chandra reported a massive cloud of hot gas at its center, some 3,000 light years across. They speculated that X-rays generated by material falling into a supermassive black hole heat the gas, which is being

---

[25] C. K. Seyfert, "Nuclear Emission in Spiral Nebulae, *ApJ*, 97 (1943), 28–40, reprinted in Abt, *AAS Centennial*, 324–40, with commentary by Donald E. Osterbrock, pp. 337–338; also reprinted in part in Lang and Gingerich, pp. 738–743. See also Dick (2013), pp. 138–140.

Fig. 14.10. Messier 77, the prototype Seyfert galaxy. Its active galactic nucleus surrounding a supermassive black hole is about 1,000 light years in diameter compared to an overall diameter of 170,000 light years for the entire galaxy. Credit: NASA, ESA, & A. van der Hoeven

blown away from the far side of the black hole at speeds exceeding 1,000 km per second. By 2010, astronomers had dubbed the central region of NGC 4151 "the eye of Sauron" after the malevolent character in Tolkien's Lord of the Rings.[26] Another Seyfert, NGC 7742 (Fig. 14.11) 72 million light years away, is famous for its appearance as a fried egg, because it is a small spiral galaxy with a yolk-yellow center surrounded by blue star-forming rings. But like all Seyferts, it is powered by a central black hole.

Seyfert galaxies were historically the first hint of active galaxies, and, we now know, the most common class of active galaxies. With the rise of radio astronomy, other galaxies were observed spewing extremely large amounts of radiation at radio wavelengths, and the term "active galaxy" came to be associated primarily with radio emission. But eventually, other parts of the electromagnetic spectrum were observed to contribute to this highly energetic behavior. Today we know that some active galaxies are radio quiet (90% of quasars and most Seyferts), while others are radio loud (radio galaxies, blazars, and the other 10% of

---

[26] Chandra Press Release, "Chandra Observed Cloud Powered by Black Hole in Distant Galaxy," June 5, 2000, online at http://chandra.harvard.edu/press/00_releases/press_060500ngc.html For more on M77 see Howard Banich, "Spotlight on a Seyfert," S&T, 136 (November, 2018), 60–65.

Fig. 14.11. NGC 7742, a Seyfert Type 2 spiral galaxy, powered by a central black hole. The thick inner ring is about 3,000 light years from the central core. Credit: Hubble Heritage Team (AURA/STScI/NASA/ESA)

quasars). These emissions were determined to be primarily from the nuclei of these active galaxies; today we refer to the "active galactic nuclei" (AGNs) hosted by active galaxies and ascribe their brightness to supermassive black holes (S 14). In short, active galaxies harbor a bright core of emission in what would otherwise be a normal galaxy. Put another way, the emission in a normal galaxy is the sum of its stars' emissions, whereas in an active galaxy the nucleus is the predominant source of emission.

Seyferts are distinguished from radio galaxies because they are usually spiral galaxies rather than ellipticals, and because they radiate in the infrared, ultraviolet, X-ray, and gamma ray regime, in addition to the optical wavelengths where they were first discovered; only about 5% of them are radio-loud. It is estimated that 1% of spiral galaxies and 10% of all galaxies are Seyferts. In the Hubble sequence, they are dominated by both barred and unbarred (Sa and SB) spirals and by S0 lenticulars. As when Seyfert first discovered them, they are usually detected today by their spectral emission lines from the ionized gas in their nuclei. Seyferts can change dramatically in brightness in periods over days to months. From their redshifts, we know they are much closer to us than quasars or blazars, normally tens of millions of light years rather than hundreds of millions or billions of light years. Seyfert galaxies are less powerful and lower luminosity than quasars (G 8). By contrast to quasars with their high-energy gamma rays up to 1Gev, Seyferts emit low energy gamma rays up to about 100 kev, probably from thermal processes similar to galactic black holes.

Today, Seyfert galaxies are categorized into types based on the strengths and widths of their emission lines. Seyfert I galaxies exhibit both narrow and broad emission lines and are believed to be young AGNs accreting mass. They show extreme variability compared to Seyfert II types. M77, the prototype Seyfert, is a class II, exhibiting only narrow emission lines, probably because of its 51-degree tilt to our line of sight. It is unclear whether these two Types are due primarily to their orientation with respect to our line of sight. Because of their X-ray activity, Seyfert galaxies have been studied by satellites such as ROSAT, XMM-Newton, and Chandra.

Astronomer and historian Donald Osterbrock has pointed out the gradual way in which the discoverer's name was conferred on a new class of objects, a rarity in astronomical classification, especially at the Class level. In 1958, 15 years after Seyfert declared the new class, Geoffrey Burbidge discussed the galaxies "which Seyfert studied," and the following year Leo Woltjer discussed "objects listed by Seyfert." Also in 1959 Geoffrey Burbidge, Margaret Burbidge, and K. H. Pendergast used the term "Seyfert galaxies" for the first time in print in their discussion of NGC 1068, the Messier object that Seyfert had singled out as the prototype in his original paper of 1943. Thereafter, the term "Seyfert galaxies" became commonplace in astronomical literature.[27]

A tour of Perseus A, courtesy of the Chandra Observatory, is at https://www.youtube.com/watch?v=ITYWqo-qvzA. For more on "the eye of Sauron" see https://www.nasa.gov/mission_pages/chandra/multimedia/11-029.html. Useful websites on Seyferts and other active galaxies are at http://www.mpe.mpg.de/xray/research/galactic_centers/index.php?lang=en and http://www.seyfertgalaxies.com/ , and on active galactic nuclei more generally at http://www.astr.ua.edu/keel/agn/.

---

[27]Donald E. Osterbrock, 'Seyfert Galaxies," in Abt, ApJ Centennial pp. 337–338.

Class G 7: Radio Galaxy

After Seyfert galaxies (G 6), the next class of active galaxy to be discovered was radio galaxies, though they were not immediately recognized as being in what we would now call the Subfamily of active galaxies. Unlike Seyferts with their intense optical emission, radio galaxies exhibit intense radio emission. They are almost always large elliptical galaxies, by contrast to the Seyfert class, which are normally spirals. Like all active galaxies, they are believed to be powered by the supermassive black holes, but in contrast to other active galaxies like blazars (G 9) and some quasars (G 8), much of the radio emission comes from large radio lobes or plumes on each side of the galaxy. Twin jets of particles are ejected from the active galactic nucleus (AGN), feeding the lobes and plumes.

According to some unification models, radio galaxies, blazars, quasars and even Seyferts are the same class of object viewed from different perspectives. Because these unification models are still uncertain, and because these classes are still used by astronomers and institutions such as NASA, we adhere to the traditional separate classes, keeping in mind that some classes of this Subfamily may be dropped in the future, as they should be if physical characteristics are the primary criteria for classification rather than arbitrary features such as viewing angle.

The discovery of radio galaxies as a new class of object owes its existence to the rise of radio astronomy. Although Karl Jansky first detected the galactic background radio noise in 1932, and Grote Reber produced maps of radio emission in the 1940s, British physicist J. Stanley Hey and his colleagues are credited with the discovery of the first discrete radio source, later known as Cygnus A, in 1946 in an extension of their wartime radar work. As Hey recalled, "We could locate the position of the source only to within about 2 degrees, and there was no obvious optical clue to its identification with visible stars. Nevertheless, we concluded that only discrete sources could produce such fluctuations." The signal was coming from the direction of the constellation Cygnus, and they believed the source must be a similar but more distant source than the Sun, which they had already observed in radio emission.[28] Only in 1952

---

[28] J. S. Hey, S. J. Parsons, J. W. Philips, "Fluctuations in cosmic radiation at radio frequencies," *Nature*, 158 (1946), 234, reprinted in part in Lang and Gingerich, pp. 774–776. For accounts see Hey's own book *The Evolution of Radio Astronomy* (New York: Science History Publications, 1973), pp. 43–46, as well as Woodruff T. Sullivan III, *Cosmic Noise: A History of Early Radio Astronomy* (Cambridge: Cambridge University Press, 2009), pp. 103–105.

did Walter Baade and Rudolph Minkowski identify Cygnus A with an optical source, an 18th magnitude galaxy, and only in 1954 did they publish their results.[29]

The energy output implications of such a distant object were astonishing. The *New York Times* described Cygnus A as a radio station at a distance of 600 million million million miles, with a power of 400,000,000,000,000,000,000,000,000,000,000,000 kilowatts. British cosmologist William H. McCrea recalled that astronomers "had to practise hard before breakfast at believing" in a radio luminosity about a million times greater than that of our Galaxy. Cygnus A, famous for its huge jet structure (Fig. 14.12) is now known to be an elliptical galaxy and one of the brightest radio sources in the sky. Because of its early discovery, it is often considered a prototype of the radio galaxy class. Its huge physical structure spans 300,000 light years, and it emits some ten million times more energy than normal galaxies. Even though it is some 800 million light years distant, it is still one of the brightest radio sources in the sky. In 2017, astronomers reported a mystery object some 1,300 light years from the center of the galaxy, clearly distinct from its nucleus, which was not there in observations three decades earlier.[30]

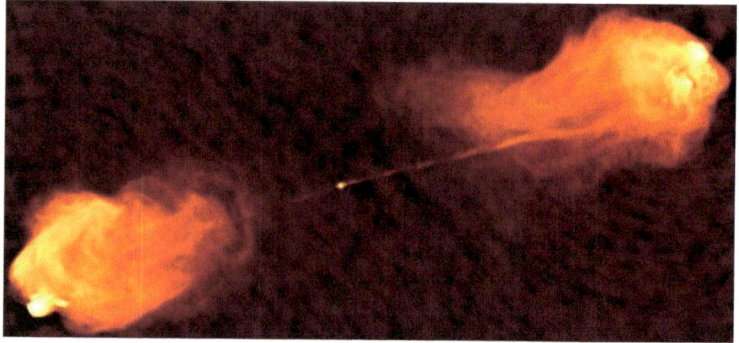

Fig. 14.12. Two jets shoot out of the radio galaxy Cygnus A. Credit: NRAO

---

[29] Sullivan, *Cosmic Noise*, pp. 335–341: 341 details the search for an optical counterpart of Cygnus A. The publication of results is W. Baade and R. Minkowski, "Identification of the Radio Sources in Cassiopeia, Cygnus A and Puppis A," ApJ, 119 (1954), 206–214, reprinted in part in Lang and Gingerich, 786–791. See also their adjacent paper "On the Identification of Radio Sources," ApJ, 119 (1954), 215–231, reprinted in Abt, *ApJ Centennial*, 538–568. This and the following paragraphs from Dick (2013), pp. 140–144.

[30] *New York Times*, 23 August, 1953, p. E9, as described in Sullivan, *Cosmic Noise*, 341; Dick (2013), pl 141, Camille M. Carlisle, "Mystery Object in Cygnus A Galaxy," January 13, 2017; https://www.skyandtelescope.com/astronomy-news/mystery-object-in-cygnus-a-galaxy-1301201623/

It took some time for astronomers to realize how common this astonishing phenomenon really was. Radio positions were not well determined in the early years of the field; optical identifications were difficult because the error box might contain thousands of optical objects. Nevertheless, over a period of years some identifications were made, so astronomers could begin to see they were dealing with an extraordinary phenomenon. In 1947, the year after Hey and his British colleagues reported the first discrete radio source, John Bolton and his Australian colleagues discovered more discrete radio sources, and in 1949 they suggested optical identifications for three of them: Taurus A, Virgo A, and Centaurus A. They identified Taurus A with the Crab Nebula within our Galaxy (Fig. 9.6), and the other two with the "nebulous objects" M87 (Fig. 14.4) and NGC 5128, respectively. But they were reluctant to identify the latter two as extragalactic, not only because "there is little definite evidence to decide whether they are true extragalactic nebulae or diffuse nebulosities within our own galaxy," but also because their identification as radio sources "would tend to favor the latter alternative, for the possibility of an unusual object in our own galaxy seems greater than a large accumulation of such objects at a great distance."[31]

We now know that Virgo A, discovered optically already in 1781, is a giant elliptical galaxy, the largest near Earth and the dominant galaxy in the Virgo Cluster (G 22). Centaurus A (NGC 5128) is now classified as an elliptical or lenticular galaxy (Fig. 14.13). At a distance 12 million light years, it is the nearest radio galaxy and one of the largest and brightest in the sky. John Herschel first called attention to this object as a peculiar galaxy in the 19th century, and it has since been studied at all wavelengths. Two lobes of radio emissions extend over very large distances (five degrees on each side) from the central source, whose likely power source is an accreting black hole 100 million times the mass of the Sun. Spitzer Space Telescope images show that it is a galaxy in collision

---

[31] The discovery of the radio sources is in J. G. Bolton, "Discrete sources of Galactic radio frequency noise," *Nature*, 162 (1948), 141–143; the optical identifications are in John G. Bolton, Gordon J. Stanley and O.B. Slee, "Positions of Three Discrete Sources of Galactic Radio-Frequency Radiation," *Nature*, 164 (1949), 101–102, reprinted in Lang and Gingerich, pp. 777–778. See also Hey, 46–47, and Sullivan, 320–324. It was Bolton who initiated the 'A' nomenclature for the brightest object in the constellation, B for the second brightest, and so on.

Fig. 14.13. Composite image of the Centaurus A radio galaxy, using instruments at three different wavelengths, showing lobes and jets emanating from the central black hole. Submillimeter data is orange, X-ray data in blue. The blue X-ray data at upper left extends about 13,000 light years from the center. Credit: ESO/WFI (optical); MPIfR/ESO/APEX/A.Weiss et al. (submillimetre); NASA/CXC/CfA/R.Kraft et al. (X-ray)

with a spiral, and in 2010 the Fermi Large Area Telescope reported gamma ray emissions from the radio lobes.[32]

In a way reminiscent of the separation of different classes of nebulae in our own Galaxy, radio astronomers were eventually able to identify different classes of radio objects. Radio surveys in the early 1950s, such as the Cambridge 1C survey and the Mills survey in Australia, identified several hundred radio objects, of which only ten or so were identified with optical counterparts by 1953. The leading theory for several years was that the majority of them were a new class of object, very nearby, dark stars termed "radio stars." But in what astronomer and historian Woodruff T. Sullivan III called the "bible" of optical identifications as of 1954, Baade and Minkowski identified four major categories of optically-identified radio sources: peculiar emission nebulosities

---

[32] A. A. Abdo et al., "Fermi Gamma-Ray Imaging of a Radio Galaxy," *Science*, 328 (7 May, 2010), pp. 725–729.

such as Cassiopeia A and Puppis A; peculiar extragalactic nebulae including Virgo A, Perseus A, Centaurus A, and Cygnus A; normal galaxies with much lower radio output than the second category; and supernova remnants such as Taurus A.[33]

As Sullivan has shown in detail, these categories of radio objects took time to separate, and in fact their first and fourth categories proved to be one and the same. The early identification of Taurus A with the Crab Nebula had already shown that not all discrete radio sources were radio galaxies, but the second half of the 20th century demonstrated that radio galaxies, with their prodigious energy outputs, were quite common. Radio astronomy had opened a new window on the universe, and our perception of the universe was forever changed. Radio galaxies and related objects are now known to number in the millions, featuring an array of characteristics including lobes, plumes, and jets (G 11). As for the radio galaxies themselves, astronomers have distinguished two morphological classes: low power and high power. These are termed Fanaroff-Riley Classes I and II, (FR I) and FR II, low-powered and high-powered respectively. In the Three Kingdom System these would be considered Types, one taxon down from Class.[34]

The existence of such prodigious amounts of energy begged the question of the energy-producing mechanism. Already in the 1950s, the favored mechanism was synchrotron radiation caused by electrons spiraling close to the speed of light in a magnetic field. Although controversial among radio astronomers until the late 1950s, this process is now known to be the most widespread type of non-thermal radio generation—that is, radiation not arising from blackbody radiation at different temperatures. [35]

More information on M87 is at http://chandra.harvard.edu/photo/2006/m87/. And more on Cygnus A, including a diagram of the jets and lobes, is at http://chandra.harvard.edu/press/00_releases/press_110600cyg.html .

---

[33] W. Baade and R. Minkowski, "Identification of the Radio Sources in Cassiopeia, Cygnus A and Puppis A," ApJ, 119 (1954), 206–214, reprinted in part in Lang and Gingerich, 786–791; Sullivan, 344–348: 346.

[34] B. M. Fanaroff and J M. Riley, "The morphology of extragalactic radio sources of high and low luminosity," MNRAS, 167 (1974), 31P-36P.

[35] "Hannes Alfven and Nicolai Herlofson, "Cosmic Radiation and Radio Stars," *Physical Review*, 78 (1950), 616, reprinted in Lang and Gingerich, 779–781.

Class G 8: Quasar

16 years after J. Stanley Hey's discovery of Cygnus A as the first discrete astronomical radio source in 1946, and 10 years after its identification as the first radio galaxy in 1952, another class of object was detected at radio wavelengths with an outpouring of radiation far more powerful than even Cygnus A. These objects appeared to be stars and so were termed "quasi-stellar objects," QSOs, or quasars for short. Quasars are now known to be the cores of extremely distant active galaxies, the most luminous of all classes of active galaxies discovered before or since, including Seyferts (G 6), radio galaxies (G 7), and blazars (G 9). Their extreme distance of billions of light years makes them the most energetic objects in the universe aside from short-lived explosive events like supernovae and gamma ray bursts. Their vast amounts of radiation, and sometimes jets (G 12) of material, originate from their centers, and their brightness may vary over time scales from months to hours. Different quasars have been observed in radio, infrared, optical, ultraviolet, X-ray, and gamma-ray regions of the spectrum, and with a variety of ground-based and space-based telescopes, including all four of NASA's Great Observatories: Hubble, Compton, Chandra, and Spitzer.

Like other active galaxies, which host active galactic nuclei just as the Seyferts and radio galaxies discovered before them, quasars are believed to be powered by matter falling into supermassive black holes (S 14), whereby gravitational potential energy is converted to kinetic energy and then thermal energy, emitting radiation. Like radio galaxies, radio emission is generated by relativistic electrons spiraling in a magnetic field, the mechanism for synchrotron radiation. Quasar formation may be driven by a process of galaxy mergers, which changes the distribution of gas around the black hole, accounting for their prodigious energy output. Despite that output quasars are extremely compact, perhaps not much larger than our solar system in diameter. They may differ from blazars and radio galaxies only in the viewing angle as seen from Earth, in which case their status as a distinct class may be in jeopardy.

The Hubble Space Telescope has revealed that, unlike Seyferts, which are associated mainly with spiral galaxies, and radio galaxies found mainly in giant ellipticals, quasars reside in a variety of host galaxy types, including spirals, ellipticals, and even

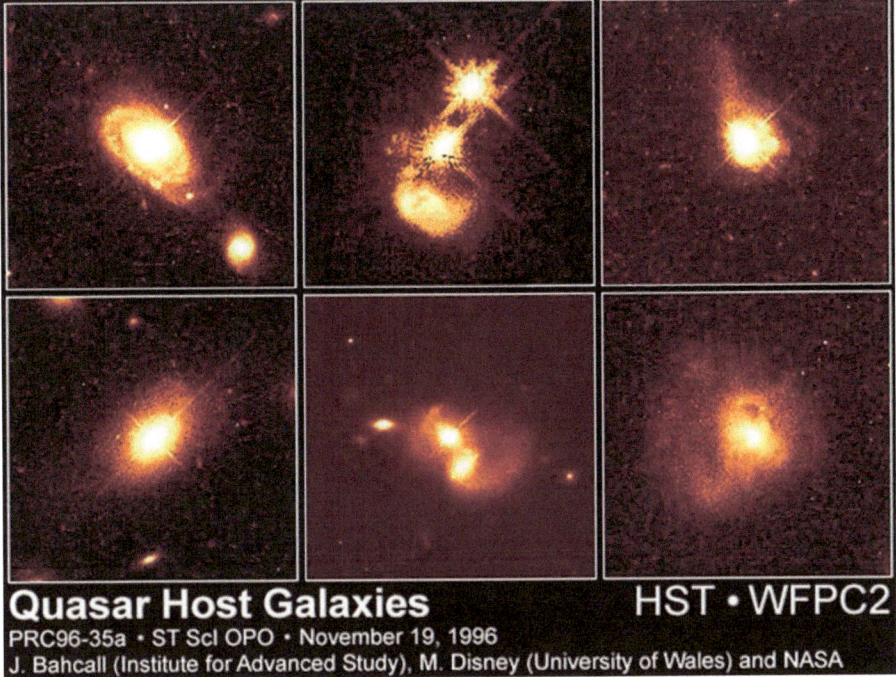

Quasar Host Galaxies   **HST • WFPC2**
PRC96-35a • ST ScI OPO • November 19, 1996
J. Bahcall (Institute for Advanced Study), M. Disney (University of Wales) and NASA

Fig. 14.14. An early survey of quasar host galaxies shows those galaxies can be spirals (top left), ellipticals (bottom left), and colliding galaxies. Credit: John Bahcall (Institute for Advanced Study, Princeton), Mike Disney (University of Wales), and NASA

colliding galaxies (Fig. 14.14). Although they were initially discovered because of their vast outpouring of radio energy, of the thousands of quasars now known, more than 90% exist in radio quiet galaxies. Nevertheless, all quasars are active galaxies because of the vast amount of radiation they emit in all wavelengths. About 50 quasars are known to emit high-energy gamma rays. Because quasars and other classes of active galactic nuclei are found only at great distances, they must become dormant as galaxies age, leaving behind a quiescent normal galaxy to live out its life. The exceptions are the radio-quiet Seyfert galaxies, which are much closer.

The story of the discovery of quasars as a new class of objects is legendary in the history of astronomy. With the birth of radio astronomy, strong discrete radio sources had been detected since the 1940s, and hundreds had been observed by the early 1960s with radio surveys of the sky such as the Third Cambridge Catalogue. They were believed to be "radio stars" within our Galaxy, but because their positions had not been accurately determined, most of their optical counterparts had not been identified, making distance determinations impossible. In 1962 Caltech astronomer Maarten Schmidt, using the 200-inch Palomar telescope, identified the optical component of an unusual radio-emitting star known as 3C273 (object 273 in the Third Cambridge Catalogue), located in the constellation Virgo. It appeared to be a blue star, but the spectrum was completely anomalous, featuring unknown broad emission lines. It took weeks until Schmidt realized these lines were actually well-known "Balmer" hydrogen lines that had an extremely large redshift, some 16%, meaning they were not simply beyond our Galaxy but at the exceedingly large distance of almost two billion light years.[36] With this Rosetta Stone decipherment, astronomers also found that 3C 48, discovered earlier by John Bolton, had a redshift amounting to 3.5 billion light years.

The interpretation of quasars, however, was not so straightforward. In his published article in 1963, Schmidt himself pointed out that an alternative explanation was a "gravitational redshift" caused by a massive star in the Galaxy. Based on a variety of considerations, he concluded that "the explanation in terms of an extragalactic origin seems most direct and least objectionable." If this was the case, the prodigious amounts of energy at these distances made them a new class of object. By 1964, the Chinese astronomer Hong-Yee Chiu suggested that the term "quasi-stellar objects" be shortened to "quasar," and this term caught on despite the fact that they were completely unrelated to stars. Quasar hunting became a favorite sport of astronomers, and their popular appeal is evident in the fact that Schmidt appeared on the cover of *Time* magazine for March 11, 1966.

Only decades later, using the Hubble Space Telescope and other instruments, were quasars proven to reside in a variety of galaxies. Astronomer John Bachall encapsulated the uncertainties that remained more than three decades after the discovery of quasars when he commented regarding their discovery in so many

---

[36] Maarten Schmidt, "3C 273: A Star-Like Object with Large Red-Shift," *Nature*, 197 (March 16, 1963), 1040, reprinted in Lang and Gingerich, 801–810, including related papers in the same issue of *Nature*; also reprinted in Bartusiak, 505–507. For a first-person account see Maarten Schmidt, "Discovery of Quasars," in K. Kellermann and B. Sheets, *Serendipitous Discoveries in Radio Astronomy*, (Green Bank: NRAO, 1983), pp. 171–174. Dick (2013), pp. 143–147.

types of galaxies, "If we thought we had a complete theory of quasars before, now we know we don't. No coherent, single pattern of quasar behavior emerges. The basic assumption was that there was only one of kind of host galaxy, or catastrophic event, which feeds a quasar. In reality we do not have a simple picture—we have a mess."[37]

A few well-known astronomers, including Margaret and Geoffrey Burbidge, Fred Hoyle, and Halton Arp, continued to believe that quasars were not at cosmological distances. Though they held this view into the early 21st century, after the detection in 1964 of the cosmic microwave background as a remnant of the Big Bang, there were fewer and fewer proponents of this renegade view.

Quasars are an example of the importance of new technology—in this case a whole new regime of the spectrum—to the process of discovery. Schmidt pointed out:

> Quasars are hard to find in the optical sky; there are as many as 3 million stars brighter than the brightest quasar, 3C 273. The situation is radically different at radio wavelengths. In the 3C catalogue 3C 273 is the sixth strongest source above galactic latitude 15 degrees. In hindsight, then, it is clear why radio astronomy was destined to lead us to the first quasars. If radio astronomy had developed much later, X-ray astronomy would have played the same role for the same reasons.[38]

More than 200,000 quasars are now known, most from the Sloan Digital Sky Survey. With redshifts between .06 and 6.5, their "co-moving" distances are between 780 million and 28 billion light years away, in terms of where they are now taking into account the expanding universe, or 600 million to 13 billion light years in terms of where they are as we see the light now observed from Earth. Most, however, are much more than three billion light years distant, between nine and 13 billion light years. Thus, they were much more plentiful in the early universe. The closest quasar is 3C 405 in the Cygnus A galaxy, some 800 million

---

[37]John Bachall, HST release, "Hubble Surveys the 'Homes' of Quasars," November 19, 1996, http://hubblesite.org/newscenter/archive/releases/1996/35/text/

[38] Maarten Schmidt, "Discovery of Quasars," in Kellermann and Sheets, *Serendipitous Discoveries*, p. 171.

Fig. 14.15. The prototype quasar 3C 273 is now known to be embedded in a galaxy with a spiral plume wound around the quasar and a red dust lane. Credit for WFPC2 image: NASA and J. Bahcall (IAS); ACS image: NASA, A. Martel (JHU), H. Ford (JHU), M. Clampin (STScI), G. Hartig (STScI), G. Illingworth (UCO/Lick Observatory), the ACS Science Team, and ESA

light years distant; although it was the first radio galaxy discovered and identified, not until 1994 did the Hubble Space Telescope find that it harbored a quasar.[39] 3C 273, the prototype quasar, is 1.5 billion light years distant and the brightest quasar in our sky at 12.8 magnitude, one of the reasons it was the first discovered. Its absolute luminosity is two trillion times that of our Sun and 100 times that of our entire Milky Way. The galaxy in which it is embedded has also been imaged (Fig. 14.15).

A gallery of quasars and active galaxies from Chandra is at http://chandra.harvard.edu/photo/category/quasars.html. A closer view of the 3C 273 jet imaged by Chandra is at http://chandra.harvard.edu/photo/2006/3c273/ .

---

[39] Hubble release, "Hubble Uncovers a Hidden Quasar in a Nearby Galaxy (Cygnus A)," September 21, 1994, http://hubblesite.org/newscenter/archive/releases/1994/42/text/

Class G 9: Blazar

Blazars are active galactic nuclei viewed head-on, made visible by their relativistic jets pointed directly at us. They vary frequently, erratically, and rapidly at an array of wavelengths, with their luminosity fluctuating by more than ten times, and by 10 to 50% in only a few hours. Like other active galaxies, this implies the central core is compact, perhaps one light day in diameter, a little larger than our Solar System. Like radio galaxies (G 7), but in contrast to the Seyfert spirals (G 6), blazars are usually associated with elliptical galaxies. Their orientation directly to our line-of-sight may enhance the continuum radiation while depressing any spectral lines. Blazars came onto the astronomical scene just when astronomers thought galactic astronomy couldn't get any stranger.

The rise of the blazars illustrates the difficulties of declaring a new class of object. The first object that eventually came to be called a blazar was discovered in 1929 by the German astronomer Cuno Hofmeister, who identified what he believed to be an irregular variable star in the constellation Lacerta, the Lizard. He dubbed it BL Lac. The object varied from 13th to 16th magnitude with fluctuations of almost a magnitude in a few days and with no discernable pattern, unusual even for a variable star. But in 1968, John Schmitt of the David Dunlap Observatory identified the BL Lac "variable star" as a powerful radio source and pointed out that its optical properties, radio polarization, and unusual featureless spectrum "make it outstandingly interesting."[40]

Only in 1972, a decade after the discovery of quasars, did Peter Strittmatter at the University of Arizona and his colleagues declare a new class of objects of the "BL Lacertae type." "Recent investigations have suggested the existence of a new class of astronomical objects with the following characteristics," they wrote in the *Astrophysical Journal*. Those characteristics included rapid variations in intensity at radio, infrared, and visual wavelengths, a high proportion of emission at infrared wavelengths, absence of emission lines in their spectra, and strong and rapidly varying polarization at visual and radio wavelengths.[41] They wrote that the objects

---

[40] Cuno Hoffmeister, "354 neue Veranderliche," *Astron. Nachr.*, 236 (1929), 233–244; John L. Schmitt, "BL Lac identified as Radio Source", *Nature* 218 (1968), 663. See also David Nakamoto, "The Enigma that is BL Lacertae," S&T, 136 (Sept, 2018), 30–35, and Dick (2013), pp. 147–148.
[41] P. A. Strittmatter, K. Serkowski, R. Carswell, R. et al., "Compact Extragalactic Nonthermal Sources," *ApJ*, 175 (1972), pp. L7–L13.

were compact sources and very similar to quasars (G 8) but exhibited no spectral lines, making a determination of their distances difficult. Reminiscent of statements made when radio galaxies and quasars were discovered, they found it "hard to understand within any reasonable physical model how a very small source of radiation can produce such a large observed flux unless the objects are not far away." But because of their similarity to quasars, they declared BL Lacertae objects likely to be extragalactic.

Two years later, in 1974, John Oke and James Gunn, both at Caltech, identified BL Lacertae itself as a galaxy rather than a star. Using the Palomar 200-inch telescope and a masking technique to examine only the region around the object rather than the object itself, they obtained a spectrum with emission lines and showed that BL Lac was actually an elliptical galaxy at a distance of 1.1 billion light years, extragalactic indeed. In concluding their otherwise staid paper, they wrote, "the same phenomena and approximately the same level of outrageousness exist for BL Lac as for higher-redshift QSOs if they are placed at the distances suggested by their redshifts." With astronomers' penchant for thrift, the shortened name "BL Lacs" caught on quickly, and by 1978 Edward Spiegel dubbed them "blazars," a term first used in the title of a paper in 1984 and that has caught on since that time.[42]

If BL Lacs were extragalactic, they had most of the characteristics of quasars except for the lack of broad emission lines, rapid and large variability, high optical polarization compared to non-BL Lac quasars, and somewhat less luminosity. In fact, many early quasars now turn out to be blazars, including the first quasar discovered, 3C 273. Why blazars vary in optical and gamma ray brightness on periods from minutes to days is still mysterious. But it soon became clear that blazars are not uncommon. The EGRET experiment on the Compton Gamma Ray Observatory in the 1990s detected 66 blazars, and 233 blazars were known by 1995. The Compton's spacecraft's successor, the Fermi Gamma Ray Telescope (formerly the GLAST) mission, discovered more than 1,100 blazars in its first four years of observations, released

---

[42]J. B. Oke and J. E. Gunn, "The Distance of BL Lacertae," ApJ Letters, 189 (1974): L5–L8. http://adsabs.harvard.edu/abs/1974ApJ...189L...5O. The first "blazar" paper title was Wardle et al. "The Radio Morphology of Blazars and Relationships to Optical Polarization and to Normal Radio Galaxies," ApJ, 279 (1984), 93–111. The paper reported on "VLA observations of BL Lac objects and highly polarized quasars ("blazars")."

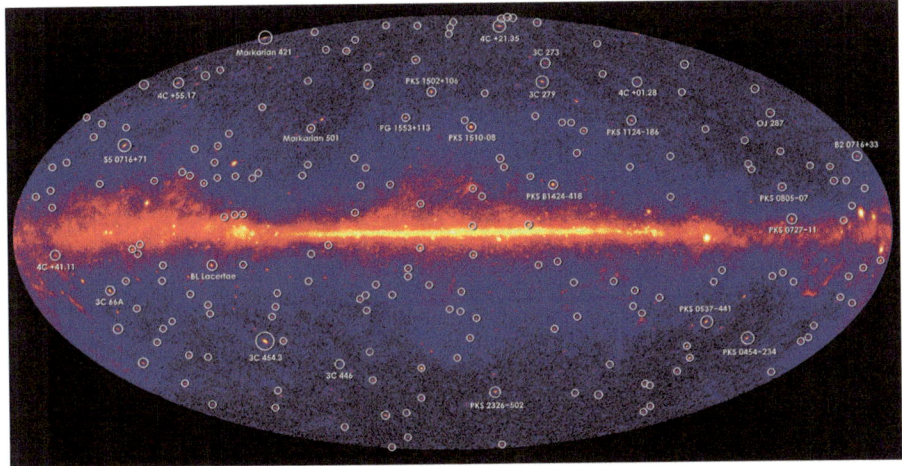

Fig. 14.16. Fermi Space Telescope plot of some of the blazars it has discovered as of 2013. The prototype BL Lac is at center left. Image credit: NASA/ DOE/Fermi LAT Collaboration

in its Third Source Catalog in 2015 (Fig. 14.16). Blazars constitute in general more than half of the gamma-ray sources detected with Fermi's Large Area Telescope.[43]

With the discovery of more members of the class, two Types of blazars have been distinguished, the BL Lac types and the optically violent variable quasars (OVVs). BL Lacs are sometimes considered weak radio galaxies, while the OVVs are considered more powerful radio galaxies. In either case, they are massive objects equal to millions or billions of Suns. They are now known to be composed of supermassive black holes (S 14) accreting stars, gas and dust, producing the usual accretion disk with an accompanying relativistic jet (G 11)—in this case aimed directly at us. The light is polarized because of an intense magnetic field whose variation causes the unpredictable variations in magnitude.

BL Lac itself changes from magnitude 12.7 to 16 and has brief flares of about one magnitude. Optical variability can be observed from night to night, and radio variations over a month. Its distance is 900 million light years. Markarian 421, a blazar 360 million light years distant, is one of the closest blazars and brightest quasars in sky; at 13.3 magnitude, it is also one of the brightest

---

[43] NASA Goddard Release, "Fermi's Five-Year View of the Gamma-Ray Sky," August 21, 2013, https://svs.gsfc.nasa.gov/11342

gamma ray sources in the sky. It has rapid variations over hours to days, and longer periodic variations of 23 years.

Much more distant blazars have also been found. W Comae Berenices, discovered by Max Wolf as a variable star, is 1.3 billion light years distant and has varied by four magnitudes over the last century. The object known as OJ 287 is at 3.5 billion light years; 3C 66A is at 4.6 billion light years; and the Fermi spacecraft has observed quasar/gamma ray blazar 3C 454.3 at a distance of 7.2 billion light years. Among other interesting blazars is PKS 2155-304, the prototype X-ray blazar. In 2017, the Fermi team announced the discovery of blazars at a distance of 12.4 billion light years, only 1.4 billion years after the Big Bang. Two of these blazars harbor black holes of a billion solar masses, challenging ideas of how black holes could have grown so large so soon in the history of the universe.

Blazars became even more famous in 2017 with the detection of the first high-energy neutrinos from the blazar designated TXS 0506+056 some four billion light years away. Its energy was an incredible 290 TeV. With this discovery, neutrinos were added to the "multimessenger" array of astronomical methods, consisting of electromagnetic radiation, cosmic rays, and gravitational waves. The combination of these methods is expected to illuminate the mechanisms energizing active galactic nuclei such as blazars.[44]

An excellent video on blazars from the Fermi Space Telescope is at https://www.nasa.gov/feature/goddard/2017/nasas-fermi-discovers-the-most-extreme-blazars-yet. Fermi celebrated its tenth anniversary in 2018, and highlights its decadal achievements at its website https://fermi.gsfc.nasa.gov/science/.

---

[44] The IceCube Collaboration et al., "Multimessenger observations of a flaring blazar coincident with high-energy neutrino IceCube170922A," Science, 361 (2018), 146–147, http://science.sciencemag.org/content/361/6398/eaat1378

15. The Circumgalactic Family

Class G 10: Satellites and Stellar Streams

Just as satellites orbit planets and planets orbit stars, so do galaxies have their own satellites and circumgalactic material orbiting them, notably in the form of smaller galaxies or their shredded debris known as rings or stellar streams. At least 59 of these galaxies orbit our Milky Way Galaxy out to about 1.5 million light years, believed to be the edge of the dark matter halo (G 12). These are dominated by irregular, elliptical, and spheroidal dwarf galaxies, although several larger irregular galaxies also orbit, including the Large and Small Magellanic Clouds. The preponderance of dwarf galaxies may be due to their origin as pieces of larger spirals pulled out by tidal interactions when galaxies collide. Stellar streams are related to these satellite galaxies because they are their shredded remains torn apart by the gravity of the Milky Way, sometimes colloquially referred to as "sky rivers" or more poetically as "gravity's rainbow."

Shredded objects as defined in this class are not to be confused with rings of molecular hydrogen (S 25) within our Galaxy, the object of research such as the Milky Way Galactic Ring Survey. Nor should they be confused with ring galaxies, a type of interacting galaxy (G 20) with a ring-like appearance likely caused by galaxy collisions, of which Hoag's object is the prototype. Avoiding this confusion is one good reason to refer to them as "stellar streams" rather than galactic rings, and this seems to be the trend among astronomers. Stellar streams as defined in this class are important not only for studying the structure of the visible halos of galaxies, but also for probing the even more enigmatic dark matter halo in which the visible halo is embedded. They can also be used for galactic archaeology to reconstruct how the Milky Way Galaxy formed, and as a method for measuring the mass of the primary galaxy. Galactic satellites and stellar streams are also found around other large galaxies.

Even some of the Milky Way Galaxy's closest satellite galaxies have only been discovered in recent years, hidden by obscuring gas, dust, and the stars of the Milky Way itself. The Sagittarius

© Springer Nature Switzerland AG 2019
S. J. Dick, *Classifying the Cosmos*, Astronomers' Universe,
https://doi.org/10.1007/978-3-030-10380-4_15

Dwarf Elliptical Galaxy discovered in 1994, not to be confused with the Sagittarius Dwarf Irregular Galaxy some 3.5 million light years away, is only 50,000 light years from the center of the Milky Way and is now being cannibalized by our own galaxy, forming one of the star streams discussed below. The long-known globular cluster M54 lies at its heart.

Even closer, in 2003 the Canis Major Dwarf Irregular Galaxy was discovered, only 25,000 light years from Earth and 40,000 light years from the galactic center. It contains perhaps a billion stars, revealed by the Two Micron All-Sky Survey (2MASS), an infrared survey able to pierce gas and dust. It too is being disrupted by the Milky Way galaxy, to the extent that it trails a stream of stars known as the Monoceros ring, which has wrapped itself three times around the Milky Way. Many of these dwarf satellite galaxies have been detected over the last decade, indicating that more will be discovered in the near future. In 2007 the Sloan Digital Sky Survey discovered two "hobbit" galaxies, the smallest of which, Leo T, is the smallest, faintest star-forming galaxy known, located at a distance of about 1.4 million light years and so at the fringes of our Galaxy's gravitational influence.[1]

In our Local Group of galaxies (G 21), at least 14 dwarf galaxies orbit the Andromeda Galaxy, dominated by M32 and M110. The Triangulum Galaxy, the third galaxy that dominates the Local Group, is also suspected to harbor satellite galaxies, although it is sometimes difficult to prove that one galaxy is a satellite of another. Nevertheless, satellite galaxies are likely to be a normal feature of large galaxies throughout the universe due to their gravitational pull. Theory predicts that these dwarf satellite galaxies should surround their primary galaxy in chaotic swarms. But in 2018 astronomers reported that 14 out of 16 of the brightest satellites of the Centaurus A galaxy (Fig. 14.13) were orbiting in the same direction and in the same plane. The reasons for this apparent order remain a mystery.[2]

Satellite galaxies are closely tied to the structures known as stellar streams or galactic rings. Like planetary rings and stellar disks and shells, galaxies show a variety of ring structures associated with the merger of galaxies and the formation of galactic halos (G 12). Just as planetary rings may be produced by disintegrated

[1] M. J. Irwin et al., "Discovery of an Unusual Dwarf Galaxy in the Outskirts of the Milky Way," ApJ, 656 (2007), L13–L16.
[2] Oliver Muller et al., "A whirling plane of satellite galaxies around Centaurus A challenges cold dark matter cosmology," Science, 359 (2018), 534–537.

satellites, galactic rings in the form of star streams may also be produced from disintegrated objects such as globular clusters and dwarf galaxies stretched by tidal forces. They are discovered by the unusually rapid velocities and chemical composition of their components, akin to extracting a needle from a haystack of billions of other stars. Once identified, however, they can be pieced together as the semi-coherent structures that they are. As mentioned above, astronomers consider the search for stellar streams as a kind of "galactic archaeology," because they are the remnants of former galaxies that reveal a great deal about the history of those galaxies they now encircle and how they were assembled.[3]

The origin of stars with unusual speeds and composition has been the subject of debate since the 1960s, and even prior to that in terms of the high-velocity subdwarfs (S 6). The first hint of star streams came in 1971 when astronomer Olin Eggen identified 53 members of a group of older disk Population I stars with the same space motion as Arcturus.[4] He dubbed these the "Arcturus Group," located some 37 light years distant and containing the bright star Arcturus. This group is now believed to be the remnant of a dwarf galaxy. At the time of their discovery the question remained whether these stars derived their motions from perturbations within the galaxy or from capture from the outside; the latter has now been confirmed in spectacular manner.

More than 20 years passed before the next stream was found in the Milky Way Galaxy, launching the modern era of galactic stream discoveries. In 1994 Rodrigo Ibata and his colleagues announced the discovery of the closest galaxy to our own, the Sagittarius Dwarf Galaxy with only $1/100^{th}$ the mass of our Galaxy. Located on the far side of the galactic center, it had eluded detection because of the large number of foreground stars. Similar to the other eight dwarf spheroidal companions of the Milky Way Galaxy known at the time, it is comparable in size and luminosity to the largest of them, the Fornax system. The Sagittarius Galaxy, however, was in an advanced stage of tidal disruption, resulting in the impressive stream of stars dubbed the Sagittarius stream.[5]

[3] For reviews see H. J. Newberg and J. L. Carlin, eds. *Tidal Streams in the Local Group and Beyond*, Astrophysics and Space Science Library, Vol. 420 (2016), and Eric Hand, "Sky Rivers," *Science*, 362 (2018), 16–21.

[4] O. J. Eggen, "The Arcturus Group," PASP, 83 (1971), 271–285.

[5] R. A. Ibata, G. Gilmore and M. J. Irwin, "A dwarf satellite galaxy in Sagittarius," *Nature*, 370 (1994), 194–196, and ibid., "Sagittarius: the nearest dwarf galaxy," MNRAS, 277 (1995), 781–800.

It stretches one million light years and has a mass of 100 million stars. Over billions of years, Sagittarius has been slowly eroded through gravitational interaction with our Galaxy, and is now in the final stages of dissolution.

Another Milky Way stream, dubbed the Monoceros Ring or Monoceros stream, is believed to have been created by the shredding of the Canis Major dwarf galaxy over billions of years. The ring, first discovered in 2002 as part of the Sloan Digital Sky Survey, is 200,000 light years long and also has the mass of 100 million Suns.[6] Numerous streams are now known to be associated with our Galaxy, including the Helmi stream (discovered in 1999), the Palomar 5 stream (2001) stretching 24 degrees, the Virgo stream (2001), and the Boötes III stream (2007). In 2006 Cambridge University astronomers Vasily Belokurov and Daniel Zucker used the Sloan Digital Sky Survey II (SDSS II), particularly useful for picking out streams due to the large number of stars it observed, to detect a "field of streams" in the area of the North Galactic Pole. The crisscrossing "field of streams" is dominated by the Sagittarius stream and includes the Monoceros stream, among numerous others.

In 2006, Caltech astronomers Carl Grillmair and Odysseas Dionatos also used the SDSS data to discover a stream stretching 63 degrees from Ursa Major to Cancer, dubbed the GD-1 stream. Because of its extreme narrowness (only about .2 degrees), it is believed to be a shredded globular cluster (S 35). The NGC 5466 stream is also believed to be the shredded remains of a globular cluster, as well as two streams discovered using the Spitzer Space Telescope in 2007 (Fig. 15.1).[7] In 2008, SDSS astronomers reported 14 more stellar streams from dwarf galaxies stretched by gravitational tides, and suggested that 1,000 streams may exist in the Milky Way. As one astronomer put it, this makes the galactic halo appear like "a jumble of pasta," a jumble that nevertheless sheds important light on the substructure of the galactic halo.

---

[6] Heidi Jo Newberg et al., "The Ghost of Sagittarius and Lumps in the Halo of the Milky Way," ApJ, 569 (2002), 2245–274; N. F. Martin, R. A. Ibata et al., "A dwarf galaxy remnant in Canis Major: the fossil of an in-plane accretion onto the Milky Way," MNRAS, 348 (2004), 12–23.

[7] C. J. Grillmair and R. Johnson, "The Detection of a 45° Tidal Stream Associated with the Globular Cluster NGC 5466," ApJ, 639 (2006), L17–L20.

Fig. 15.1. Three star streams discovered with the Spitzer Space Telescope in 2007. The two smaller streams are likely shredded globular clusters, while the larger one is a shredded dwarf galaxy. Credit: NASA/JPL-Caltech/R. Hurt (SSC/Caltech)

New streams continue to be discovered around our Galaxy. In 2018 astronomers working with the Dark Energy Survey, undertaken with the 4-meter Victor Blanco Telescope at Cerro Tololo in Chile, announced 11 new stellar streams. And with the release of positions and motions of a billion stars from the Gaia satellite in April 2018, the floodgates were opened to refine past discoveries and to discover even more streams in our Galaxy. Already, data from Gaia indicates that the GD1 stream may have been affected by one of the dark matter clumps in the galactic halo, one of the few ways to study dark matter. Even more streams will be discovered once the Large Synoptic Survey Telescope (LSST) becomes operational in Chile in 2022.[8]

---

[8] V. Belokurov et al., "The Field of Streams: Sagittarius and Its Siblings," ApJ, 642 (2006), 137–140; the 2008 discovery is Kevin Schlaufman et al., "Insight into the Formation of the Milky Way Through Cold Halo Substructure. I. The ECHOS of Milky Way Formation," ApJ, 703 (2009), 2177–2204, reported in Andrea Thompson, "Milky Way's Halo Loaded with Star Streams," August 16, 2008, at http://www.space.com/scienceastronomy/080816-milky-way-map.html; N. Shipp et al., "Stellar Streams Discovered in the Dark Energy Survey," https://www.darkenergysurvey.org/wp-content/uploads/2018/01/StellarStreams.pdf

Star streams are also known to exist around other galaxies. In 1974, the first extragalactic stream was discovered as a neutral hydrogen (H I) feature, a filament of gas and stars connecting the Large and Small Magellanic Clouds and named the Magellanic stream.[9] In 2001, stellar streams associated with the Andromeda Galaxy (M31) were reported, and in 2005 Sandra Chapman and colleagues used the streams to determine the mass of Andromeda. A panoramic survey of M31 in 2009 showed stars and coherent structures that are "almost certainly" the remnants of dwarf galaxies destroyed by the tidal field of M31; there was also evidence that its brightest companion, Triangulum (M33) had a recent encounter with M31. "This panorama of galaxy structure directly confirms the basic tenets of the hierarchical galaxy formation model," the authors wrote, "and reveals the shared history of M31 and M33 in the unceasing build-up of galaxies." In 2010 even more streams were discovered in Andromeda.[10] More than a dozen star streams are now known around galaxies beyond our Local Group, and more are being uncovered every year.

A list of satellite galaxies of the Milky Way is at https://en.wikipedia.org/wiki/Satellite_galaxies_of_the_Milky_Way, and an incomplete list of star streams is at https://en.wikipedia.org/wiki/List_of_stellar_streams.

---

[9] D. S. Mathewson and M. N. Cleary, "The Magellanic Stream," ApJ, 190 (1974), 291–296. For the latest see HST Release, "Hubble Solves Cosmic 'Whodunit' with Interstellar Forensics," March 22, 2018, http://hubblesite.org/news_release/news/2018-15/4-galaxies

[10] R. Ibata et al., "A giant stream of metal-rich stars in the halo of the galaxy M31," Nature, 412 (2001), 49; S. C. Chapman et al., "A Kinematically Selected, Metal-poor Stellar Halo in the Outskirts of M31," ApJ, 653 (2006), 255–266; Alan McConnachie et al., "The remnants of galaxy formation from a panoramic survey of the region around M31," Nature, 461 (2009), 66–69.

Class G 11: Galactic Jet

Galactic jets are energetic emissions of matter associated with supermassive black holes in the nuclei of active galaxies (G 6–G 9). As accreting matter is drawn toward and falls into the black hole, superheated gas, energetic electrons, and other subatomic particles are ejected from the poles in the form of narrow collimated columns in opposite directions. They may be viewed from Earth at many angles. The exact composition of jets and their mechanism are unknown, but the mechanism is believed to be related to the rotation of the black hole in conjunction with magnetic forces that eject materials. In contrast to stellar jets (S 20) that may stretch several astronomical units equal to a small fraction of a light year, galactic jets may be hundreds of thousands of light years in extent, and some can span more than a million light years in one direction from the center of the black hole. Such a jet in our own Galaxy would extend half way to the Andromeda galaxy. These structures betray the extremely high energies that power them.

The first to observe a galactic jet was Lick Observatory astronomer Heber D. Curtis, who in 1918 noticed an unusual feature around the elliptical galaxy M87 (Fig. 14.4) and wrote, "A curious straight ray lies in the gap in the nebulosity in p.a. [position angle] 20 degrees, apparently connected to the nucleus by a thin line of matter."[11] Because Curtis did not know the distance to M87, he could not fathom the true nature of the object, also known as NGC 4486.

More than three decades later, Walter Baade and Rudolph Minkowski studied NGC 4486 in more detail as a "peculiar extragalactic nebula" with radio emission, the object radio astronomer pioneers had labeled Virgo A. "NGC 4486 has a unique peculiarity which has been known for a long time," they wrote, referring to Curtis's observation. "In the center of the nebula is a straight jet, extending from the nucleus in position angle 290 degrees...Several strong condensations are in the outer parts of the jet, which extends about 20 arcseconds from the nucleus and has an average width of about 2 arcseconds." They calculated that the linear length of the jet is about 300 parsecs (1,000 light years) with a width about 100 light years. Noting an oxygen emission line, they ventured,

---

[11] H. D. Curtis, "Descriptions of 762 Nebulae and Clusters Photographed with the Crossley Reflector," *Publications of the Lick Observatory*, 13 (1918), 31.

"The interpretation which suggests itself is that the jet was formed by ejection from the nucleus and that the [O II] line is emitted by a part of the material which forms the jet and is still very close to, if not still inside, the nucleus ... No possibility exists at this time of forming any hypothesis on the formation of the jet, the physical state of its material, and the mechanism which connects the existence of the jet with the observed radio emission."[12] Two years later, Baade suggested the synchrotron mechanism was responsible for the radio emission.

We now know that M87 is the largest galaxy in the Virgo Cluster, 55 million light years distant. Its jet extends at least 5,000 light years, with lobes of matter extending 250,000 light years beyond that. The nearest radio galaxy, the elliptical galaxy Centaurus A (Fig. 14.13), has a relativistic jet extending some 13,000 light years from its center. Approximately 13 million light years distant, the jet is traveling outward at a speed estimated at 10 to 45% the speed of light.[13] These jets are hardly the record-holders, however. Observations by radio telescopes of the elliptical galaxy CGCG 049-933 in 2007 revealed a jet extending 1.5 million light years in one direction. The jet emanates from a supermassive black hole and is unusual because of its asymmetry in the sense that, while it is visible in the optical on both sides, it has a radio lobe on only one side.[14]

Many more jets have been subsequently discovered emanating from other active galaxies. Jets associated with these galaxies have been seen in the optical by the Hubble Space Telescope, in the radio by arrays such as the Very Large Array, and in the X-ray by Chandra. In the years since the launch of Chandra in 1999, it has become apparent that many radio galaxies and BL Lac objects have X-ray jets extending thousands of light years. When jets are seen head-on, as in the BL Lac objects also known as blazars (G 9), they may appear to be moving faster than the speed of light, an illusion

---

[12] W. Baade and R. Minkowski, "On the Identification of Radio Sources," ApJ, 119 (1954), 215–231: 221–222. Baade, "Polarization in the Jet of Messier 87," ApJ, 123 (1956), 550–551.

[13] M. J. Hardcastle, "Radio and X-Ray Observations of the Jet in Centaurus A," ApJ, 593 (2003), 169; Chandra PR, April 1, 2003, "New View of X-ray Jet Blasting Through Nearest Radio Galaxy," http://chandra.harvard.edu/press/03_releases/press_040103.html

[14] Joydeep Bagchi et al., "A Giant Radio Jet Ejected by an Ultramassive Black Hole in a Single-lobed Radio Galaxy," ApJ, 670 (2007), L85–L88.

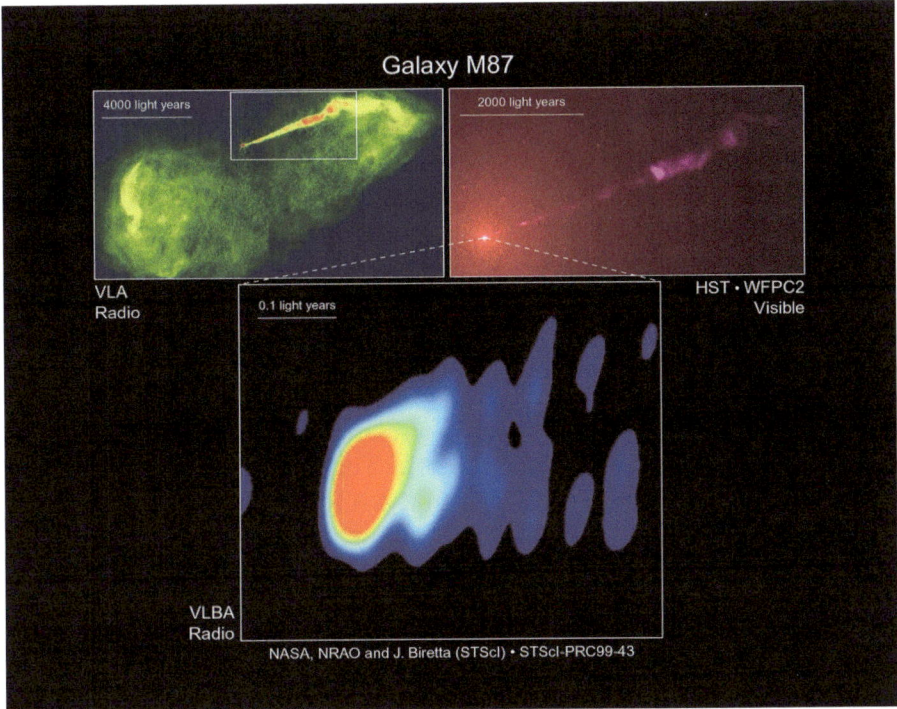

Fig. 15.2. Radio (top left and bottom) and visible (top right) images of the M87 jet, consisting of subatomic particles coming from the galaxy's central three-billion-solar mass black hole. The radio images show the jet near the black hole. Credit: NASA, National Radio Astronomy Observatory/National Science Foundation, John Biretta (STScI/JHU), and Associated Universities, Inc.

caused by the geometry of the situation. The quasar 3C 279 was the first superluminal jet discovered in 1973 at radio wavelengths. In 1999, the Hubble Space Telescope detected apparent superluminal motion in the M87 jets for the first time (Fig. 15.2).[15] The first jet from a Seyfert galaxy (G 6) was observed in the year 2000 in the compact galaxy III Zw 2, so called because it is the second object in Zwicky's third list of galaxies, produced in 1964.

A radio survey of jets led by astronomer Ken Kellermann and published in 2004 catalogued 208 distinct features in 110 quasars and other types of active galaxies. In addition to the underlying continuous flow of individual jets at relativistic velocities, the

---

[15] J. A. Biretta et al., "Hubble Space Telescope Observations of Superluminal Motion in the M87 Jet," ApJ, 520 (1999), 621–626,

survey also showed significant bends and twists in the jets deviating from outward motion. It found that sources emitting gamma rays tended to have higher velocity jets.[16]

Jets have been known to be gamma ray sources since the EGRET detections on the Compton Gamma Ray Observatory in the 1990s. Its successor, the Fermi telescope, has detected bursts in objects ranging from the relatively "close" radio galaxy Perseus A at 233 million light years to blazars and quasars 10 billion light years away. "We don't know what the jets are made of or how they are produced. It is one of the biggest unsolved mysteries of astrophysics. But jets are the link between the activity of the supermassive black hole and the AGN's surrounding environment in intergalactic space," said Peter Michelson of Stanford University, the Principal Investigator of the Fermi Telescope's primary science instrument, the Large Area Telescope (LAT).[17] Chandra X-ray observations of nine elliptical galaxies have shown a relation between accretion rate and jet power; the higher the accretion rate, the greater the power.[18] Despite progress, however, many aspects of galactic jets remain enigmatic, including the mechanism whereby jets form from accreting matter.

Another view of the M87 jet taken with the Hubble Space Telescope is at http://hubblesite.org/newscenter/archive/releases/2000/20/image/a/. In 2007, multiple ground and space-based telescopes imaged a jet from one galaxy punching into another galaxy. The offending galaxy was dubbed "the death star galaxy:" http://chandra.harvard.edu/press/07_releases/press_121707.html.

---

[16] K. I. Kellermann et al., 'Sub-Milliarcsecond Imaging of Quasars and Active Galactic Nuclei. III. Kinematics of Parsec-scale Radio Jets," ApJ, 609 (2004), 539–563.

[17] Fermi release, March 11, 2009, "Top Ten Gamma Ray Sources from the Fermi Telescope," http://www.universetoday.com/26831/top-ten-gamma-ray-sources-from-the-fermi-telescope/; Peter Michelson, quoted in http://www.nasa.gov/mission_pages/GLAST/science/blazers.html

[18] S. W. Allen et al., "The relation between accretion rate and jet power in X-ray luminous elliptical galaxies," MNRAS, 372 (2006), 21–30.

Class G 12: Halo

A halo is a slightly flattened spherical population of stars, globular clusters (S 35), and diffuse ionized gas surrounding disk galaxies, including spirals (G 4) and lenticulars (G 3), as well as some types of massive elliptical galaxies. This visible halo is believed to be the stellar component of a much larger dark matter halo, which is inferred by its gravitational influence and dominates the mass of the galaxy. For spirals, the visible halo is one of three basic structural components, including the disk and the nucleus, or central bulge. In the case of our Milky Way Galaxy, the visible halo has a radius of about 150,000 light years from the galactic center, though some objects, including globular clusters, have been found much farther out. Unlike the galactic disk, star formation is not taking place in the halo, and its stars have been shown to be old and metal-poor, lacking elements more complex than hydrogen and helium. Halo stars move at high velocities in eccentric orbits compared to the more regular circular orbits and relatively slower velocities of disk stars.

The visible halo is likely to be among the first objects to have formed in the Galaxy after the much more extensive dark halo, whose gravitational attraction may have initiated the formation of the Galaxy. As such, the visible halo records the early dynamics and chemical evolution of the Galaxy. The sparse nature of the halo by contrast to the disk, and its general lack of gas and dust, make it difficult to observe around other galaxies; nevertheless, halos have been observed around nearby galaxies such as the Andromeda Galaxy and M81. The dark halo may extend 250,000 light years, one tenth of the way to the Andromeda galaxy. Because of their relevance to protogalaxies (G 1), in particular cold dark matter (CDM) theory, detailed studies of galactic halos are an important part of research on galaxy formation.

The existence of the visible halo around our own Galaxy was discovered only gradually. A first step was Harlow Shapley's demonstration in the 1920s that globular clusters were arranged in a sphere, and it was natural to infer that such a spherical distribution of matter orbited around the galactic center, which he showed was nowhere near the Solar System. A more important step was Walter Baade's delineation in 1944 of two populations of stars (S 36) in the Andromeda Galaxy, wherein Population II objects consist of older stars found in globular clusters and in the outer regions of the galaxy as distinct from Population I stars in the disk. The structure of the halo was gradually derived by observations of globulars, population II objects, and subdwarfs (S 5), both in our Galaxy and in others.

Perhaps the first use of the word "halo" as a component of the Galaxy came in 1938 when D.W. Rosebrugh, describing a lecture by Harlow Shapley on the structure of the Milky Way from his observations of Cepheid variables, wrote that "it is believed that the Milky Way is a discoid 35,000 parsecs in diameter and 300 parsecs thick. Surrounding this flattened discoid is a globular halo about 40,000 parsecs in diameter the density of which falls to about 1/10,000 of that in the galactic plane at a distance of 12,500 parsecs from the galactic plane."[19] (A parsec is 3.26 light years). Whereas standard astronomy textbooks of the 1920s, 30s, and 40s did not speak of a halo, from 1955 to 1960 11 papers used the word "halo" in their titles in connections with galaxies. In his 1958 lectures *Evolution of Stars and Galaxies*, Baade divided the structure of the Milky Way into "the disk, the surrounding halo, and the central nucleus," and this description rapidly became commonplace thereafter.[20]

In 2007, observations of 20,000 stars in the Sloan Digital Sky Survey showed that the visible halo of the Milky Way Galaxy consists of two distinct structural components rotating in opposite directions, an inner halo rotating the same direction as the disk stars at 50,000 mph, and an outer halo speeding along at 100,000 mph (Fig. 15.3). The inner halo stars also have a different chemical composition, containing three times more heavy atoms such as iron and calcium than the outer halo stars. This may mean that the inner halo formed first from the collision of massive galaxies, while the outer halo formed later from smaller galaxies orbiting the Milky Way in the opposite direction.[21] The orbits of halo stars

---

[19] D. W. Rosebrugh, "Spring Meeting of the American Association of Variable Star Observers," JRASC, 32 (1938), 340–342.

[20] Walter Baade, *Evolution of Stars and Galaxies*, Cecilia Payne-Gaposchkin, ed., (Cambridge, MA: MIT Press, 1958), p. 267. Among the important earlier articles are "Overall Structure: Nuclear Region and Halo," Conference on Co-ordination of Galactic Research, IAU Symposium no. 1 (Cambridge: Cambridge University Press, 1955), pp. 4 ff; G. Larsson-Leander, Nancy Roman, et al., "The Galactic Halo," in Second Conference on Co-ordination of Galactic Research, IAU Symposium no. 7, held in 1957, A. Blaauw, ed., (Cambridge: Cambridge University Press, 1959), pp. 22–27. See also Sidney van den Bergh, "The Halo Phase of Galactic Evolution," PASP, 73 (1961), 135–142.

[21] Daniela Carollo, Timothy Beers et al., "Two Stellar Components in the Halo of the Milky Way," *Nature*, 450 (2007), 1020–1025.

Fig. 15.3. Structure of the Milky Way's halo. The age of the inner halo stars is about 11.5 billion years, while the outer halo stars may be as much as 13.5 billion years—formed shortly after the Big Bang. Credit: NASA, ESA, and A. Field (STScI)

are so different from disk stars that at times they pass through the disk. Kapteyn's star and Groombridge 1830 are thought to be red dwarf halo stars now doing just that. In 1998, astronomers using the Compton Gamma Ray Observatory reported gamma rays associated with the halo, evidence of high energy processes still not understood.[22] In 1990, J.M. Dickey and F.J. Lockman used radio observations to show that H I gas (S 23) was also located in the galactic halo. In 2002, Lockman reported further the discovery of H I clouds in the Galactic halo. Lockman estimated that as much as half the mass of the neutral halo may be in the form of hydrogen clouds.[23]

---

[22] D. D. Dixon, D. H. Hartmann et al, "Evidence for a Galactic gamma-ray halo," *New Astronomy*, 3 (1998), 539–561.

[23] J. M. Dickey and F. J. Lockman, "H I in the Galaxy," ARAA, 28 (1990), 215–261; Felix J. Lockman, "Discovery of a Population of H I Clouds in the Galactic Halo," ApJ, 580 (2002), L47–L50.

The ability to observe the halos of several other galaxies sheds light on galaxy formation.

The bright giant stars in the halo of the Andromeda Galaxy (Fig. 15.4), a Local Group galaxy located at a distance of some 2.5 million light years, were seen by ground-based telescopes using deep CCD technology as early as 1986.[24] Andromeda is now known to harbor approximately 500 globular clusters, about three times more than the Milky Way, many of them in the halo. In 2003 the Hubble Space Telescope observed some 300,000 fainter normal stars in the Andromeda Galaxy halo and showed that at least one third of its stars formed six to eight billion years ago, compared to 11 to 13 billion years for the stars of the Milky Way halo. Because the younger stars in Andromeda's halo are richer in

Andromeda Galaxy Halo Details
Hubble Space Telescope • Advanced Camera for Surveys

NASA, ESA and T. Brown (STScI) • STScI-PRC03-15b

Fig. 15.4. The stars and globular cluster (lower right) in these images are found in the Andromeda halo, whereas the galaxies are in the background. Credit: NASA, ESA, and T.M. Brown (STScI)

---

[24] J. Mould and J. Kristian, "The stellar population in the halos of M31 and M33," ApJ, 305 (1986), 591–599.

heavier elements than the stars in our Galaxy's halo, this suggests a collision with a relatively massive galaxy whereby those stars must have either come from the disk of the invading galaxy or from the disk of Andromeda itself. Alternatively, the young stars could have been formed during the collision itself.[25]

In 2006, Sandra Chapman and colleagues found the Andromeda halo to be similar in many respects to the Milky Way's. But in the same year, Rodrigo Ibata and his colleagues published detailed observations of the halos of Andromeda and Triangulum, another member of the Local Group of galaxies. For Andromeda, they found a halo component with predominantly metal-poor stars extending 450,000 light years from the galactic center. They found that "This vast, smooth, underlying halo is reminiscent of a classical 'monolithic' model [of galaxy formation] and completely unexpected from modern galaxy formation models, in the sense that those models predict structures like arc, shells and streams rather than a smooth underlying halo of stars."[26]

Triangulum (M33), the third brightest galaxy in the Local Group, has a mass ten times lower than either Andromeda or the Milky Way Galaxy. In 2002 Rupali Chandar and colleagues found evidence for a halo, and this was confirmed by the Keck telescope. Beyond these relatively nearby galaxies, the halo has been extremely difficult to detect, although in 2010 astronomers using the Subaru telescope reported possibly sighting the halo of M81, a member of one of closest galaxy groups to our own Local Group, but at a distance of 11.7 million light years, more than four times the distance of Andromeda.[27] The halo seemed to contain

---

[25] HST Release, "Deepest View of Space Yields Young Stars in Andromeda Halo," May 7, 2003, at http://hubblesite.org/newscenter/archive/releases/2003/15/text/ states that "The ACS is the first astronomical camera to combine ultra-sharp vision and sensitivity to ferret out M31's faint halo population." T. Brown et al., "Evidence of a Significant Intermediate-Age Population in the M31 Halo from Main-Sequence Photometry," 592 (2003), L17–L20; and Brown et al., "RR Lyrae Stars in the Andromeda Halo from Deep Imaging with the Advanced Camera for Surveys," AJ, 127 (2004), 2738–2752.

[26] S. C. Chapman et al., "A Kinematically Selected, Metal-Poor Stellar Halo in the Outskirts of M31," ApJ, 653 (2006), 255–266; R. Ibata et al., "The Haunted Halos of Andromeda and Triangulum: A Panorama of Galaxy Formation in Action," ApJ, 671 (2007), 1591–1623.

[27] R. Chandar et al., "Kinematics of Star Clusters in M33: Distinct Populations," ApJ, 564 (2002), 712–735; McConnachie et al., "The Stellar Halo and Outer Disk of M33," ApJ, 647 (2006), L25–L28; For the M81 halo see Subaru Telescope, "M81's 'Halo' Sheds Light on Galaxy Formation," April 13, 2010, http://www.physorg.com/news190360095.html

more metal-rich stars by comparison with the Milky Way and Andromeda and should help in understanding galaxy formation.

In light of their study of M31 and M33, Ibata and his colleagues concluded somewhat poetically:

> Halos are truly misnamed: they are in reality dark galactic graveyards, full of the ghosts of galaxies dismembered in violent clashes long ago. Other, even more ancient remnants have lost all memory of their original form, and in filling these haunted halos with the faintest shadow of their former brilliance, they follow faithfully the dark forces to which they first succumbed. The true nature of this most somber of galactic recesses is finally beginning to be revealed.[28]

Several views of the Andromeda Galaxy halo are at http://hubblesite.org/newscenter/archive/releases/2003/15/text/ and http://hubblesite.org/image/1338/news_release/2003-15. Some schematic views of the Milky Way galaxy, including the halo, are illustrated at http://chandra.harvard.edu/resources/illustrations/milkyWay.html.

---

[28] Ibata et al, 2007, 1622.

16. The Subgalactic Family

Class G 13: Subgalactic Objects

The term "subgalactic" is most often used in astronomical literature in connection with protogalaxies (G 1), in which subgalactic objects are the building blocks of galaxies as part of the "hierarchical clustering" process in the cold dark matter (CDM) theory.[1] In theory, subgalactic objects could also be failed galaxies, analogous to the failed stars known as brown dwarfs (S 22) in the substellar Family (Chapter 10), or they could be remnants of galaxy formation analogous to dwarf planets (P 9) and the small bodies of the solar system (P 11, 12, and 13) in the subplanetary Family (Chapter 4). Because little evidence exists as yet for either failed galaxies or remnants of galaxy formation that are not incorporated into another galaxy as galactic rings (G 10), we define the class of subgalactic clumps to be the building blocks of galaxies that are not yet part of a protogalaxy in formation.

Observational evidence does exist for subgalactic objects, but their identity depends on definition. What appear to be subgalactic clumps (Fig. 16.1) have been found to exist early in the history of the universe at high redshift, and, more surprisingly, much closer to our own Galaxy. Moreover, though dwarf galaxies are usually considered to be galaxies in their own right, an increasing body of evidence indicates they are part of the hierarchical clustering that forms larger galaxies at later stages of galaxy formation still underway even in our own Galaxy. While the definition and criteria for galaxy status has not been the subject of major controversy as it has for dwarf planets, dwarf galaxies might therefore also

---

[1] For example, S. M. Pascarelle et al., "A Cluster of Lyman-α Emitting Candidates at $z \simeq 2.39$ in Deep WFPC2 Images: Galaxy Formation from Subgalactic Clumps?," *Science with the Hubble Space Telescope – II*, Piero Benvenuti, F.D. Macchetto, and Ethan J. Schreier, eds. (Baltimore: Space Telescope Science Institute, 1996), p.109 ff.; online at http://www.stsci.edu/stsci/meetings/shst2/pascarelle.html

© Springer Nature Switzerland AG 2019
S. J. Dick, *Classifying the Cosmos*, Astronomers' Universe,
https://doi.org/10.1007/978-3-030-10380-4_16

be viewed as subgalactic objects, at least in relation to the larger galaxies like our own that swallow them. In the end, like "dwarf planets," the subgalactic family remains a matter of definition, and one to which other classes may be added in the future.

Detailed analysis of the Hubble Deep Fields has revealed structures at high redshift that cannot be classified in the traditional Hubble scheme. One study in 2000 using the Hubble Deep Field North concluded that ordinary Hubble sequence spiral and elliptical galaxies reached maturity between redshifts of 1 and 2 (three to six billion years after the Big Bang), beyond which so-called Lyman break galaxies appear compact and irregular.[2] In 2003 and 2004, the Hubble Ultra Deep Field uncovered more than 10,000 new galaxies, many with redshifts greater than 3, corresponding to the epoch less than two billion years after the Big Bang (Fig. 13.1). Astronomers singled out objects believed to be galaxy building blocks. As with the Hubble North and South Deep

Galaxy Building Blocks   HST · WFPC2
PRC96-29b · ST ScI OPO · September 4, 1996 · R. Windhorst (Arizona State University), NASA

Fig. 16.1. Faint blue subgalactic clumps 11 billion light years from Earth, each containing a few billion stars, could be the building blocks of galaxies. Each image is about 2,000 to 3,000 light years wide, covering an area about two million light years wide. They were found in this image from Hubble released in 1996. Credit: Rogier Windhorst and Sam Pascarelle (Arizona State University) and NASA

---

[2] Mark Dickinson, "The first galaxies: structure and stellar populations," PTRAS, 358 (2000), 2001.

Fields taken in 1995 and 1998, these high-redshift galaxies, or galaxy building blocks, are smaller and less symmetric in shape than galaxies at lower redshift. Hubble Ultra Deep Field astronomers argued, "these results confirm the conclusion of the first HDF observations that typical galaxies in the early universe look markedly different than galaxies today, showing that galaxies evolved rapidly in the first few billion years after the big bang."[3] Among these structures are the so-called chain galaxies with their knotty components. Simulations show that in a cold cloud medium, the gaseous disk fragments and develops several massive clumps of gas and dust, which eventually merge to build a massive bulge as part of the hierarchical clustering process.[4]

Galaxies may also still be forming from subgalactic clumps. Although it seems logical that such clumps would have been swept up and incorporated into galaxies long ago, when astronomers Michael Corbin and William Vacca examined a sample of dwarf galaxies chosen for their small size and young stars, they found surprising results. Beginning in 2000, they reported finding an ultracompact blue dwarf galaxy, POX 186, 70 million light years away, only 100 million years old, and less than 1000 light years across, making it one of the smallest known galaxies. Its small size, asymmetric shape, and youthful age were consistent with the idea that it is composed of two partly coalesced subgalactic clumps, each about 300 light years in extent, in the act of forming a new galaxy. The astronomers theorized that because it was located in one of the voids (G 24) of space far from other galaxies, these particular subgalactic clumps remained undisturbed until a chance and very rare encounter. They speculated further that "clumps of stars this small may represent the building blocks required by hierarchical models of star formation."[5]

In 2005, Corbin, Vacca, and their colleagues reported finding a similar ultracompact blue dwarf galaxy assembling subgalactic components, and the following year reported more observations of similar events in several of the nine ultracompact blue dwarfs

[3] Steven Beckwith et al., "The Hubble Ultra Deep Field," AJ, 132 (2006), 1729–1755: 1754;

[4] Andreas Immeli et al., "Subgalactic Clumps at High Redshift: A Fragmentation Origin?" APJ, 611 (2004), 20–25.

[5] Michael Corbin and William D. Vacca, "Pox 186: A Dwarf Galaxy in the Process of Formation?", ApJ, 581 (2002), 1039–1046.

observed with the Hubble Space Telescope. Because these dwarf galaxies contain stars about 10 billion years old, they concluded that they are not protogalaxies, forming their first generation of stars, but rather dwarf galaxies "in the process of assembling from clumps of stars intermediate in size between globular clusters and objects previously classified as galaxies." In light of these observations, the authors also speculated that a population of subgalactic clumps may exist in the voids of space.[6]

The study of galaxy formation from subgalactic clumps begs the question of when a galaxy actually becomes a galaxy. According to one definition, a galaxy is "a vast, contiguous collection of stars, gas, dust, and other matter, totaling at least a few million times the mass of the Sun, all held together by mutual gravitational attraction."[7] Under this definition, a protogalaxy would be a galaxy forming its first generation of stars, whereas subgalactic clumps would precede the protogalaxy stage. On the other hand, in 2005 astronomers reported finding the first "dark galaxy," a cloud of gas in the Virgo cluster that consists of hydrogen gas and dark matter, with no detectable stars.[8] Also likely related are the so-called Lyman alpha blobs (G15), large regions of hydrogen in which star formation has begun its early stages.

Results from the Hubble Deep Fields from 1996 to the present are available at http://hubblesite.org/news/14-deep-fields. The Hubble Ultra Deep Field observations continue today with the observing campaign known as Beyond Ultra-deep Frontier Fields And Legacy Observations (BUFFALO). Its object is to identify galaxies less than 800 million years after the Big Bang: https://buffalo.ipac.caltech.edu. On Pox 186 see http://hubblesite.org/newscenter/archive/releases/2002/16/text/. For recent news on dwarf galaxies as building blocks of massive galaxies see https://public.nrao.edu/news/2017-dwarf-galaxy-groups/.

---

[6] Michael R. Corbin, William D. Vacca et al., "Hubble Space Telescope Imaging of the Ultracompact Blue Dwarf Galaxy HS 0822+3542: An Assembling Galaxy in a Local Void?," ApJ, 629 (2005), L89–L92, and Corbin, Vacca et al., "Ultracompact Blue Dwarf Galaxies: Hubble Space Telescope Imaging and Stellar Population Analysis," ApJ, 651 (2006), 861–873.

[7] Charles Liu, "Let's Make a galaxy: astronomers have identified a cosmic infant 'nearby' 70 million light years from Earth," Natural History (March, 2003),

[8] Ken Crosswell, "The First Dark Galaxy?" Astronomy online, February 18, 2005, http://astronomy.com/news/2005/02/the-first-dark-galaxy

17. The Intergalactic Medium Family

Class G 14: Warm Hot Intergalactic Medium (WHIM)

We now move into the Family of the intergalactic medium, consisting of both gas and dust. Intergalactic gas is a hot, highly rarefied plasma consisting primarily of ionized hydrogen, along with traces of other elements such as helium and oxygen. It has been detected at temperatures ranging from 300,000 to five million degrees Celsius. This so-called Warm Hot Intergalactic Medium (WHIM) constitutes the bulk of the intergalactic medium (IGM). As with the interstellar medium, there is also a cooler neutral hydrogen component of the intergalactic medium in regions where there is not enough energy for ionization. Intergalactic molecules have not been discovered, but dust (G 16) also often coexists with the gas components.

It is well known that the universe is made of three major components: normal "baryonic" matter (5%), dark matter (26%), and dark energy (69%). Until recently, only about two thirds of the normal matter has been identified in the form of planets, stars, galaxies, and black holes. The diffuse IGM has a density of only ten to 100 hydrogen atoms per cubic meter, and perhaps 1,000 times that in rich galaxy clusters. Yet because of the immensity of intergalactic space, it is estimated that intergalactic gas nevertheless defines the cosmic landscape, likely accounts for a good part of the "missing mass" of normal matter, and may help astronomers trace invisible dark matter through its gravitational influence.

The first hints about the intergalactic medium began close to home with the discovery of the cool medium. In the 1950s, Jan Oort proposed that the Milky Way's galactic halo (G 12) might contain cold clouds of neutral hydrogen, analogous to H I

© Springer Nature Switzerland AG 2019
S. J. Dick, *Classifying the Cosmos*, Astronomers' Universe,
https://doi.org/10.1007/978-3-030-10380-4_17

regions (S 23) in the interstellar medium. Such clouds were discovered in 1963 based on their 21-cm radio emission.[1] These so-called "high-velocity clouds," or HVCs, are clumps of hydrogen as much as ten million times the mass of the Sun and 10,000 light years across, moving up to 400 km per second through the outer regions of the Galaxy—much faster than the rotation of the Galaxy itself.

What is now known as the Warm Hot Intergalactic Medium (WHIM) was first detected by a very different method—absorption lines in spectra produced at optical wavelengths when the relatively cooler gas is seen against the background of distant quasars. Like shining a flashlight through the fog, an analysis of the quasar light yields information about the intervening medium, which imprints its chemical fingerprints onto the quasar light in the form of absorption lines. This is the so-called "Lyman alpha forest" of neutral hydrogen, produced by photoionized gas at temperatures of about 10,000 K, which traces the hotter components of the gas. Conversely, a cooler foreground object such as a galaxy could cause the absorption of X-rays from the hottest IGM gas components. Such intergalactic absorbers along line-of-sight of quasars or galaxies therefore provide the chief means for studying the otherwise invisible IGM.

Spacecraft have been especially useful for this kind of study. In 1992 the Hubble Space Telescope reported absorption lines in the ultraviolet spectra of a quasar they interpreted as due to intervening objects such as Lyman alpha blobs (G 15) and galaxy clusters. Similar absorption lines had been seen for two decades by ground-based telescopes.[2] By 2000, astronomers used Hubble's ultraviolet capabilities to detect for the first time highly ionized oxygen between low redshift galaxies by analyzing the light of a

---

[1] Bart P. Wakker and H. van Woerden, "High Velocity Clouds, ARAA, 35 (1997), 217–266; Wakker and Philipp Richter, "Our Growing Breathing Galaxy," SciAm (January, 2004), 38–47.

[2] John N. Bahcall et al., "The Ultraviolet Absorption Spectrum of the Quasar H1821 +643 (z = 0.297)," ApJ, 397 (1992), 68–80, reported at HSTPR, January 13, 1992, "NASA's Hubble Space Telescope Observations Indicate Nearby Hydrogen Clouds May Be Associated with Galaxies," http://hubblesite.org/news_release/news/1992-04

quasar.[3] The highly ionized oxygen (O VI) is believed to trace large quantities of hot ionized hydrogen, itself invisible because it is too hot to be seen in visible light, but too cool to be seen in X-rays. The oxygen tracer likely came from exploding stars in galaxies that spewed the oxygen back into intergalactic space, where it mixed with hydrogen via a shockwave that heated the oxygen to very high temperatures. Subsequent surveys reported in 2005 and 2008 detected ionized oxygen (O VI) and neutral H I absorption lines in numerous systems in the low-redshift IGM, within about four billion light years of Earth, suggesting there are both hotter and warm components to the IGM.[4] It was these observations using ionized oxygen to trace ionized hydrogen that were touted as finding the missing mass, confirming cosmological models. "This is a successful, fundamental test of cosmological models," said Todd Tripp of Princeton University, the lead author on the first paper to find the ionized oxygen. "This provides strong evidence that the models are on the right track."

In 2001, astronomers using the Far Ultraviolet Spectroscopic Explorer (FUSE) spacecraft reported the detection of intergalactic helium in the early universe by observing a quasar 10 billion light years distant and showing how the intervening helium gas absorbed the quasar's light at certain wavelengths. By comparing hydrogen and helium absorption, FUSE astronomers were able to shed light on the power source of the reionization phase in the early universe, concluding that it was likely a mix of quasars powered by supermassive black holes and the light from newly formed stars.[5]

---

[3] Todd Tripp et al., "Intervening O VI Quasar Absorption Systems at Low Redshift: A Significant Baryon Reservoir," ApJ, 534 (2000), L1–L5; HSTPR, May 3, 2000; "Lost and Found: Hubble Finds Much of the Universe's Missing Hydrogen," https://www.spacetelescope.org/images/opo0018a/

[4] Charles Danforth and J. Michael Shull, "The Low-z Intergalactic Medium. I. O VI Baryon Census," ApJ, 624 (2005), 555–560; Charles Danforth and J. Michael Shull, "The Low-z Intergalactic Medium. III. H I and Metal Absorbers at z < 0.4," ApJ, 679 (2008), 194–219; HST PR, May 20, 2008, "Hubble Survey Finds Missing Matter, Probes Intergalactic Web," http://hubblesite.org/newscenter/archive/releases/2008/20/full/

[5] HST PR, August 9, 2001, "New View of Primordial Helium Traces the Structure of Early Universe," http://hubblesite.org/newscenter/archive/releases/2001/27.

Over the last two decades, claims of finding some or all of the missing mass of normal matter have proliferated. The hotter component of the IGM is best observed in the X-ray spectrum, and already in 2002 Joel Bregman and Jimmy Irwin, using the Chandra X-ray Observatory, observed the absorption of X-rays from the hot gas by a foreground galaxy, NGC 891.[6] In the same year, four independent teams used Chandra to detect the gas by observing dimming of X-rays from the quasars PKS 2155-304 and H 1821+643, caused by oxygen and other elements in the gas. This dimming allowed them to measure the temperature, density, and mass of an absorbing gas cloud ranging from a local filament to distances of a few billion light years. They described the gas as "like a fog in channels carved by rivers of gravity," hidden from view since the galaxies formed. These studies represent the first X-ray detections of absorption from the warm/hot intergalactic medium.[7]

Even more distant clouds of hydrogen gas may also exist at high redshifts, even before the epoch of reionization, and methods have been proposed for observing the 21-cm radiation of such neutral (H I) hydrogen. If such gas was observed before it was heated above the cosmic background radiation temperature, it could be observed in absorption against the background radiation. Once the IGM is heated above the temperature of the cosmic background radiation, it would be rendered invisible in absorption. The transition from the neutral IGM to a fully ionized one must have been due to early ionizing sources such as quasars, very young galaxies or Population III stars. Study of the IGM is therefore very important for probing the epoch of reionization and perhaps detecting the onset of the first generation of stars.[8] The epoch of reionization is expected to have taken place when the universe was 3 to 7% of its current age, about half a billion years after the Big Bang, corresponding to redshifts 6 to 12.

---

[6] Joel N. Bregman and Jimmy A. Irwin, "A Shadow of the Extragalactic X-Ray Background," ApJ, 565 (2002), L13–L16.

[7] Fabrizio Nicastro et al., 'Chandra Discovery of a Tree in the X-Ray Forest Toward PKS 2155-304: The Local Filament?" ApJ, 573 (2002), 157–167; Chandra PR, "Hot Intergalactic Gas: Chandra Discovers 'Rivers of Gravity' That Define Cosmic Landscape," at http://chandra.harvard.edu/photo/2002/igm/;

[8] Piero Madau et al., "21 Centimeter Tomography of the Intergalactic Medium at High Redshift," ApJ, 475 (1997), 429–444.

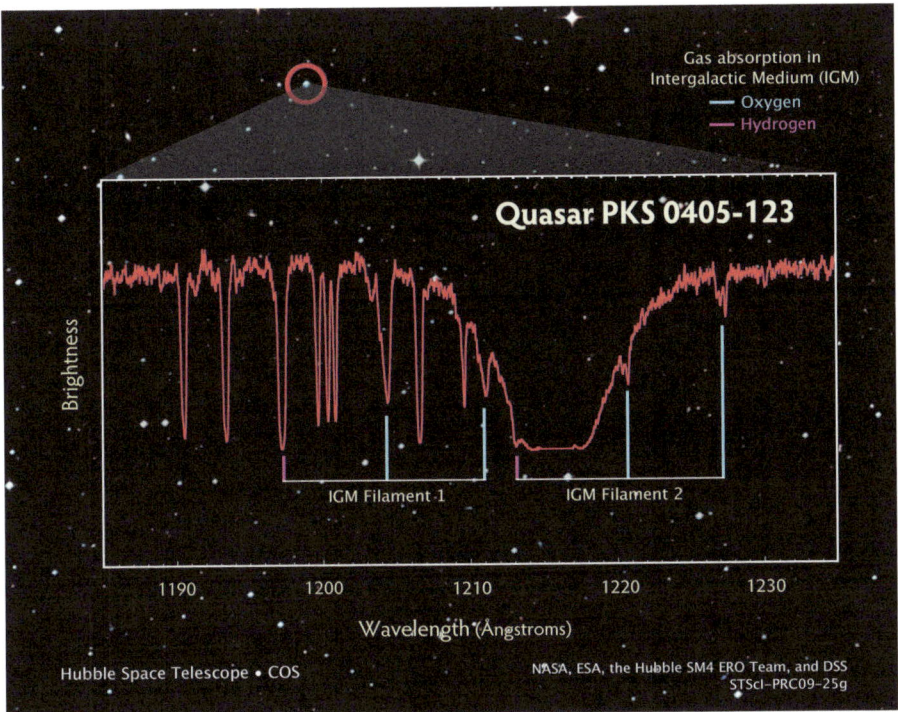

Fig. 17.1. Evidence of warm hot intergalactic matter from the Hubble Space Telescope. Credit: NASA, ESA, the Hubble SM4 ERO Team, and Digitized Sky Survey observations

    With cosmological worldviews at stake, research on the IGM proceeds apace. In 2010, astronomers used the new Cosmic Origins Spectrograph on the Hubble Space Telescope to probe a string of intergalactic clouds using the background light from quasar PKS 0405-123, located eight billion light years distant (Fig. 17.1). The resulting absorption lines revealed many previously unseen clouds.[9] Similarly, the Chandra Telescope used X-rays from a blazar passing through the Sculptor Wall, one of the great filaments of galaxies (G 24) located some 400 million light years distant, to study the medium. The WHIM in the wall yielded its characteristic absorption features, indicating a temperature of about one

[9] HST PR, "September 9, 2009, "Fingerprinting the Distant Universe Using the Light from Quasar PKS 0405–123," at http://hubblesite.org/newscenter/archive/releases/2009/25/image/ak/

million degrees Kelvin and a density for the cloud of six hydrogen atoms per cc, compared to one million per cc typical of the interstellar medium. In 2018, astronomers using the XMM-Newton X-Ray Space Telescope reported that they had determined that the WHIM is made of ionized oxygen gas at temperatures of about 1.8 million degrees Fahrenheit, about 1 million Celsius. They claimed that all of the missing normal baryonic mass could now be accounted for, either in the IGM or in the filamentary structures (G 24) that connect the cosmic web.[10] Using thousands of quasars and galaxies in this manner, astronomers are well on their way to mapping the gossamer cosmic web between the galaxies.

On Chandra's discovery of "rivers of gravity" see http://chandra.harvard.edu/photo/2002/igm/.

---

[10] Chandra PR, May 11, 2010, "X-Ray Discovery Points to Location of Missing Matter,"   http://www.nasa.gov/mission_pages/chandra/news/10-048.html; L. Zappacosta et al., "Studying the WHIM Content of the Galaxy Large-Scale Structures along the Line of Sight to H 2356-309," ApJ, in press; T. Fang et al., "Confirmation of X-ray Absorption by Warm-Hot Intergalactic Medium in the Sculptor Wall," ApJ, 714 (2010), 714 ff.; F. Nicastro et al., "Observations of the missing baryons in the warm–hot intergalactic medium," *Nature*, 558 (2018), 406–409; Taotao Fang, "Missing matter found in the cosmic web," *Nature News*, June 20, 2018, https://www.nature.com/articles/d41586-018-05432-2; Daniel Strain, University of Colorado Press Release, June 20, 2018, "Researchers find last of universe's missing ordinary matter," https://www.colorado.edu/today/2018/06/20/missing-baryons

Class G 15: Lyman Alpha Blobs

Lyman alpha blobs are large concentrations of hydrogen gas in the intergalactic medium (IGM), spanning several hundred thousand light years and emitting light at the "Lyman alpha" wavelength produced when electrons recombine with ionized hydrogen. These wavelengths, discovered in the laboratory in 1906 by Harvard physicist Theodore Lyman, are normally in the ultraviolet region of the spectrum, but in the case of extremely distant objects, they are redshifted to longer wavelengths in the optical. Many such structures have been found in the early universe, and they span an area up to 400,000 light years across, some of the largest structures in the universe. They are usually found as a byproduct of searches for Lyman alpha emitting galaxies at high redshift. They may be the building blocks of galaxies and thus related to subgalactic objects (G 13), but in a more advanced stage where galaxy formation has begun. If this is true, they may be a type of protogalactic object (G 1). Their relationship to the warm hot intergalactic medium (G 14) is unclear.

Astronomers found the first Lyman alpha blobs serendipitously around 1999 during surveys for so-called Lyman break galaxies. Highlighting the uncertainties of first discovery, the authors of one of the teams, led by Charles Steidel, wrote, "We have discovered two objects, which we call 'blobs,' that are very extended (>15"), diffuse, and luminous Ly alpha nebulae, with many properties similar to the giant Lyman alpha nebulae associated with high-redshift radio galaxies...we have considered several explanations for the blobs, with no firm conclusion emerging...it is unclear whether or not this phenomenon is unusual." They went on to say that "While we refrain from making too much of this result at present, given the many uncertainties involved, it does suggest that the blobs may represent a different class of rare object that, like QSOs [quasars], are preferentially found in rich environments at high redshift."[11] Their true nature remains unknown, but they are likely related to the birth to galaxies.

Among possible energy sources for Lyman alpha blobs are hidden quasars, star formation, superwinds from starburst galaxies, and cold gas accretion onto a dark matter halo.[12] Some or all of

---

[11] Charles C. Steidel et al., "Lyα Imaging of a Proto-Cluster Region at <z>=3.09," ApJ, 532 (2000), 170–182: 172, 181–182.

[12] K. K. Nilsson et al, "A Lyman-α blob in the GOODS South field: evidence for cold accretion onto a dark matter halo," A&A, 452 (2006), L23–26.

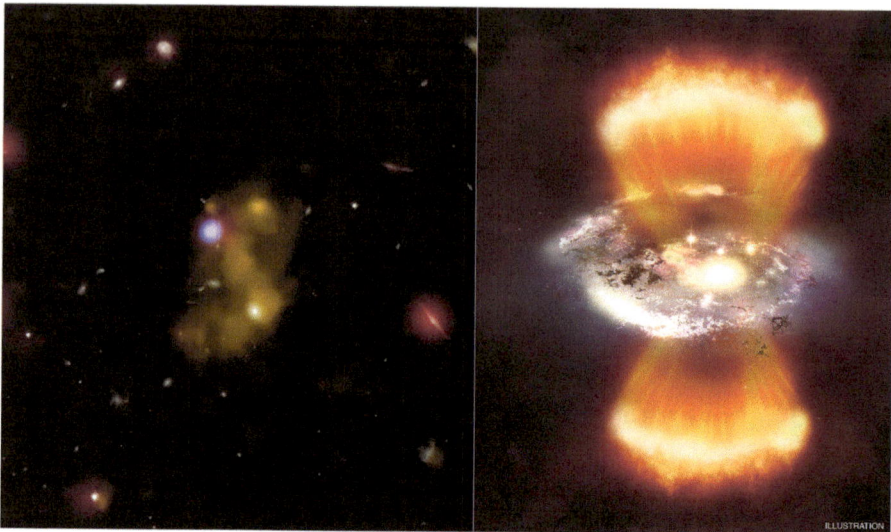

Fig. 17.2. One of the largest Lyman Alpha blobs in a survey of 29 such clouds by NASA's Chandra X-ray Observatory, is shown on the left panel. The right panel is an artist's rendering of a growing galaxy inside the blob, with outflows powered by a supermassive black hole. Credit: Left panel: X-ray (NASA/CXC/Durham Univ./D.Alexander et al.); Optical (NASA/ESA/ STScI/IoA/S.Chapman et al.); Lyman-alpha Optical (NAOJ/Subaru/Tohoku Univ./T.Hayashino et al.); Infrared (NASA/JPL-Caltech/Durham Univ./J. Geach et al.); Right, Illustration: NASA/CXC/M.Weiss

these mechanisms may apply. Peering inside the clouds as part of a survey of 29 blobs, in 2009 astronomers using the Chandra X-ray Observatory showed that two of the sources of energy were growing supermassive black holes within the blobs, as well as remarkable levels of star formation in galaxies inside the blobs (Fig. 17.2). In this scenario, the blobs may represent the outflows of heated gas from galaxies and their black holes. Such outflows of gas, detected in this case in blobs about two billion years after the Big Bang (redshift 3), likely prevent galaxies from becoming larger. Blobs in an earlier stage of evolution may represent the gaseous materials that fed their galaxies and black holes before the outflows began.[13]

---

[13] J. E. Geach et al., "The Chandra Deep Protocluster Survey: Lyα Blobs are Powered by Heating, Not Cooling," ApJ, 700 (2009), 1–9; Chandra Press release, June 24, 2009, "Lyman Alpha Blobs: Galaxies Coming of Age in Cosmic Blobs," at http://chandra.harvard.edu/photo/2009/labs/

Fig. 17.3. Himoko, a giant bubble of intergalactic gas, is shown in the two right panels from Hubble (top) and (bottom) a combination of Hubble, Subaru, and Spitzer telescopes. On the left panel, the position of Himoko is marked with a square. Credit: NASA, ESA, ESO, NRAO, NAOJ, JAO, M. Ouchi University of Tokyo), R. Ellis (California Institute of Technology), Y. Ono (University of Tokyo), K. Nakanishi (The Graduate University for Advanced Studies (SOKENDAI) and Joint ALMA Observatory), K. Kohno and R. Momose (University of Tokyo), Y. Kurono (Joint ALMA Observatory), M. Ashby (Harvard-Smithsonian Center for Astrophysics), K. Shimasaku (University of Tokyo), S. Willner and G. Fazio (Harvard-Smithsonian Center for Astrophysics), Y. Tamura (University of Tokyo), and D. Iono (National Astronomical Observatory of Japan)

Most Lyman alpha blobs have been found to exist when the universe was two to three billion years old. But in 2009 astronomers using a suite of telescopes reported detection of a huge Lyman alpha blob, dubbed Himiko (Fig. 17.3) for a legendary Japanese queen, at a distance of 12.9 billion light years (redshift 6.595), which means that it existed only 800 million years after the Big Bang. This places it in the so-called "reionization epoch" of the universe, a time when the formation of stars and galaxies had begun. About 55,000 light years in diameter, Himiko has an estimated mass equivalent to some 40 billion suns. Alan Dressler, a member of the discovery team, remarked, "If this was the discovery of a class of objects that are ancestors of today's galaxies, there should be many more smaller ones already found—a continuous distribution. Because this object is, to this point, one-of-a-kind, it makes it very hard to fit it into the prevailing model of how normal galaxies were assembled. On the other hand, that's what makes it interesting!"[14] Astronomers remain unsure whether it is a primordial galaxy, a collision of galaxies, or some other class of object.

However, instruments such as the Atacama Large Millimeter/submillimeter Array (ALMA), together with ground-based telescopes such as the European Southern Observatory's Very Large Telescope and large-scale simulations, are shedding more light on the blobs. These observations give credence to the hypothesis that star-forming galaxies embedded in the blobs are illuminating them, like a street light on a foggy night. "For a long time the origin of the extended Lyman-alpha light has been controversial," said Jim Geach, one of the ALMA astronomers. "But with the combination of new observations and cutting-edge simulations, we think we have solved a 15-year-old mystery: LAB-1 is the site of formation of a massive elliptical galaxy that will one day be the heart of a giant cluster...We are seeing a snapshot of the assembly of that galaxy 11.5 billion years ago."[15] (LAB-1 refers to the original Lyman Alpha Blob discovered by Steidel et all in 1999). Lyman Alpha Blobs are therefore important to our understanding of galaxy formation.

The 2016 European Southern Observatory press release on Blob observations with ALMA, including an informative diagram of the structure of the blobs, is at https://www.eso.org/public/news/eso1632/. A popular press report is at http://www.sci-news.com/astronomy/lyman-alpha-blobs-04208.html.

---

[14] Carnegie Institution for Science Press Release April 22, 2009, "Mysterious Space Blob Discovered at Cosmic Dawn," http://www.ciw.edu/news/mysterious_space_blob_discovered_cosmic_dawn; Masami Ouchi et. al., "Discovery of a giant Lyα Emitter near the Reionization Epoch," ApJ, 696 (2009), 1164–1175.

[15] The quotation is reported at Science News, "Astronomers Unravel Mystery of Gigantic Lyman-Alpha Blobs," http://www.sci-news.com/astronomy/lyman-alpha-blobs-04208.html., based on the press release from the European Southern Observatory September 21, 2016, "ALMA Uncovers Secrets of Giant Space Blob," at https://www.eso.org/public/news/eso1632/. The research is published at J. Geach et al. 2016. "ALMA observations of Lyman-alpha Blob 1: halo sub-structure illuminated from within," ApJ, Volume 832, Issue 1, article id. 37, 7 pp. (2016).

Class G 16: Dust

With the Subfamily of intergalactic dust, we enter a regime distinct from intergalactic gas such as the Warm Hot Intergalactic Medium (G 14) and Lyman Alpha Blobs (G 15). Intergalactic dust consists of small particles found between galaxies, believed to originate in dying stars and driven out of galaxies, along with gas, by stellar and galactic winds (S 29 and G 17) as well as radiation pressure. Because the light of distant objects such as supernovae and quasars are seen through this dust, they may cause dimming and an "intergalactic reddening" analogous to interstellar reddening. A similar effect occurs when dust grains in the Earth's atmosphere at sunset block blue light and result in red sunsets. Objects such as supernovae seen through intergalactic dust may be closer than they seem due to the dimming of their light, with cosmological ramifications, especially for the Type Ia supernovae (S 11) used as standard candles to demonstrate an accelerating expansion of the universe.[16] This in turn may affect investigations of concepts such as dark energy. While interstellar dust and its effects have been known since the 1930s, intergalactic dust has only recently been confirmed by observations. Interstellar dust is not to be trifled with: approximately half the metals (those elements other than hydrogen and helium) in the Milky Way and other local galaxies is locked up in dust grains.

Fritz Zwicky was the first seriously to suggest intergalactic dust when he argued, "The apparent non-uniformities in the apparent distribution of clusters of galaxies are due to the effect of intergalactic (and interstellar) obscuration." More specifically, referring to his studies of the Coma and Virgo clusters, among others, he proposed that this non-uniformity "naturally suggests that a local concentration of *intergalactic* dust accounts for the relatively large fluctuations in the distribution of distant clusters of galaxies, a conclusion greatly strengthened by the analogous results that we have discussed regarding the relative numbers of very faint galaxies in regions covered by rich nearby clusters and regions not so covered." Zwicky did not believe he had observed the dust, only its effects, and that the distribution of distant clusters of galaxies

---

[16] Simone Bianchi and Andrea Ferrara, "Intergalactic medium metal enrichment through dust sputtering," MNRAS, 358 (2005), 379 ff; for general material see Mark Bailey, ed., *Dust in the Universe* (Cambridge: Cambridge University Press,1988).

"are satisfactorily accounted for by the assumption of locally concentrated intergalactic dust between the member galaxies of the
large clusters." Moreover, he believed the apparent lack of new
observed clusters with the Palomar 200-inch telescope as compared to its 48-inch Schmidt was due to the existence of intergalactic dust. Finally, he believed the observation of some luminous
intergalactic formations and intergalactic neutral hydrogen made
it "extremely probably that there exists also intergalactic dust."[17]

Astronomers did not immediately take Zwicky's inference
into account, in part because it was just that—an inference. But by
the late 1980s, observing techniques had evolved to the extent that
the comparison the colors of background objects could be used as a
technique for probing dust outside the galactic disk using the intergalactic reddening technique. In 1994, Dennis Zaritsky and colleagues reported dust in galactic halos for two nearby spirals, NGC
2835 and NGC 3521, as seen in the color change in background
galaxies whose light passed through these halos (G 12). This photometric technique provided the first suggestion that dust surrounds
galaxies to distances extending to 200,000 light years.[18] Such dust
could have formed in situ during the early phases of galaxy formation, or it might have been carried out of the disks by galactic
winds (G 17). In 1999 and 2005, astronomers provided evidence of
the latter, reporting that galactic winds triggered by supernovae
would expel some of the galactic dust into intergalactic space.[19]

In 1997, the European Space Agency's Infrared Space
Observatory first detected dust between galaxies. Observing the
Coma cluster of galaxies about 450 million light years distant,
they found dust concentrated toward the center of the cluster.
This dust, about -250° C at the very cold end of the temperature
scale, coexists with gas at a temperature of 80 million degrees,

---

[17] F. Zwicky, "Non-Uniformities in the Apparent Distribution of Clusters of
Galaxies," PASP, 69 (1957), 518–529: 522 and 525; and Zwicky, "New
Observations of Importance to Cosmology," in *Problems of Extra-Galactic
Research* (New York: MacMillan, 1962), p. 347. See also K. H. Schmidt,
"Existence and amount of intergalactic dust," *Astrophysics and Space
Science* 34 (1975) 23–31.

[18] Dennis Zaritsky, " Preliminary Evidence for Dust in Galactic Halos," AJ,
108 (1994), 1619–1626.

[19] Anthony Aguirre, "Intergalactic Dust and Observations of Type IA
Supernovae," ApJ, 525 (1999), 583; Simone Bianchi and Andrea Ferrara,
"Intergalactic medium metal enrichment through dust sputtering," MNRAS,
358 (2005), 379 ff

also observed in the Coma Cluster by the ROSAT X-ray observatory. It is believed such hot gas will eventually destroy the dust. "We think that the intergalactic dust in the Coma Cluster has been ejected from galaxies during the past 100 million years," astronomer Manfred Stickel commented. "The two largest galaxies in the middle of the Coma Cluster do not show up in our infrared scan. They've lost their dust, either in collisions between galaxies or more probably in the merger of the Coma Cluster with another cluster. Fierce cosmic winds generated in such an event can blow the dust right out of the galaxies and into the surrounding space. That may be typical of the way in which intergalactic dust clouds arise, throughout the Universe." In 2002, observations with Hubble showed massive dust clouds in the ultra-luminous infrared galaxy Arp 220 (Fig. 17.4), located 250 million light years away in the constellation Serpens. The observations revealed hundreds of massive star clusters, but astronomers inferred that vast dust clouds obscured hundreds more. Claims have also been made for dust clouds near our own Milky Way Galaxy.[20]

Fig. 17.4. Intergalactic dust is evident in this image of Arp 220, which collided with another galaxy about 700 million years ago. Credit: NASA, ESA, and C. Wilson (McMaster University, Hamilton, Ontario, Canada)

---

[20] ESA Press Release, November 6, 1997, "ISO Proves that Intergalactic Space is Dusty;" M. Stickel et al., "Far-infrared emission of intracluster dust in the Coma galaxy cluster," A&A, 329 (1998), 55–60. HST Press Release, "Super Star Clusters in Dust-Enshrouded Galaxy," http://hubblesite.org/image/1940/news/9-active-galaxies-quasars. On dust clouds near the Milky Way Galaxy see Bogdan Wszolek et al., "Far-Infrared Emission from an Intergalactic Dust Cloud?," Astrophysics and Space Science, 152 (1988), 29–34.

The advent of large CCD imaging arrays permitted ever-larger numbers of galaxies to be detected for use with the reddening technique. In a 2006 article entitled "Abundant Dust Found in Intergalactic Space," astronomers used such an array on the Isaac Newton Telescope and reported a systematic shift in the color of distant background galaxies viewed through the intergalactic medium of the nearby M81 group of galaxies. In addition to M81, the major members of the group are M82 and NGC 3077, all of which are interacting. They found that the reddening coincided with the cool H I atomic gas (S 23) already known to exist between the M81 members, and that the 1:20 dust-to-H I mass ratio was high compared to that of the solar neighborhood (1:120). They concluded that M82 (Fig. 14.9) was the likely source of most of the dust, through its starburst-drive galactic wind (G 17), and that the intergalactic medium is a major repository for galactic dust.[21]

A similar technique was used on a much larger scale in 2009 when astronomers using data from the Sloan Digital Sky Survey II analyzed the light of 85,000 distant quasars as it passed through some 20 million foreground galaxies. Because dust grains block blue light more effectively than red light, they used the "reddening" effect and were able to find dust hundreds of thousands of light years beyond any galaxy. "Putting together and analyzing this huge dataset required cutting-edge ideas from computer science and statistics," said team member Gordon Richards of Drexel University. "Averaging over so many objects allowed us to measure an effect that is much too small to see in any individual quasar." The search for dust in the high-redshift universe is now undertaken using a variety of techniques. These techniques have shown that almost all types of high-redshift sources, including not only quasars but also Lyman Break Galaxies and Lyman Alpha Galaxies, contain dust. All these involve star formation, so there is a high correlation between dust and star formation. [22]

---

[21] E. Xilouris et al., "Abundant Dust Found in Intergalactic Space," APJ, 651 (206), L107–L110.

[22] Brice Menard et al., "Measuring the galaxy-mass and galaxy-dust correlations through magnification and reddening," MNRAS 405 (2009), 1025–1039, reported in "Colors of Quasars Reveal a Dusty Universe," SDSS News Release, Feb 26, 2009; G. R. Meurer, "Dust in the High Redshift Universe," in Astrophysics of Dust, A. N. Witt et al., eds (San Francisco: ASP Conference series, 2003).

Class G 17: Galactic Wind

Galactic winds are streams of charged particles, gas and dust that blow off galaxies, analogous to solar wind (P 16) and stellar wind (S 29), creating filaments and bubbles of multi-million-degree gas clouds that may extend tens of thousands of light years from the galaxies themselves. Galactic winds, sometimes called "super-winds," are believed to be the chief mechanism for recycling matter and energy within the Galaxy and the intergalactic medium (IGM). A study of 10 galaxies between 20 and 900 light years from Earth showed that galactic winds blow between 300 and 3,000 km per second, and that if they do not have enough speed to escape the gravitational pull of the galaxies, they "rain" back down upon the galaxy.[23] Galactic winds are believed to originate with massive stars, supernovae, and the black holes of active galactic nuclei. They are an important factor for the evolution of galaxies because they determine how much material is available for star formation.

The origin of interest in galactic winds may be traced to the discovery in 1963 by C.R. Lynds and Allan Sandage of an explosion at the center of the galaxy M82 (Fig. 14.9), an explosion that drew attention to the possibility of huge outflows of gas. In 1968 J. A. Burke first suggested the presence of a galaxy-scale wind, and by 1971 astronomers were speculating that the lack of an interstellar medium in elliptical galaxies was due to galactic winds that had swept them free of gas. The analogy to stellar winds was explicit; the first of two papers using "Galactic Winds" in their titles began by saying, "The large-scale flow of gas in an elliptical galaxy can be described as a galactic wind analogous to the solar wind and stellar winds...The essential difference between the two cases is that for a galactic wind the gravitating mass and the source of the gas are distributed instead of being entirely concentrated in a central body of negligible size."[24] The search for the engines of galactic winds had begun.

---

[23] S. Villeux et al., "A Search for Very Extended Ionized Gas in Nearby Starburst and Active Galaxies," AJ, 126 (2003), 2185–2208.

[24] C. R. Lynds and A. R. Sandage, "Evidence for an Explosion in the Center of the Galaxy M82," AJ, 137 (1963), 1005–1021; J. A. Burke, "Mass flow from stellar systems-I. Radial flow from spherical systems," MNRAS, 140 (1968), 241 ff. The quote is from the first paper with "Galactic Winds" in its title, H. E. Johnson and W. I. Axford, "Galactic Winds," ApJ, 165 (1971), 381–390, which appeared in April of that year. A second paper, William G. Mathews and James C. Baker, "Galactic Winds," ApJ, 170 (1971), 2241–259 appeared in December. This history is reviewed in Sylvain Veilleux et al., "Galactic Winds," ARAA, 43 (2005), pp. 769–826.

Three decades later, galactic-scale outflows of gas were well-established as a ubiquitous phenomenon in the most active star-forming galaxies in the local universe, and as the likely source of metals and dust in the intergalactic medium. Prior to 2000, most of the information about these winds came from observations of X-ray emission produced by hot gas, or optical line emission produced by warm gas. Those approaches remain very important but over the last decade have been supplemented by the observation of interstellar absorption lines seen against the starlight background.[25]

The study of galactic winds received a boost with the launch of the Chandra X-ray Observatory in 1999. In 2003, Sylvain Veilleux and colleagues reported unexpectedly large galactic winds from ten galaxies. Based on X-ray data from Chandra and optical data from the Anglo-Australian Observatory and the William Herschel Telescope in the Canary Islands, Veilleux reported, "We are seeing that these galactic winds are blowing off of galaxies on a very large scale...We have detected these winds in both visible light and X-ray light on scales that are sometimes much larger than the galaxies themselves." Today, galactic winds are studied in both nearby and distant galaxies. These observations "strongly implicate wind-related feedback processes as key to the chemical and thermal evolution of galaxies and the IGM."[26]

The results of many studies at optical, radio, and infrared wavelengths lead to the conclusion that galactic winds are dependent on starburst strength for a given galaxy, confirming that one of the sources of galactic winds are stars, as stellar winds blow off massive stars during their youth and during supernovae explosions. The other source is believed to be black holes at the centers of active galaxies, so-called AGN-driven winds. Outflows from starburst-driven winds tend to have a bipolar distribution perpendicular to the disk; this is not the case with AGN driven winds.

In 2003 Kurt Adelberger and colleagues, reporting on observations of star-forming galaxies through which the light of background quasars passed, found that "a large fraction of the gas in starburst galaxies at low and high redshift appears to be flowing

[25] Timothy Heckman et al, "Absorption-Line Probes of Gas and Dust in Galactic Superwinds," ApJ Supplement, 129 (2000), 493–516.
[26] Sylvain Veilleux, quoted in Fraser Cain, "Galactic Winds Connects Galaxies," http://www.universetoday.com/9056/galactic-wind-connects-galaxies/, November 21, 2003; and AJ (2003), above, p. 3.

outward rapidly enough to escape the galaxies' gravitational pull; galaxies that experienced intense bursts of star formation in the past now contain little interstellar gas; and metals produced by stars can be found far from known galaxies."[27] In general, galaxy outflows exist if the radiation and mechanical momentum from the starburst or AGN exceeds the gravity of the host galaxy.

Our own Galaxy and the Large Magellanic Cloud (LMC) exhibit processes that shed light on more distant galactic winds. The LMC contains the largest H II region in the Local Group, 30 Doradus, with a mini-starburst region of 50 massive stars, generating shells and compact knots moving at about 200 km per second, perhaps forming the basis for a large-scale wind that can escape the Large Magellanic Cloud. In our own Galaxy, powerful activity in the galactic center, focused on the four-million-solar mass black hole, seems to be the driver of galactic winds.

Observation of galactic winds, especially at high redshifts, remains fraught with difficulties. Astronomers are able to detect galactic winds because of the energy emitted when particles that make up the wind collide with other particles. "We can detect these galactic winds because collisions among the charged particles create electromagnetic energy emissions in the form of X-rays, visible light and radio waves," Veilleux noted. "These emissions are not uniform in the regions around the galaxies. Rather, they are clumpy, being most notable in the regions where hot gas in the wind collides with colder material from the galaxies themselves or from the intergalactic medium."[28] The result is filaments of emissions surrounding galaxies in irregular bubble-shaped regions out to at least 65,000 light years from the galaxy centers. In many ways, the field of galactic winds is still in its infancy.

For ALMA observations of galactic winds 12 billion light years distant see https://phys.org/news/2018-09-galactic-stifling-star-formation-distant.html.

---

[27] Kurt Adelberger, Charles Steidel et al., "Galaxies and Intergalactic Matter at Redshift z~3: Overview," ApJ, 584 (2003), 45–75: 46.

[28] Sylvain Veilleux, quoted in http://www.universetoday.com/9056/galactic-wind-connects-galaxies/; J.S. Spilker el al., "Fast molecular outflow from a dusty star-forming galaxy in the early Universe," *Science* (2018).

Class G 18: Extragalactic Cosmic Rays

Extragalactic cosmic rays (ECRs) are ultra-high energy particles, thought to originate in active galactic nuclei, in star-forming galaxies, and from sources still unknown. They are the highest energy particles in the universe, with speeds very near the speed of light. In contrast to anomalous cosmic rays (P17) in our Solar System with energies in the tens to hundreds of MeV range, and galactic cosmic rays (S 30) with energies in the 100 MeV to ten GeV range, extragalactic cosmic rays can have energies up to or even exceeding billions of GeV (10^{20} eV)—one hundred million times more energy than the particle beams at the Large Hadron Collider at CERN. The energy of such a tiny subatomic particle is about the same as that of a baseball thrown at 50 miles per hour.

Particles with energies greater than 10^{17} eV are often referred to as ultra-high energy (UHE) particles. They are thought to emanate predominantly from black-hole-powered active galactic nuclei. Very-high energy (VHE) cosmic rays, defined as 100 GeV to one or more TeV (10^{11} to 10^{12} eV), are also found in extragalactic regions. Because high-energy cosmic rays may produce gamma rays when they interact with matter and radiation under certain conditions, at these energies they have been observed indirectly by gamma-ray observatories such as Compton, Fermi, and the ground-based High Energy Stereoscopic System (HESS, see S 30). In addition, in 2018 scientists reported the first high-energy neutrinos, emanating from a blazar. Because neutrinos as well as gamma rays are thought to be produced when cosmic rays accelerated in a jet interact with gas or photons, neutrinos are also thought to indicate the presence of high-energy cosmic rays.

Ultra-high energy cosmic rays that impact Earth are exceedingly rare, amounting to only a few dozen per square kilometer per century. Building on the work detecting galactic cosmic rays, ever larger arrays of detectors had to be built to capture the extended air showers produced by the rare ultra-high energy cosmic rays. The first of the large arrays to tackle the detection of ultra-high energy particles was led by Bruno Rossi's group from MIT, where Rossi had established a Cosmic Ray Group in 1946 following his work at Los Alamos. Led by the physicist John Linsley, the team built an array in a remote area of New Mexico consisting of 19 detectors, each with an area of 3.3 square meters, spread over eight square kilometers. With this facility, the so-called Volcanic Ranch array, in 1962 Linsley and his team first made the astonishing discovery of an ECR above 10^{20} eV.[29] The facility operated for three years in this form, observing 1,000 showers with energies above 10^{18} eV. The ECRs appeared isotropic, not emanating from any particular direction, so the source remained mysterious and would for many decades. The working hypothesis, proposed as early as 1960 by Bernard Peters of the Tata Institute in Bombay, suggested that lower-energy cosmic rays are produced predominantly inside our own galaxy, whereas those of higher energy come from more distant sources.[30]

After these pioneering efforts, arrays built in the United Kingdom, the Soviet Union, and Australia also reported cosmic ray detection in the 10^{20} eV range. The array near Leeds, England was completed in 1968, covered 12 square kilometers, and used 600 tons of water to detect ECRs via the Cerenkov effect. Using photomultiplier tubes, it could detect Cerenkov light lasting 20 billionths of a second. The Soviet project covered 20 square kilometers in the mid-1970s. The Australian array was the largest built at the time, spanning 70 square kilometers. In 1967, the first of a new type of cosmic ray detector was built: a 25-sided building in Utah dubbed the "Fly's Eye" that pointed in all directions and gathered fluorescent light from cosmic ray showers. Subsequent Fly's Eye detectors were built, including a very successful one at the University of Utah in the 1970s. Its descendant, the High Resolution Fly's Eye, or HiRes, operated from 1996 to 2006. Another giant array, the Akenko Giant Air Shower Array (AGASA), operated by the Institute for Cosmic Ray Research at the University of Tokyo, eventually spanned 100 square kilometers. Over seven years it recorded nearly 4,000 particles with energies above 10^{18} eV, 461 above 10^{19} eV, and 6 above 10^{20} eV.[31]

---

[29] John Linsley, "Evidence for a Primary Cosmic-Ray Particle with Energy 10^{20} eV," *Phys Rev Letters*, 10 (1963), 146–148. On the context see Roger Clay and Bruce Dawson, *Cosmic Bullets: High Energy Particles in Astrophysics* (Reading, Mass.: Addison-Wesley, 1997), 122–125.

[30] James W. Cronin, Thomas K. Gaisser and Simon P. Swordy, "Cosmic Rays at the Energy Frontier," SciAm, Jan 1997, 62–67.

[31] Michael Freidlander, *A Thin Cosmic Rain: Particles From Outer Space* (Cambridge, Mass.: Harvard University Press, 2000), p. 118.

Since 2004 the Pierre Auger Observatory, located in the high plains of Argentina near the Andes, has been searching for these elusive cosmic rays using 1,600 detectors at separations of 1.5 km, spread over 3,000 square km. It is named after Pierre Auger, the pioneer cosmic ray researcher who discovered "air showers" of cosmic rays in 1938. In 2010, the Observatory reported that at energies above a few times 10^{18} eV, the cosmic ray flux changes from one dominated by protons to one dominated by iron nuclei.[32] Conflicting data from other sources indicates the cosmic ray flux at those energies is still dominated by protons. In 2007, the Auger project announced these highest energy cosmic rays were correlated with active galactic nuclei.[33]

In addition to their observation by air showers in the Earth's atmosphere, both galactic and extragalactic cosmic rays are known to produce gamma rays, and extragalactic cosmic rays may also produce high energy neutrinos. In 2018, a new model was published showing how all three particles can result from the acceleration of cosmic rays by jets emanating from supermassive black holes. Also in 2018, scientists reported that the Ice Cube Neutrino Observatory had detected high-energy neutrinos from blazar TXS 0506+056, located four billion light years from Earth, where high energy gamma rays had already been reported by Fermi. The neutrino energy was an incredible 290 TeV. They concluded, "The energies of the γ-rays and the neutrino indicate that blazar jets may accelerate cosmic rays to at least several PeV. The observed association of a high-energy neutrino with a blazar during a period of enhanced γ-ray emission suggests that blazars may indeed be one of the long-sought sources of very-high-energy cosmic rays, and hence responsible for a sizable fraction of the cosmic neutrino flux observed by IceCube."[34]

---

[32] Bertram Schwarzschild, "The highest-energy cosmic rays may be iron nuclei," *Physics Today* (May 2010), 14–18.

[33] The Pierre Auger Collaboration, "Correlation of the Highest-Energy Cosmic Rays with Nearby Extragalactic Objects," *Science* **318**, 938 (2007), and J. Abraham et al., Auger collaboration, "Measurement of the energy spectrum of cosmic rays above 10^{18} eV using the Pierre Auger Observatory," Physics Letters B, 685, (2010), 239–246.

[34] The IceCube Collaboration et al., "Multimessenger observations of a flaring blazar coincident with high-energy neutrino IceCube170922A," *Science*, 361 (2018), 146–147, http://science.sciencemag.org/content/361/6398/eaat1378.

In our own Galaxy, the diffuse gamma ray emission is dominated by gamma rays produced by cosmic rays interacting with interstellar gas and radiation. A much fainter component, the diffuse extragalactic gamma ray background (EGB), was first detected by the second Small Astronomy Satellite (SAS-2) in 1978 and confirmed in the 1990s by analysis of the EGRET instrument data aboard the Compton Gamma Ray Observatory. The extragalactic diffuse gamma ray background is likely the combined result of many extragalactic processes, including interactions of high-energy cosmic rays with photons.[35] Active galactic nuclei were believed to be the major source of the diffuse extragalactic gamma-ray background, and indeed 569 out of 1,451 sources in the Fermi Telescope's first catalogue of sources were seen to be those AGNs known as blazars. Nevertheless, in 2010 astronomers reported that for the extragalactic gamma ray background, at energies from 0.1 to 100 GeV, active galaxies were only minor players. "Active galaxies can explain less than 30 percent of the extragalactic gamma-ray background Fermi sees," noted Fermi scientist Marco Ajello. "That leaves a lot of room for scientific discovery as we puzzle out what else may be responsible."[36] Leading contenders are particle acceleration in star-forming galaxies, or during the final assembly of the large-scale structure of the universe where clusters of galaxies (G 22 and G 23) are merging into filaments and voids (G 24). However, gamma rays from clusters of galaxies have not been detected thus far.

On Fermi observations of gamma ray sources see https://www.nasa.gov/mission_pages/GLAST/news/cosmic-rays-source.html. And http://www.nasa.gov/mission_pages/GLAST/news/gamma-ray-dragons.html. For the latest results from the Pierre Auger Observatory see http://www.auger.org/cosmic_rays/faq.html.

---

[35] C. E. Fichtel et al, "Diffuse Gamma Radiation," ApJ, 222 (1978), 833–849; P. Sreekumar et al., "EGRET Observations of the Extragalactic Gamma-Ray Emission," ApJ, 494 (1998), 523–534, including historical references.

[36] A. A. Abdo and the LAT Collaboration, "Spectrum of the Isotropic Diffuse Gamma-Ray Emission Derived from First-Year Fermi Large Area Telescope Data," Phys. Rev. Lett., 104 (March 12, 2010); Fermi PR, March 2, 2010, "NASA's Fermi Probes 'Dragons' of the Gamma-ray Sky," at http://www.nasa.gov/mission_pages/GLAST/news/gamma-ray-dragons.html

18. The Galactic Systems Family

Class G 19: Binary Galaxies

From the intergalactic medium Family we now move to the Family of systems of galaxies, ranging from binary galaxies to clusters, superclusters, and filaments and voids, the largest gravitationally bound structures in the universe. A binary galaxy, sometimes called a double galaxy, consists of two galaxies either orbiting one another in bound orbits or involved in close encounters. They are distinguished from chance alignments of galaxies in the line-of-sight. Binary galaxies are of interest because, analogous to binary stars, they can in principle be used to determine the masses of the two galaxies involved, as well as the masses of halos of dark matter (G 12). Studies have shown that binary galaxies constitute about 10% of all galaxies not in clusters, and that a large fraction of these have disturbed morphologies and enhanced optical and far infrared emission, indicating they are interacting galaxies (G 20).[1] The famous Whirlpool Galaxy M51 (Fig. 14.7) is an example of an interacting binary. Simulations of interacting binary galaxies have shown similarities to the process of mass transfer in semi-detached binary stars.

In practice, mass determination for binary galaxies is difficult because the orbital times are so long, but this has not prevented astronomers from trying. The Swedish astronomer Knut Lundmark was the first to point out the importance of studying double galaxies for determining their masses, and as early as 1937 another Swedish astronomer, Erik Holmberg, published a study of 827 double and multiple galaxies for this purpose, which also showed how satellite galaxies move in certain orbits, now called the "Holmberg effect." Holmberg's study of binary galaxies was followed by that of Thornton Page 15 years later, a tradition that has continued to the present. In 1976, Edwin Turner produced an

---

[1] Cong Xu and Jack Sulentic, "Infrared emission in paired galaxies. II - Luminosity functions and far-infrared properties," ApJ, 374 (1991), 407–430.

© Springer Nature Switzerland AG 2019
S. J. Dick, *Classifying the Cosmos*, Astronomers' Universe,
https://doi.org/10.1007/978-3-030-10380-4_18

influential thesis on a selection of 156 binary systems selected from Zwicky's *Catalog of Galaxies and Clusters of Galaxies.*[2]

Binary galaxies are best studied in isolation from galaxy groups, but galaxy groups may also contain binaries. Our own Local Group has at least two binaries with components of approximately equal mass. The Large and Small Magellanic Clouds, both irregular galaxies (though weak spirals have recently been detected), may be considered binary galaxies. The spheroidal galaxies NGC 147 and NGC 185, both studied in detail by Walter Baade in 1944, probably form a stable binary system. They are located on the near side of the Andromeda subgroup of the Local Group. Sidney van den Bergh has noted that these galaxies, the only two Local Group binaries with similar masses, also have similar morphologies.[3]

Binary galaxies of different types may be useful for numerous studies, including galaxy evolution, since binary galaxies may evolve differently than isolated galaxies. I.D. Karachentsev's *Catalogue of Isolated Pairs of Galaxies in the Northern Hemisphere*, containing 603 galaxy pairs, found that about 25% consisted of a spiral or irregular paired with an elliptical or lenticular galaxy, indicating that they were not paired by random capture. Such pairs are simpler to study for star formation and other characteristics because they contain a single gas-rich component, as opposed to spiral pairs where both components are gas-rich and rapid rotators. About half of these pairs are sources of far infrared radiation, enabling studies with spacecraft such as IRAS and the Infrared Space Observatory.[4] Another study showed that a substantial fraction were late-type dwarf galaxies, with luminosities lower than the Magellanic Cloud and separations of only about

[2] Erik Holmberg, "A Study of Double and Multiple Galaxies," *Annals of the Observatory of Lund*, 1937; Thornton Page, "Radial Velocities and Masses of Double Galaxies," ApJ, 116 (1952), 63; Edwin Turner, *Binary galaxies and groups of galaxies.* Dissertation, California Institute of Technology, 1976, summarized in Edwin Turner, "Binary Galaxies I: A Well-Defined Statistical Sample," ApJ, 208 (1976), 20.

[3] Walter Baade, "NGC 147 and NGC 185, Two New Members of the Local Group of Galaxies," ApJ, 100 (1944), 147–150; Sidney van den Bergh, "The Binary Galaxies NGC 147 and NGC 185," AJ, 116 (1998), 1688–1689.

[4] D. L. Domingue et al, "Multiwavelength Insights into Mixed-Morphology Binary Galaxies. I. ISOCAM, ISOPHOT, and Hα Imaging," AJ, 125 (2003), 555–571.

100,000 light years. In 1993, a study showed the existence of wide pairs with separations up to three million light years.[5]

Binary galaxies may also involve active galaxies. In 2007, NASA astronomers reported a galactic jet (G 11) from a supermassive black hole in the binary galaxy 3C 321 was blasting its smaller companion galaxy, separated by only 20,000 light years (Fig. 18.1). They dubbed it the "Death Star Galaxy." It is located at a distance of about 1.4 billion light years from Earth, a distance that required an array of telescopes to determine what was going on. The Chandra, Hubble and Spitzer Space Telescopes were part of the effort, as well as the Very Large Array (VLA) and MERLIN radio telescopes. Chandra's X-ray images showed that each galaxy harbors a black hole, while Hubble revealed the stars in each galaxy, and VLA and MERLIN radio observations reveal the jet striking the smaller galaxy. The jet began impacting the smaller galaxy about a million years ago. "We've seen many jets produced by black holes, but this is the first time we've seen one punch into another galaxy like we're seeing here," said Dan Evans, the leader of the study. "This jet could be causing all sorts of problems for the smaller galaxy it is

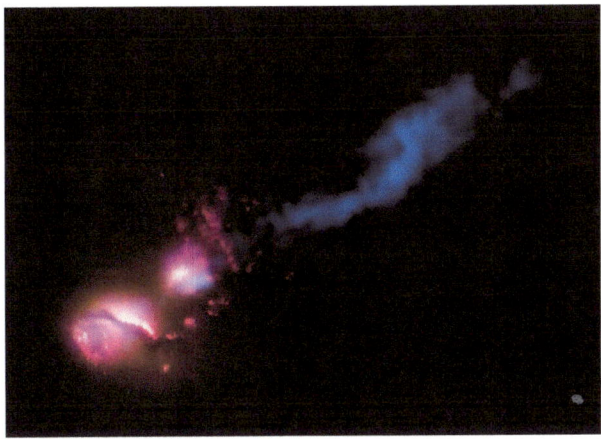

Fig. 18.1. The so-called "death star" binary galaxies comprising 3C 321 are interacting in an aggressive way as an energetic jet (blue) emanating from the black hole of the lower red galaxy strikes the smaller red galaxy and is diverted to the upper right. Credit: X-ray: NASA/CXC/CfA/D.Evans et al.; Optical/UV: NASA/STScI; Radio: NSF/VLA/CfA/D.Evans et al., STFC/JBO/ MERLIN

---

[5]I. D. Karachentsev and D. I., Makarov, "Binary Galaxies in the Local Supercluster," Astrophysical Bulletin, 63 (2008), 299–345. J. N. Chengalur et al, "Dynamics of Binary Galaxies. I. Wide Pairs", ApJ, 419 (1993), 30–46.

pummeling." While the impinging jet could have disastrous effects on any planets in its path, astronomers speculate it could also trigger the formation of new stars and planets. 3C 321 appears to be a rare example of two merging active galaxies, similar to NGC 6240, though the latter is in a much more advanced stage of merger.[6]

In 2010, the first binary quasar was reported within a pair of merging galaxies. Labeled SDSS J1254+0846, it was initially detected by the Sloan Digital Sky Survey. Observations from the Chandra, Kitt Peak, and Palomar observatories were subsequently used to show the object was likely a binary quasar in the midst of a galaxy merger.[7]

A recent Hubble image of the binary pair NGC 4302 and 4298, both at a distance of 55 million light years, is at http://hubblesite.org/news_release/news/2017-14/4-galaxies, and a zoom video at https://www.youtube.com/watch?v=pxyHbAFpfLU. See also the frontispiece of Part III.

---

[6] Daniel Evans et al, "A Radio through X-Ray Study of the Jet/Companion-Galaxy Interaction in 3C 321," ApJ, 675 (2008), 1057–1066, reported in NASA release, December 18, 2007, "NASA Announces Discovery of Assault by a Black Hole," http://science.nasa.gov/science-news/science-at-nasa/2007/18dec_assault/; quote from Chandra Release, Dec. 17, 2007, "'Death Star' Galaxy Black Hole Fires at Neighboring Galaxy," http://www.nasa.gov/mission_pages/chandra/news/07-139.html

[7] Paul Green et al, "SDSS J1254+0846: A Binary Quasar Caught in the Act of Merging," ApJ, 710 (2010), 1578–1588.

Class G 20: Interacting Galaxies

Interacting galaxies are two or more galaxies substantially disturbing each other due to mutual gravitational attraction. Interactions range from minor disturbances, such as a disruption of spiral arms, to full-on galaxy collisions. Such interaction occurs more often than one might think, since galaxies are much closer together as a function of their size than are stars; whereas the diameter of a star compared to the distances between them are normally a billion to one, for galaxies the ratio is 25 to 1. Even during major galaxy interactions, it is not the stars that collide, but the interstellar gas, often setting off a burst of star formation and creating what are known as "starburst galaxies." Computer simulations indicate that such interactions may last one to two billion years. All of this interaction is courtesy of the far-reaching force of gravity. As one poetic astronomer wrote, "Like majestic ships in the grandest night, galaxies can slip ever closer until their mutual gravitational interaction begins to mold them into intricate figures that are finally, and irreversibly woven together. It is an immense cosmic dance, choreographed by gravity."[8]

Galaxy interactions may be divided into early, intermediate, and advanced stages. The Whirlpool Galaxy M51 (Fig. 14.7) is an example of galaxy interaction at an early stage, the case of a large primary galaxy interacting with its satellite, NGC 5195. Located at 23 million light years, it is the prototype spiral galaxy (G 4), discovered by Messier in 1774 but only recognized as a spiral by Lord Rosse in 1845, and the photogenic subject of ground- and space-based telescopes ever since. It is known to harbor a supermassive black hole at its center. The satellite galaxy is believed to have passed through M51 a half billion years ago, producing the interaction seen today.

Two galaxies in an intermediate stage of collision are the Mice Galaxies, both spiral galaxies with a single catalogue number, NGC 4676, designated A and B (Fig. 18.2). Located about 300 million light years away, they are part of the huge Coma Cluster of galaxies. This interacting binary system was first described in some detail in 1957 by Boris A. Vonontsov-Velyaminov, who suggested

---

[8] Lars Lindberg Christensen, Davide de Martin and Raquel Yumi Shida, *Cosmic Collisions The Hubble Atlas of Merging Galaxies* (New York: Springer, 2009), p. 53.

Fig. 18.2. The Mice galaxies, an interacting binary system. The blue areas indicate star forming regions precipitated by the interacting galaxies. Streams of material flow between the two galaxies. Credit: NASA, H. Ford (JHU), G. Illingworth (UCSC/LO), M.Clampin (STScI), G. Hartig (STScI), the ACS Science Team, and ESA

calling the object the "playing mice," after which their designation as "the Mice" became universally accepted. Their motions were modeled already as part of the first computer interaction simulations by the Estonian astronomers Alar and Juri Toomre in 1972, and they became part of the "Toomre sequence" showing various stages of spiral merging over 500 million years.[9] They are characterized by extremely long tails produced by tidal action in which star formation is taking place. They will eventually merge into a single galaxy. The Hubble Space Telescope, among others, has studied the Toomre Sequence of many other merging galaxies.[10]

Merging galaxies are a subset of interacting galaxies, so termed when not enough momentum exists for the invading galaxy to penetrate the other galaxy. Such mergers may be responsible for both elliptical and irregular galaxies (G 2 and G 5). The Antennae galaxies (Fig. 18.3) are the prototype for merging galaxies in an advanced stage of interaction. Discovered by William Herschel in 1785, and

---

[9] N. Ya. Sotnikova nd V. P. Reshetnikov, "Star Formation in the NGC-4676 System (the Mice," *Astronomy Letters*, 24 (1998), 73–83; A. Toomre and J. Toomre, "Galactic Bridges and Tails," ApJ, 178 (1972), 663. Some of the stages of the Toomre sequence are illustrated at http://astronomy.swin.edu.au/cosmos/T/Toomre+Sequence.

[10] S. Laine et al., "HST Observations of the Nuclear Regions of the Toomre Sequence of Merging Galaxies," in *The Central Kiloparsec of Starbursts and AGN: the La Palma Connection, ASP Conference Proceedings* (San Francisco, Calif.: Astronomical Soc. of the Pacific, 2000) 249: 179.

Fig. 18.3. The Antenna galaxies, an example of advanced interaction precipitating star formation and a "starburst" galaxy. Two long streamers of stars, gas, and dust extend from the crash site. Credit: NASA, ESA, and the Hubble Heritage Team (STScI/AURA)-ESA/Hubble Collaboration; Acknowledgment: B. Whitmore (STScI)

also known as NGC 4038 and 4039, they are located at a distance of about 45 million light years, twice the distance of the Whirlpool Galaxy. More than a billion years ago they were separate spiral galaxies; they are now characterized by two long tails of stars, gas, and dust ejected from the interaction. They will likely eventually produce an elliptical galaxy. The Antenna galaxies are an example of a starburst galaxy, a type of galaxy which star formation is taking place at an extremely high rate. This is believed to be due to high concentrations of cool molecular gas (S 25) generated by the collisions of galaxies. The irregular radio galaxy M 82 (Fig. 14.9), located only 12 million light years away, is the prototype of a starburst galaxy, believed to be undergoing high rates of star formation because of its interaction with the nearby spiral M 81. Galaxy mergers are high stakes events involving extremely high energies. In 2010, astronomers observed for the first time merging galaxies that created a binary quasar.[11]

As we will see in the next entry, the compact galaxy group Stephan's Quintet is an example of the interaction of multiple galaxies. Four of the five galaxies form a physical group undergoing mutual interaction that will probably result in a merger. Ring galaxies, characterized by a ring of dust or star formation in the plane

---

[11] Space Daily, Feb. 4, 2010, "Merging Galaxies Create A Binary Quasar," http://www.spacedaily.com/reports/Merging_Galaxies_Create_A_Binary_Quasar_999.html

of rotation of the central object, constitute yet another type of galaxy resulting from interactions. Sometimes classed as a starburst galaxy, Hoag's object, discovered in 1950, is the prototype. The Cartwheel Galaxy is another famous example. In 2010, astronomers released an *Atlas of Images of Nuclear Rings* (AINUR) that included 107 objects. Six are dust rings in elliptical galaxies, and the remainder are star-forming rings in disk galaxies. They range in size from 500 to 3,000 light years.[12]

Research over several decades indicates galaxy evolution is likely driven by galaxy interactions in all these various stages. The Hubble Ultra-Deep Field demonstrated that galaxy interactions were even more common early in the history of the universe. Unlike Edwin Hubble's original sequence from ellipticals to spirals, the early universe appears to be dominated by spirals, which may have merged into ellipticals according to the currently favored theory. As described in the entry on galactic satellites and star streams (G 10), our own Milky Way Galaxy is swallowing at least two small galaxies, the Sagittarius and Canis Major dwarf ellipticals, as evidenced by the existence of shredded galaxies in the form of galactic rings (G 10). In billions of years our Milky Way may merge with the Andromeda Galaxy (another spiral) to produce a giant elliptical.

Many interacting galaxies are represented in Halton Arp's *Atlas of Peculiar Galaxies,* consisting mainly of images taken with the Palomar 200-inch and the 48-inch Schmidt and first published in 1966. It contains 338 "peculiar" galaxies, many earning that name because of their interactions. It was followed two decades later by *A Catalogue of Southern Peculiar Galaxies and Associations.*[13] An even earlier catalogue, Vorontsov-Velyaminov's *Atlas and Catalogue of Interacting Galaxies*, had been published

---

[12] S. Comeron et al., "AINUR: Atlas of Images of Nuclear Rings," MNRAS, 402 (2010), 2462.

[13] Jeff Kanipe and Dennis Webb, *The Arp Atlas of Peculiar Galaxies: A Chronicle and Observer's Guide* (Willmann-Bell, 2006); H. Arp, F. Madore and W. Roberton, *A Catalogue of Southern Peculiar Galaxies and Associations (Cambridge: CambridgeUniversityPress, 1987).* A version of the original 1966 Atlas is online at http://nedwww.ipac.caltech.edu/level5/Arp/frames. html, The 1987 Southern catalogue is at http://ned.ipac.caltech.edu/level5/ SPGA_Atlas/frames.html. The earlier Vorontsov-Velyaminov catalogue, and its second part published in 1976 is at http://www.sai.msu.su/sn/vv.

in 1959 but was not well-known in the West. It is now available online at http://nedwww.ipac.caltech.edu/level5/VV_Cat/frames. html.

A Hubble gallery of interacting galaxies, released on Hubble's 18th anniversary in 2008, can be found at http://hubblesite.org/ images/news/release/2008-16. A simulation of the Mice galaxies collision is at http://hubblesite.org/video/285/news. A composite image of the antenna galaxies from Hubble, Spitzer, and Chandra is at http://hubblesite.org/image/2755/category/19-interacting-galaxiesa and a zoom video is at http://hubblesite.org/video/123/ news_release/1997-34. A video of the interacting galaxy Arp 273: http://hubblesite.org/video/657/news. A video zoom of Stephan's quintet is at http://hubblesite.org/image/2575/category/15-gal-axy-clusters.

Class G 21: Galaxy Group

Most galaxies do not exist in either isolation or in large clusters, but rather in smaller groups. Beyond binary galaxies (G 19), a galaxy group has usually been considered a collection of between three and 50 gravitationally bound galaxies spanning less than 10 million light years. However, the discovery of many dwarf elliptical and spheroidal satellite galaxies (G 10), including 59 now known to be orbiting the Milky Way itself, has expanded this definition to potentially one hundred or more if those dwarf galaxies are included. Groups in turn may form galaxy clusters (G 22), containing up to several thousand members and spanning 10 to 30 million light years. Superclusters (G 23) are the largest aggregations of galaxies, spanning several hundred million light years. The dividing line between groups, clusters, and superclusters of galaxies is somewhat arbitrary but nevertheless useful. Conferring on them distinct class status is therefore also somewhat arbitrary, but also useful and further justified by physical differences other than scale.

The discovery that galaxies tend to form groups not only awaited large telescopes that could detect fainter galaxies, but also required the determination of galaxy distances so they could be confirmed as close enough to each other to be gravitationally bound. "Surveys of the sky show that nebulae are scattered singly and in groups of various sizes up to the occasional great clusters," Edwin Hubble wrote in 1936. "The small-scale distribution resembles that of stars in the stellar system. Analogies among the nebulae [galaxies] are readily found with individual stars, doubles, triples, multiples, sparse clusters, and open clusters. Globular clusters alone seem to have no counterpart in the realm of the nebulae."

Not surprisingly, the observational confirmation of galaxy groups began close to home when it was discovered that the Milky Way Galaxy is part of a "Local Group" of galaxies. Hubble began detailed studies of nearby galaxies in 1925, and by 1936 he spoke of nine definite and three possible members of the Local Group in his classic book *The Realm of the Nebulae*. He also identified about 20 other galaxy clusters, and we now know that many Messier objects are actually members of galaxy groups.[14] Today we recognize that

---

[14] Edwin Hubble, *The Realm of the Nebulae*, 1936, pp. 77, 124–151. Walter Baade appears to have been the first to use the term "Local Group" in a 1935 paper, where he noted "our galaxy and the nearest extra-galactic systems

our Local Group may contain more than a hundred galaxies, if we include the 59 members orbiting our own Galaxy within 1.5 million light years. The Local Group spans nearly 10 million light years. Put another way, the Local Group occupies a volume of space with a radius of about five million light years. The Local Group is in turn part of the Virgo supercluster, sometimes called the Local Supercluster. The Virgo Supercluster contains at least 100 galaxy groups and clusters spanning 110 million light years. Our Local Group is an outlying member.

As the Group of galaxies nearest us, the Local Group is important for the study of galaxy Groups in general. Aside from the Milky Way Galaxy itself, the Local Group includes two other large spirals, Andromeda (M31) and Triangulum (M33). These three galaxies dominate the Local Group, with about 400 billion, one trillion, and 40 billion stars respectively. Andromeda (Fig. 14.6), at a distance of 2.5 million light years, is most likely a spiral barred galaxy, though the bar is difficult to observe because we view it along its axis. Like the Milky Way Galaxy, Andromeda contains numerous globular clusters (about 460 detected thus far) and appears to harbor a supermassive black hole at its center, which is an X-ray source observed with the Chandra X-ray telescope. The Andromeda Galaxy is likely on a collision course with the Milky Way, with contact expected in about 2.5 billion years. The much smaller Triangulum spiral, at a distance of about three million light years, has numerous star-forming regions as well as its accompaniment of globular clusters. Unlike the two larger spirals of the Local Group, Triangulum contains a stellar-mass, rather than a supermassive, black hole; at about 15 solar masses, however, it is one of the largest stellar black holes. Satellite galaxies (G 10) tend to form around these three dominant spirals.

The closest members of the Local Group were the first to be discovered by naked eye observers, including the eponymous Ferdinand Magellan during his 1519 Southern Hemisphere circumnavigation of the Earth. The Large Magellanic Cloud (LMC) is now known to be at a distance of 160,000 light years, and the Small Magellanic Cloud (SMC) at about 200,000 light years (Fig. 18.4). Both have long been classified as irregulars, though weak spiral features have recently

---

form what may be termed a local group of nebulae," Baade, "The Globular Cluster NGC 2419," ApJ, 82 (1935), 396–412: 412. Hubble had an entire chapter titled "The Local Group" in his 1936 volume. The first paper with the term in its title was Walter Baade, "NGC 147 and NGC 185, Two New Members of the Local Group of Galaxies," ApJ, 100 (1944), 147-150.

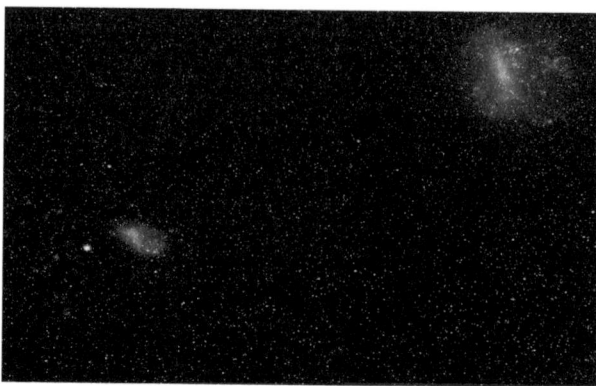

Fig. 18.4. The Magellanic Clouds, two members of the Local Group of galaxies, visible with the naked eye from the Southern Hemisphere. They are classified as irregular, though they have notable bar features across their central disks. Credit: ESO/S. Brunier

been discovered. The Large Magellanic Cloud harbors about 20 billion stars, one-tenth the number in the Milky Way. They are connected to the Milky Way Galaxy by a bridge of hydrogen gas, and to each other by a bridge of neutral hydrogen and a thin line of stars. Because of their proximity the Magellanic Clouds are closely studied. The Large Magellanic Cloud, for example, is known to contain at least 60 globular clusters, 400 planetary nebulae, and 700 open clusters, and when Supernova SN1987 A exploded (S 11) in the Large Magellanic Cloud it was immediately noticed.

The Local Group provides an excellent opportunity to study galaxy groups up close, using an admittedly expansive definition of "close." Looking beyond the Local Group, the nearest galaxy groups are Maffei (several dozen galaxies at 10 million light years), Sculptor (13 galaxies at nine million light years) and the M81 group (about 35 galaxies at 12 million light years). Not far beyond (Fig. 18.5) are the M101 group (seven galaxies at about 20 million light years, including the Pinwheel Galaxy) and M51 group (seven galaxies at 23 million light years). Paolo Maffei recognized Maffei 1 and 2 as galaxies in 1968 (they were previously believed to be H II regions), and most of the members of the Maffei group were discovered only in the last 30 years. Thousands of galaxy groups are now known to exist, and they are undoubtedly spread throughout the universe back to the era when galaxies first formed.

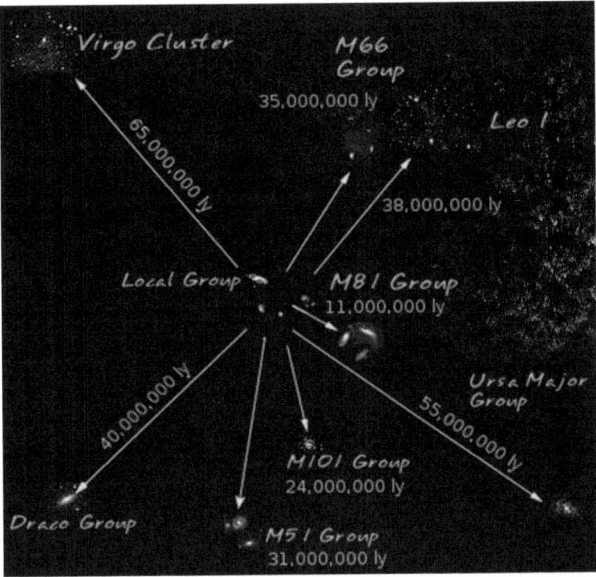

Fig. 18.5. Galaxy groups and the Virgo cluster. All distances are approximate. Credit: NASA

One distinct type of galaxy group is the compact group. In the 1980s, Canadian astronomer Paul Hickson compiled a catalogue of 100 of these compact groups, typically consisting of four or five galaxies isolated in the general field. Hickson considered them an extreme example of galactic systems having a wide range of densities and populations.[15] A high proportion of these groups have active galaxies, contain large quantities of gas, and have motions dominated by dark matter. Such groups serve as good objects for studying galaxy interaction and mergers. The first compact group detected was Stephan's quintet, also known as Hickson 92 (Fig. 18.6), discovered by Edouard Stephan from the Marseille Observatory in 1877. Three of the five galaxies in this group show strong tidal distortions. Stephan's quintet is one of the most studied compact galaxy groups, and was one of the first images released taken by the Wide Field Camera 3 after Hubble's

---

[15] P. Hickson, "Compact Groups of Galaxies," ARAA, 35 (1997), 357–88, online at http://nedwww.ipac.caltech.edu/level5/Sept01/Hickson/Hickson_contents.html; P. Hickson, *Atlas of compact groups of galaxies*, (Basel, Gordon & Breach, 1994). Compact Group literature as of 1998 is given at http://www.astro.ubc.ca/people/hickson/hcg/refs.html

Fig. 18.6. Stephan's Quintet, a very active galaxy group. The image in the center is actually two galaxies, and the spiral galaxy at top left is a foreground object not physically associated with the others. Credit: NASA, ESA, and the Hubble SM4 ERO Team

last servicing mission in 2009. It has also been observed in the infrared by the Spitzer Telescope. The second compact group discovered was Seyfert's sextet (Hickson 79), found in 1948 and one of the densest known. Other well-known groups include Hickson 40, at a distance of 300 million light years, incorporating three spirals, an elliptical, and a lenticular galaxy.

A list of Local Group members may be found at http://www.ast.cam.ac.uk/~mike/local_members.html. A list of members of the Maffei group may be found at http://www.atlasoftheuniverse.com/galgrps/maffei.html. A list of nearby galaxy groups within 100 MLY may be found at http://www.atlasoftheuniverse.com/galaclus.html, although some of the data is out of date. A Chandra look at Stephan's Quintet is at http://spitzer.caltech.edu/images/1605-ssc2006-08a-A-Shocking-Surprise-in-Stephan-s-Quintet. See also http://www.raycash.us/deepsky/besthick.htm for images of some of the best Hickson groups.

Class G 22: Cluster

A cluster of galaxies is a gravitationally bound aggregation of galaxies intermediate in size between a Galaxy Group (G 21) and a Supercluster (G 23). Galaxy clusters are given the name of the constellation in which they are found, and typically contain between 50 and 1,000 galaxies, though some may contain up to 2,000 galaxies. They span five to 30 million light years. Nearby prominent galaxy clusters include the Coma Cluster and the Virgo Cluster, the latter spanning eight degrees on the sky. "Nearby" is relative in this case, since the Virgo Cluster is 60 million light years away. Others are much more distant, including the often-studied Centaurus, Perseus, Leo and Hercules clusters. The Fornax and Ursa Major clusters are closer than the Virgo and Coma clusters, but not nearly as rich. Clusters are also distinguished from galaxy groups in that the orbital velocities of the latter are lower—between 70 and 150 miles per second, ten times lower than for galaxies in clusters, as the higher density of clusters whips their galaxies around faster.

Clusters of galaxies, not to be confused with galactic clusters of stars (S 34) within our own Galaxy, are sometimes divided into two types: regular clusters and irregular clusters. Regular clusters have a spherical structure and concentrated core and appear to be populated mainly by elliptical galaxies. Irregular clusters have no well-defined center and often include all galaxy types. Many galaxy clusters have an "Abell" number attached to them, after the American astronomer George Abell, who catalogued the richest of the clusters (those with more than 30 members) using the National Geographic-Palomar Sky Survey. Abell's original "Northern Survey" of 1958 contained 2,712 rich clusters, and its southern extension in 1989 added another 1,361 clusters for a total of 4,073, giving an idea of the abundance of galaxy clusters. Named clusters also have Abell numbers associated with them (the Coma Cluster is Abell 1656, for example), but where they exist the names are more often used.

The recognition of this class of object came slowly and was highly dependent on the improvement of telescopes and methods for distance determination, especially redshifts via spectroscopy. Already in the 18th century, telescopic observers noticed a concentration of "nebulae" in certain regions of the sky. In his catalogue of nebulae and clusters published in 1781, Charles Messier wrote, "The constellation of Virgo, & especially the northern Wing, is one of the constellations which encloses the most Nebulae."[16] Although Messier knew neither the true nature of the nebulae nor their distances, today we recognize that 15 out of the 110 objects in Messier's extended catalogue are actually galaxies in the Virgo Cluster.

The modern idea of clustering galaxies began in the 1930s, shortly after the extragalactic nature of some of the nebulae became known. By this time Edwin Hubble had already been led to the idea of a "Local Group" of galaxies, but clusters were much larger. Hubble and Milton Humason listed eight galaxy clusters in their study of 1931: Virgo, Pegasus, Pisces, Cancer, Perseus, Coma, Ursa Majoris, and Leo. In 1933, Harlow Shapley listed 25 clusters, and the decade of the 1930s saw the landmark studies of Shapley and Adelaide Ames in producing a catalogue of 1246 bright galaxies down to 13th magnitude, which advanced the identification of more clusters. To their eyes, the plot of the 1025 galaxies brighter than 13th magnitude (Fig. 18.7) clearly indicated clustering.[17] Also in the 1930s, Fritz Zwicky began his work on clusters and cluster

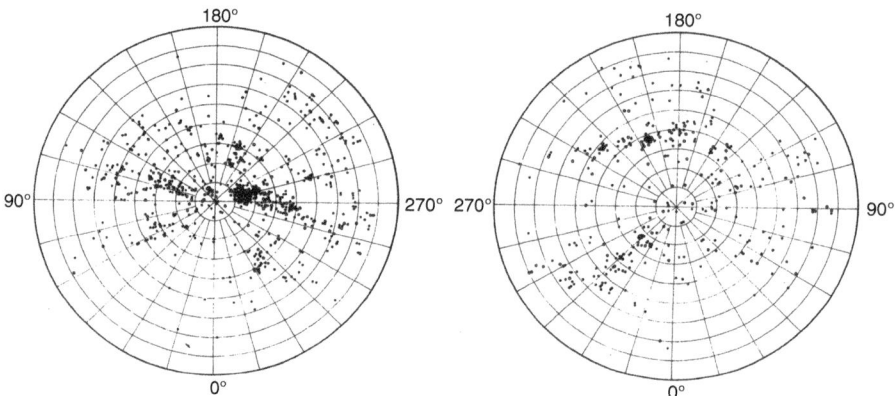

Fig. 18.7. Early indications of galaxy clustering, according to Harlow Shapey and Adelaide Ames, from the Shapley-Ames Catalogue, 1932. The plots show the distribution of galaxies brighter than 13th magnitude in the northern (left) and southern (right) galactic hemispheres

---

[16] Messier's description follows the entry for M91 in the catalogue he published in the *Connaissance des Temps* for 1784.

[17] E. Hubble and M. Humason, "The Velocity-Distance Relation among Extra-Galactic Nebulae," ApJ, 74 (1931), 43–80, especially pp. 60 ff; Harlow Shapley, "Luminosity Distribution and Average Density of Matter in Twenty-five Groups of Galaxies," PNAS (1933), 19, 591–596. The Shapley-Ames Catalogue was published in 1932; Allan Sandage and G. A. Tammann published a revised version in 1981, online at http://nedwww.ipac.caltech.edu/level5/Shapley_Ames/frames.html. See also Dick (2013), pp. 154–165.

masses and found from the orbital motions of galaxies within the Coma cluster a mass for the cluster that implied dark matter must be holding it together in addition to the matter that was visible. This was the first use of the "virial theorem" method, whereby random motions of the members of a cluster can be used to yield the mass of the entire cluster. Beginning in 1938, Zwicky published his work on the distribution of galaxy clusters over different regions of the sky.[18]

As late as 1949, only a few dozen clusters of galaxies were known. This changed dramatically in the 1950s with the advent of large-scale photographic programs, including Lick Observatory's proper motion survey and, most especially, the National Geographic Society-Palomar Observatory Sky Survey, which revealed tens of thousands of galaxy aggregates. It was the latter that George Abell used to determine the 2,700 richest of these in his catalogue of clusters published in 1958, extended to more than 4,000 objects in 1989, six years after Abell's untimely death.[19]

From early on, galaxy clusters were used to probe the extent and age of the universe. In 1956 Humason, Nicholas Mayall, and Allan Sandage used 474 of the brightest members of clusters of galaxies to determine a Hubble recession factor, now called the Hubble constant, of 180 km per second per megaparsec.[20] Much controversy ensued over the next 50 years about the true value of the Hubble constant, with most determinations found to be between 50 and 100, corresponding to an age of the universe between 10 and 20 billion years. In 1994, a team led by Wendy Freedman used the Hubble Space Telescope to determine a distance of 17 megaparsecs (about 56 million light years) to the Virgo Cluster spiral galaxy known as M100. The team determined a Hubble constant of 72, corresponding to an age of the universe of 13 billion years, with an uncertainty of a billion years. The current widely accepted value is 71 km per second per megaparsec,

---

[18] Fritz Zwicky, "On the Masses of Nebulae and of Clusters of Nebulae," ApJ, 86 (1937), 217–246, reprinted in part in Lang and Gingerich, pp. 729-107; Zwicky, "On the Clustering of Nebulae," PASP (1938), 50, 218–220. In his 1937 paper Zwicky describes the gravitational lens effect for the first time; Lang and Gingerich, 734.

[19] George Abell, "The Distribution of Rich Clusters of Galaxies," ApJ Supplement, 3, (1958), 211; G. Abell, H. G. Corwin, and R. P. Olowin, *A Catalog of Rich Clusters of Galaxies*. ApJ Supplement, 70 (1989). pp. 1–138.

[20] Milton L. Humason, Nicholas U. Mayall and Allan R. Sandage, "Redshifts and Magnitudes of Extra-Galactic Nebulae," AJ, 61 (1956), 97–162, reprinted in part in Lang and Gingerich, 753–762.

yielding a universe 13.8 billion years old. But as of 2018, the two major methods for determining the Hubble constant—one using the cosmic distance ladder as determined by the Hubble and Gaia spacecraft, the other using the cosmic background radiation as measured by the Planck spacecraft—continue to give incompatible results for the expansion rate of the nearby universe as compared to the distant universe shortly after the Big Bang.[21]

The Virgo Cluster (Fig. 18.8), at a distance of approximately 60 million light years, is a dominant example of a nearby irregular cluster of galaxies. Its eight-degree span as seen from Earth is almost 20 times the diameter of the full Moon, equivalent to 15 million light years at that distance. It contains about 2,000 member galaxies and is much denser than our Local Group with its 40 members spanning five million light years. It is estimated to contain 100,000 billion solar masses; some 90% of its galaxies are

Fig. 18.8. The Virgo Cluster. M87 is the large galaxy at lower left. The dark spots indicate where bright stars were removed from the image. Credit: Chris Mihos (CaseWestern Reserve University)/ESO

---

[21] G. A. Tammann, in *The Virgo Cluster, ESO Workshop proceedings No. 20,* ed. O.-G. Richter and B. Binggeli, (Garching: ESO 1985), p. 3. For the current controversy see HST Release, March 22, 2018, "Hubble Solves Cosmic 'Whodunit' with Interstellar Forensics," http://hubblesite.org/news_release/ news/2018-15/4-galaxies, and HST Release, July 12, 2018, "Hubble and Gaia Team Up to Fuel Cosmic Conundrum," http://hubblesite.org/news_release/ news/2018-34/4-galaxies. The Hubble and Gaia data yield a value for the Hubble constant of 73.5 kilometers per second per megaparsec, while the final Planck results are 67 kilometers per second per megaparsec.

dwarfs. The huge elliptical galaxy known as M87 is the king of the Virgo Cluster. Observational evidence indicates that the Virgo cluster may be pulling our Local Group in its direction. The Virgo cluster is in turn the dominant cluster of the local supercluster, also known as the Virgo supercluster (G 23).

Another giant cluster is the Coma Cluster, also known as Abell 1656. At a distance of about 300 million light years, it spans more than 20 million light years and has perhaps 3,000 members. Because it has the densest galaxy environment in the local universe, it is the object of a project known as "The Hubble Space Telescope Treasury Survey of the Coma Cluster of Galaxies."[22] Abell 1689 (Fig. 18.9) is also one of largest galaxy clusters, but located 2.2 billion light years distant. Its immense gravitational field has been used as a "gravitational lens" to map dark matter in the cluster, which is about 80% of its mass. Closer to home are many smaller clusters, including the Centaurus cluster of 100 galaxies at 155 million light years; the Perseus cluster with

Fig. 18.9. Abell 1689 cluster, showing gravitational lensing. Credit: NASA, ESA, the Hubble Heritage Team (STScI/AURA), J. Blakeslee (NRC Herzberg Astrophysics Program, Dominion Astrophysical Observatory), and H. Ford (JHU)

---

[22] "The Hubble Space Telescope Treasury Survey of the Coma Cluster of Galaxies," http://astronomy.swin.edu.au/coma/project-overview.htm

190 galaxies at 230 million light years, observed by the Chandra Telescope as a bright X-ray source; the Leo cluster of about 100 objects at 330 million light years; and the Hercules cluster of 100 galaxies at 650 million light years.

All of this begs the question of when galaxy clustering began. As more and more distant galaxy clusters have been discovered, observations indicated at first that protoclusters may have started forming about 10 billion years ago. In 2001, astronomers reported that a cluster named 3C 294 was located at about 10 billion light years. Four years later, astronomers using the ESA's XMM-Newton spacecraft reported an object with the unwieldy name XMMXCS J2215.9-1738 at about the same distance.[23] In 2006, astronomers undertaking a deep sky survey with the UK Infrared Telescope detected the farthest ever galaxy cluster known at the time, JKCS041, at 10.2 billion light years (redshift 1.9). In 2009, Chandra Observatory X-ray observations confirmed the detection. "This discovery is exciting because it is like finding a *Tyrannosaurus rex* fossil that is much older than any other known," said Ben Maughan of the University of Bristol in Britain, co-author of the paper describing the cluster in *Astronomy & Astrophysics*. "One fossil might just fit in with our understanding of dinosaurs, but if you found many more, you would have to start rethinking how dinosaurs evolved. The same is true for galaxy clusters and our understanding of cosmology." Astronomers believe the object formed about as early in the universe as it could. "We don't think gravity can work fast enough to make galaxies much earlier," team member Stefano Andreon was quoted as saying.

In an indication of how fast-paced astronomy (and gravitational processes) can be, by 2018 astronomers reported clustering and possible merging of massive galaxies already 1.5 billion years after the Big Bang. Using the ALMA array, they reported 14 starburst galaxies congregating into a protocluster they dubbed SPT 2349–56 (because it was first detected by the South Pole Telescope). It is the most distant protocluster detected thus far. "The fact that

---

[23] A. C. Fabian et al., "Chandra detection of the intracluster medium around 3C 294 at z=1.7863," MNRAS, 322 (2001), L11–L15. The XMM claim detection is S. A. Stanford et al., "The XMM Cluster Survey: A Massive Galaxy Cluster at z = 1.45," ApJ, 646 (2006), L13–L16; *Space Daily*, June 7, 2006, "Astronomers Find Most Distant Galaxy Cluster Yet," http://www.spacedaily.com/reports/Astronomers_Find_Most_Distant_Galaxy_Cluster_Yet.html

this is happening so early in the history of the universe poses a formidable challenge to our present-day understanding of the way structures form in the universe," said Scott Chapman. Another protocluster consisting of 10 galaxies forming only 200 million years later has also been announced.[24]

Although galaxy clusters are generally widely separated, in 2009 the Hubble and Chandra telescopes detected collisions among three galaxy clusters. The clusters are located at a distance of about 5.4 billion light years. Such mergers are among the most energetic events in the universe.[25]

An incomplete list of the Abell catalogue of galaxy clusters is at https://en.wikipedia.org/wiki/List_of_Abell_clusters. A Hubble view of the Coma cluster is at http://hubblesite.org/image/2357/category/15-galaxy-clusters. A zoom video of the Abell 370 cluster is at http://hubblesite.org/video/1171/news_release/2018-39. For a zoom and visualization of the members of the Virgo cluster see http://www.atlasoftheuniverse.com/galgrps/vir.html. See also http://hubblesite.org/video/948/news/91-astronomical. On the collision of galaxy clusters see http://hubblesite.org/newscenter/archive/releases/2009/17/image/a/.

---

[24] S. Andreon, B. Maughan et al., "JKCS 041: a colour-detected galaxy cluster at z_{phot} ~ 1.9 with deep potential well as confirmed by X-ray data," A&A, 507 (2009), 147–157; original detection is A. Lawrence, et al. "The UKIRT Infrared Deep Sky Survey (UKIDSS)," MNRAS, 379 (2007), 1599. Maughan and Andreon are quoted in http://www.wired.com/wiredscience/2009/10/farthest-galaxy-cluster-ever-detected/. Chandra Press release, Oct 22, 2009, "Galaxy Cluster Smashes Distance Record," is at http://www.nasa.gov/centers/marshall/news/news/releases/2009/09-086.html. The 2018 observations are reported at T. B. Miller et al. "A massive core for a cluster of galaxies at a redshift of 4.3," https://arxiv.org/abs/1804.09231.

[25] Cheng-Jiun Ma et al., "An X-Ray/Optical Study of the Complex Dynamics of the Core of the Massive Intermediate-Redshift Cluster MACSJ0717.5+3745," ApJ, 693 (2009), L56–L60. http://hubblesite.org/pubinfo/pdf/2009/17/pdf.pdf; HST release, April 16, 2009, "Galaxy Cluster MACS J0717," http://hubblesite.org/newscenter/archive/releases/2009/17/

Class G 23: Supercluster

Through the relentless reach of gravitation, clusters of galaxies may form superclusters spanning a hundred million light years or more, ten times more extensive than clusters (G 22). Like clusters, superclusters are distinguished by their gravitational interaction at these much larger scales, which tends to pull them together, overcoming some of the Hubble flow due to the expansion of the universe. About 50 superclusters are known, 17 of them within a billion light years of our Local Supercluster, also known as the Virgo Supercluster. They are usually named after their dominant member; thus, the Virgo Supercluster includes the huge Virgo Cluster with its 2,000 galaxies. Whereas the Virgo Cluster itself spans about eight million light years, the Virgo Supercluster, with its 100 galaxy groups and clusters in addition to the Virgo Cluster, spans 110 million light years.

The discovery of a higher order of clustering came as a surprise and was even slower to be accepted than the idea of lower-order clustering. Speaking of already well-known clusters such as Coma, Perseus, and Pegasus, Fritz Zwicky wrote in 1937, "Clusters of nebulae [galaxies] are the largest known characteristic aggregations of matter and their investigation provides the stepping stone for the investigation of the accessible fraction of the universe as a whole. For instance, the counts in the Coma cluster have furnished the first and, so far, only proof that in the first approximation Newton's law of gravitation adequately describes the interactions among nebulae."[26] Zwicky vigorously denied higher-order clustering throughout his career into the 1970s, partly as a question of nomenclature because he considered clusters to consist of anything ranging from small groups of galaxies to those spanning 150 million light years. This highlights the important point that higher-order clustering is different from one large cluster and is in fact one of the distinguishing features of a cluster from a supercluster.[27]

---

[26] Fritz Zwicky, "On the Clustering of Nebulae," PASP (1938), 50, 218–220.

[27] De Vaucouleurs, "The Local Supercluster of Galaxies," *Astronomical Society of India Bulletin*, 9 (1981), 1–23: 6. De Vaucouleurs here gives a detailed history, from his point of view, of the idea of superclustering.

Aside from early general suggestions of higher-order clustering, our Local Supercluster was naturally the first to be supported with solid observational evidence. In 1953, the American astronomer Gerard de Vaucouleurs suggested that the Virgo cluster was a dominant member of a large aggregate of galaxies that formed what he called a "supergalaxy." By 1958 he referred to it as the "Local Supercluster." As he put it in his 1958 paper, "Preliminary evidence has been presented...that the majority of the brighter galaxies in the Harvard Survey [Shapley and Ames, 1932] and a good many of the fainter ones listed in the NGC, whether members of recognized groups or clusters or so-called 'field nebulae,' belong to a flattened super-cluster of galaxies, the 'Local Supergalaxy,' of which the Local Group of galaxies is merely an outlying condensation."[28] He estimated the diameter of the supercluster at 60 to 100 million light years, depending on the uncertain distance to the Virgo cluster. Because of the dominance of the Virgo cluster, the supercluster is sometimes called the Virgo Supercluster. By 1974 de Vaucouleurs had identified 54 clusters forming the Local Supercluster. The general idea of superclusters, however, remained controversial. While George Abell had argued already in his 1959 catalogue that superclusters exist, some astronomers, including Zwicky, found no higher order clustering. Many others dismissed superclusters as accidental alignments or sheer speculation into the 1970s.[29]

In the 1970s, however, Princeton University astronomer James Peebles and his colleagues, using galaxy counts and a tool called the "galaxy correlation function," showed that galaxy clustering appeared to occur on a very wide range of scales, from small groups to superclusters. Using the Lick Observatory counts of galaxies first carried out by Donald Shane and C.A. Wirtanen, they produced a famous "Map of a Million Galaxies," widely published both in science publications and in popular culture venues such as posters and Stewart Brand's *Whole Earth Catalogue*, in order to provide a perspective on Earth's place in the universe.[30]

---

[28] Gerard de Vaucouleurs, "Evidence for a Local Supergalaxy," AJ, 58 (1953), 30–32; de Vaucouleurs, "Further evidence for a local super-cluster of galaxies: rotation and expansion," AJ, 63 (1958), 253–266.

[29] De Vaucouleurs, 1981, p. 6.

[30] C. D. Shane and C. A. Wirtanen, "The Distribution of Galaxies," *Publications of the Lick Observatory*, 22 (1957), 1–60; M. Seldner, B. Siebars, E. J. Groth and P. J. E. Peebles, "New Reduction of the Lick catalog of galaxies, AJ, 82 (1977), 249–256.

The map showed a very inhomogeneous distribution of galaxies, but even Peebles was skeptical of the tendency of the human eye to find patterns. However, at the same time the Estonian astronomer Jaan Einasto and his colleagues were using redshift data to show beyond doubt that structure did indeed exist beyond galaxy clusters.[31] Such redshift data, which gave three-dimensional information including distance, would prove to be the key to determining the large-scale structure of the universe even beyond superclusters, the so-called filaments and voids (G 24).

The existence of superclusters is now widely accepted. After our Local Supercluster and within 200 million light years, is the Hydra Supercluster, following the convention of being named after its brightest member, the Hydra cluster (Abell 1060). It is about 100 million light years long. Stretching between 150 and 400 million light years away is the Pavo-Indus Supercluster, not particularly rich but identifiable nonetheless. By contrast, the Perseus-Pisces Supercluster is one of the largest structures of the universe, at a distance of about 250 million light years. Next out is the Coma Supercluster, located at about 350 million light years distant, some five or six times further away than the Virgo cluster. The Shapley Supercluster, at 650 million light years, was only named after Shapley when Somak Raychaudhury rediscovered it in 1989 and referred to it as the "Shapley Concentration," a concentration that Shapley had noted already in 1930. Raychaudhury concluded that the Shapley Supercluster is responsible for less than 10% of the acceleration of our Local Group of galaxies, the rest being due to nearer galaxies.[32]

---

[31] M. Joeveer and J. Einasto, "Has the Universe the cell structure?" in *The Large Scale Structure of the Universe*, eds. M.S. Longair and J. Einasto (Dordrecht: Reidel, 1978), 241–251; J. Einasto, "Dark Matter and large scale structure," in *Historical Development of Modern Cosmology*," in PASP Conference series, vol. 252, eds. V. J. Martinez, V. Trimble and M. J. Pons-Bordeia (San Francisco: ASP, 2001), 85–107.

[32] Harlow Shapley, "Note on a Remote Cloud of Galaxies in Centaurus," *Harvard Coll. Obs. Bull.* 874 (1930), 9–12; Somak Raychaudhury, "The distribution of galaxies in the direction of the 'Great Attractor'," *Nature*, 342 (1989), 251–255.

The search for the cause of the motion of our Local Group of galaxies, however, continued. By determining the velocities of thousands of galaxies, Alan Dressler and his colleagues proposed in 1987 a huge concentration of dark matter that they dubbed "The Great Attractor." It existence was based on the observed motion of the Virgo Supercluster toward Hydra-Centaurus at a velocity of some 630 km/second, 1.4 million miles per hour. If it exists the Great Attractor would be a large supercluster beyond 200 million light years. In 2006, however, astronomers used X-ray observations to peer beyond the gas and dust of the Milky Way's "zone of avoidance" and announced that the Milky Way was being pulled to an even more massive Supercluster beyond the Great Attractor, none other than the Shapley Supercluster.[33]

A visualization of the Virgo Supercluster is at http://www. atlasoftheuniverse.com/virgo.html. Superclusters to one billion light years are visualized at http://www.atlasoftheuniverse.com/ superc.html.

---

[33] Alan Dressler, *Voyage to the Great Attractor* (New York: Vintage Books, 1994); Dale Kocevski, "On the Origin of the Local Group's Peculiar Velocity," ApJ, 645 (2006), 1043–1053; Institute for Astronomy, University of Hawaii Press Release, "X-rays Reveal What Makes the Milky Way Move," January 11, 2006, at http://www.ifa.hawaii.edu/info/press-releases/kocevski-1-06/

Class G 24: Filaments and Voids

The existence of structures larger than superclusters, on the scale of hundreds of millions to billions of light years, was not suspected until four decades ago, when large surveys of galaxy redshifts began to reveal the coherent large-scale structure of the universe. "Many astronomers had imagined roughly spherical galaxy clusters floating amongst randomly scattered field galaxies, like meatballs in sauce," two galactic astronomers wrote looking back from the vantage point of the year 2000. "Instead, they saw galaxies concentrated into enormous walls and streamers, surrounding huge voids that appear largely empty."[34] These structures have variously been called "sheets," "walls," "bubbles" and "holes," but the terms most used are "filaments" and "voids." By whatever name, they constitute the "cosmic web" channeling intergalactic gas (G 14) into galaxies and clusters. One might well argue whether these filaments and voids constitute an object, much less a new class of object. On the other hand, one might also argue that they actually constitute two classes, one of voids and one of filaments. Recognizing that we are stretching the definition of "object" to "coherent structure," as indeed we have been for clusters and superclusters, we believe filaments do qualify for class status. And because voids could not exist without the filaments that enclose them, we designate them here as a single class rather than two classes, in fact the largest class of objects in the universe in terms of their size.

The discovery of filaments and voids depended on accurate distances for large numbers of galaxies, not an easy task. Edwin Hubble and Vesto M. Slipher had shown in the 1920s that redshifts in the absorption or emission lines of galactic spectra are a measure of recessional velocity, and that these velocities are in turn a measure of distance, Hubble's velocity-distance relationship published in 1929 being one of the landmarks of 20th century cosmology. Lists of galactic redshift measurements were gradually and painstakingly built up over the decades. Some 600 were known by 1956, 2,700 by 1976, 5,000 by 1980, and 30,000 by 1989.

During the 60-year period from 1930 to 1990, measurement time for a single spectrum went from several hours with a spectrograph on the Mt. Wilson 100-inch telescope to a few minutes with digital detectors and image intensifiers. Still, as the authors of the first large redshift survey emphasized in recounting this history and announcing the first large-scale structure of filaments and

[34] Linda S. Sparke and John S. Gallagher, *Galaxies in the Universe: An Introduction* (Cambridge: Cambridge University Press, 2000), p. 281.

voids, these advances allowed the mapping of only 1/10,000[th] of the volume of the visible universe, equivalent to the surface of Rhode Island compared to the surface of the Earth.[35] The mapping process has since accelerated with projects like the Sloan Digital Sky Survey, which has confirmed large-scale structure of filaments and voids out to 2.5 billion light years and continues to push the frontiers. It is striking that, like its smallest structures, the large-scale structure of the universe is predominantly determined by the work of gravity. In this case, those structures likely have cosmological significance, reflecting the enhanced density regions of visible and dark matter in the early universe as measured over the last several decades by the COBE, WMAP, and Planck spacecraft. Results are also consistent with the Dark Energy Survey's map of dark matter.[36]

An early hint of large-scale, low-density regions came in two papers published in 1978, and in 1981 with the discovery of the "Boötes Void," a region 300 million light years in diameter where the density of galaxies was observed to be less than 20% of the average. In other words, voids are not completely void: once over the shock that voids exist, the next question is why there are any galaxies at all in a void. This was the largest of several other isolated regions previously noticed, and one of several dozen voids or "supervoids" since discovered, including the Local Void identified in Brent Tully and Richard Fisher's *Nearby Galaxies Atlas* in 1987, which we now know rivals the Boötes Void in extent. It starts about four million light years from Earth. True to form, even the Local Void has a few galaxies, including the dwarf irregular KK 246, some 25 million light years away but with nothing else surrounding it in any direction for 10 million light years.[37]

---

[35] M. J. Geller and J. P. Huchra, "Mapping the Universe," *Science*, 246 (1989), 897–903: 897, reprinted in Bartusiak, pp. 585–590.

[36] Siyu He et al., "The detection of the imprint of filaments on cosmic microwave background lensing," *Nature Astronomy*, vol. 2 (2018), 401–406; Dark Energy Collaboration, T. M. C. Abbott et al., "Dark Energy Survey Year 1 Results: Cosmological Constraints from Galaxy Clustering and Weak Lensing, https://arxiv.org/abs/1708.01530

[37] R. P. Kirshner et al., "A million cubic megaparsec void in Boötes," ApJ, 248 (1981), L57–L60; Ia. B. Zedlovich et al, "Giant voids in the universe," *Nature*, 300 (1982), 407–413; precursors to the Boötes Void include S. A. Gregory and L. A. Thompson, "The Coma/A1367 supercluster and its environs," ApJ, 222 (1978), 784–799. On early filament and void hints see R. Giovanelli and M. P. Haynes, "The Lynx-Ursa Major supercluster," AJ, 87 (1982), 1355, and Laird A. Thompson and Stephen A. Gregory, "An Historical View: The Discovery of Voids in the Galaxy Distribution," preprint at https://arxiv.org/abs/1109.1268 On the Local Void see Ken Crosswell, "The Void Next Door," S&T, 136 (October, 2018), 12–19.

Meanwhile, another component of large-scale structure was revealed in 1982 with the discovery of long sheets, walls, or "filaments" of galaxies. But the first large redshift survey to reveal a pattern in the large-scale structure of the universe was undertaken by the Harvard Center for Astrophysics (CfA), led by Margaret Geller and John Huchra. Their survey, begun in the early 1980s using the 1.5-meter telescope operated by the Smithsonian Institution on Mt. Hopkins, Arizona, consisted of thousands of galaxies in a six by 117 degree slice through the Coma cluster out to 700 million light years. When their graduate student, Valerie de Lapparent, first plotted 1057 of these galaxies in the summer of 1985 as part of her dissertation, the "slice map" revealed chains and sheets of galaxies separated by giant empty regions they called voids (Fig. 18.10). "Several features of the results are striking," they wrote in their paper published the following year. "The distribution of galaxies in the redshift survey slice looks like a slice through the suds in the kitchen sink; it appears that the galaxies are on the surfaces of bubble-like structures with diameter 25-50 h^{-1} Mpc. This topology poses serious challenges for current models for the formation

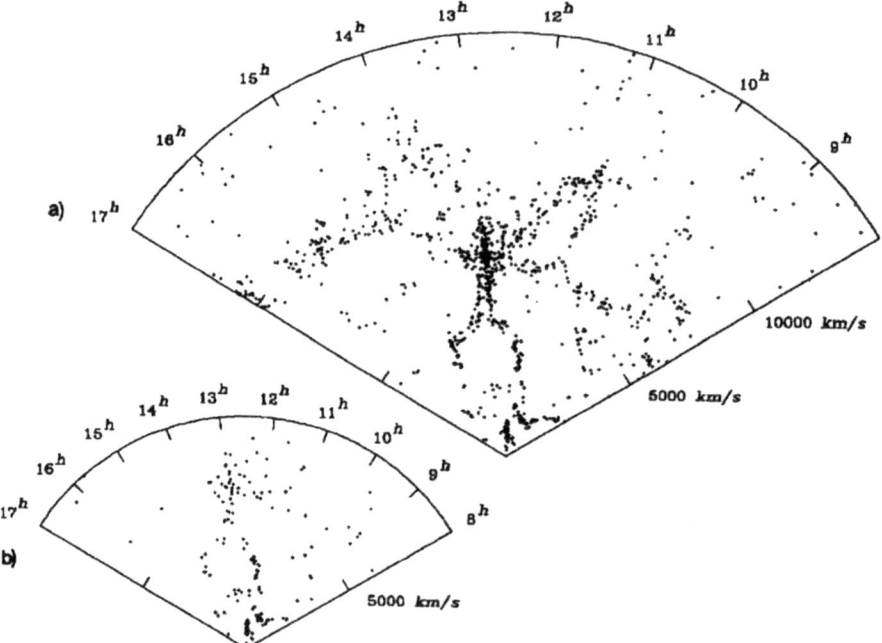

Fig. 18.10. A slice of the universe out to 700 million light years, evidence for large-scale structure of filaments and voids. From Lapparent, Geller and Huchra, "A Slice of the Universe," ApJ, 302 (1986), 12

of large-scale structure." In the lingo of astronomers 50 h^{-1}, where "h" is 1/100th of the Hubble constant now known to be around 71 (km/sec)/megaparsec, is equivalent to about 225 million light years. The following year astronomers reported a thin "filament" of galaxies with a narrow redshift range, known as the Perseus-Pisces filament.[38]

When the Harvard map was completed in 1989 incorporating 14,000 bright galaxies, it revealed the "Great Wall," a sheet of galaxies 500 million light years long, 200 million light years wide, and 15 million light years thick, surrounded by great voids 200 million light years in diameter.[39] In 1986, J. Richard Gott and his colleagues at Princeton showed that the large-scale structure was "sponge-like," with the galaxies representing the material of the sponge and the voids representing the holes in the sponge. Others prefer to describe it as bubble-like. Ever-deeper surveys continued to show the same structure at increasing distances. In 1996, the Las Campanas Redshift Survey, employing data from 26,418 galaxies, showed the cellular structure extending at least four times further than the original Harvard survey. In 2001, astronomers at the Anglo-Australian Telescope announced that their survey of redshifts of 180,000 galaxies in the Southern Hemisphere, the so-called Two Degree Field Survey (2dF for short), showed the same cell structure out to five times the distance of the Huchra-Geller survey. A follow-up Six Degree Field Survey (6dF) expanded these results.[40]

Finally, the Sloan Digital Sky Survey (SDSS) measured redshifts of a million galaxies out to 2.5 billion light years. Like the nearer Harvard survey, in 2003 the Sloan survey announced the existence of voids and filaments, in this case out to two billion light years, including a 1.37 billion-light-year structure dubbed the "Sloan Great Wall." Three times further away and 80% longer than

---

[38] Valerie de Lapparent, Margaret J. Geller, and John P. Huchra, "A slice of the universe," ApJ, 302 (1986), L1–L5. R. P. Kirshner et al, ApJ, "A Survey of the Boötes Void," 314 (1987), 493; Dick (2013), p. 166. For her personal story on this research see Geller, "The Large-Scale Structure of the Universe, in David DeVorkin, ed., *Beyond Earth: Mapping the Universe* (Washington, DC: National Geographic, 2002), pp. 182–185.

[39] M. J. Geller and J. P. Huchra, "Mapping the Universe," *Science*, 246 (1989), 897–903.

[40] N. Cross et al., "The 2dF Galaxy Redshift Survey: the number and luminosity density of galaxies," MNRAS, 324 (2001), 825. Dick (2013), pp. 167–168.

the Great Wall discovered by Geller and Huchra in 1989, it was at the time the largest observed structure in the universe. That title is perhaps now superseded by the "Huge Large Quasar Group" and the "Hercules-Corona Borealis Great Wall," both reported in 2013 by separate teams. The latter contains billions of galaxies and may extend over 10% of the observable universe. Other well-studied filaments include the Sculptor Wall, the Centaurus Wall, and the Coma Wall.

Redshift surveys continue to be undertaken, shedding light on galaxy evolution and further confirming the filament/void structure. Today, several dozen voids have been identified, ranging from the "Local Void" to the "Giant Void" more than a billion light years in diameter, the largest void in the northern galactic hemisphere.[41] New filaments are also being discovered, and galaxies seem to flow out of voids into filaments. Some of these structures may be seen in a panoramic view of the combined data from many redshift surveys compiled by the 2 Micron All-Sky Survey (2MASS) team (Fig. 18.11).

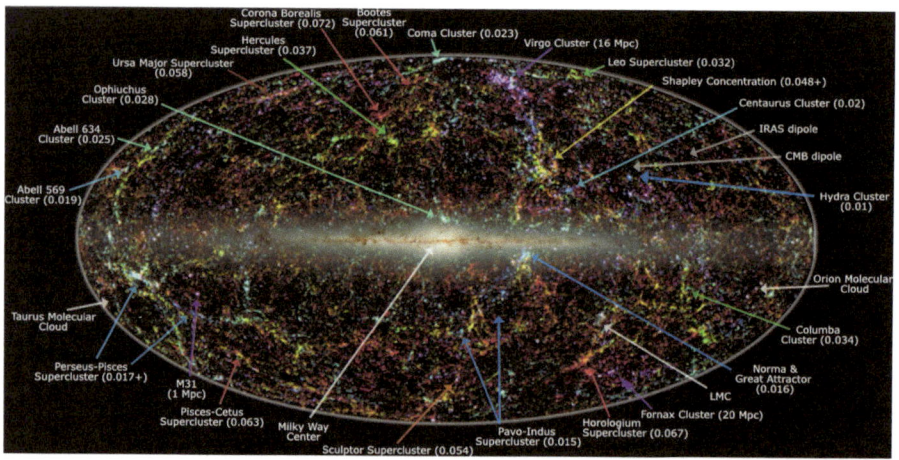

Fig. 18.11. Panoramic view of the infrared sky showing large scale structure based on the distribution of galaxies, clusters and superclusters of galaxies for the "local universe" out to about one billion light years. From the 2 Micron All-Sky Survey (2MASS) Extended Source Catalog incorporating 1.5 million galaxies, and using redshift data from many sources including the SDSS. Blue and purple indicate the nearest galaxies, green are at moderate distances, and red are most distant. Graphic created by T. Jarett (IPAC/Caltech)

[41] J. Richard Gott III et al., "A Map of the Universe," ApJ, 624 (2005), 463-484; Dick (2013), pp. 167–168.

Theory, simulation, and observation indicate this large-scale structure of the universe was built on a framework of invisible dark matter believed to constitute by far the majority of matter in the universe. To end where we began Part III, observations from the Great Observatory Origins Deep Survey (GOODS) program, incorporating the Hubble, Spitzer, and Chandra space telescope data, is consistent with the cold dark matter theory that normal matter was attracted to dark matter regions to build star clusters, galaxies, and clusters of galaxies from the bottom up, resulting in that small portion of luminous matter we see in our universe today.

A list of galaxy redshift surveys is at http://www.astro.ljmu. ac.uk/~ikb/research/galaxy-redshift-surveys.html. Many of the redshift surveys describe their work on their own websites such as http://www.6dfgs.net, http://www.gama-survey.org, and http:// vipers.inaf.it. The latest work from the SDSS, including a video of large scale structure, is at https://www.sdss.org/science/. The Dark Energy Survey dark matter map is at http://news.fnal.gov/2017/08/ dark-energy-survey-reveals-accurate-measurement-dark-matter-structure-universe/, and further results are at https://www.dark-energysurvey.org/the-des-project/overview/. The online Atlas of the Universe has excellent visualizations ranging from the nearest stars to superclusters and large-scale structure at http://www. atlasoftheuniverse.com. A 3-D video simulation of large-scale structure is at https://www.youtube.com/watch?v=FFlzyxSQhTc. A list and map of some of the well-known voids and filaments can be found at https://en.wikipedia.org/wiki/Void_(astronomy) and http://en.wikipedia.org/wiki/Galaxy_filament.

About the Author

Steven J. Dick is the former NASA Chief Historian and Director of the NASA History Office. He was the 2014 Baruch S. Blumberg NASA/Library of Congress Chair in Astrobiology at the Library of Congress's John W. Kluge Center. In 2013, he testified before the United States Congress on the subject of astrobiology. From 2011 to 2012, he held the Charles A. Lindbergh Chair in Aerospace History at the National Air and Space Museum. Prior to that, he was an astronomer and historian of science at the US Naval Observatory for more than two decades. He is the author or editor of 22 books, including most recently *Discovery and Classification in Astronomy: Controversy and Consensus (Cambridge, 2013), The Impact of Discovering Life Beyond Earth (Cambridge, 2015), and Astrobiology, Discovery, and Societal Impact (Cambridge, 2018)*. In 2006, Dick received the LeRoy E. Doggett Prize from the American Astronomical Society for a career that has significantly influenced the field of the history of astronomy. In 2009, minor planet 6544 Stevendick was named in his honor.

© Springer Nature Switzerland AG 2019
S. J. Dick, *Classifying the Cosmos*, Astronomers' Universe,
https://doi.org/10.1007/978-3-030-10380-4

Index

© Springer Nature Switzerland AG 2019
S. J. Dick, *Classifying the Cosmos*, Astronomers' Universe,
https://doi.org/10.1007/978-3-030-10380-4